What Happened to Governance in Kashmir?

What Happened to Governance in Kashmir?

Aijaz Ashraf Wani

OXFORD
UNIVERSITY PRESS

OXFORD
UNIVERSITY PRESS

Oxford University Press is a department of the University of Oxford.
It furthers the University's objective of excellence in research, scholarship,
and education by publishing worldwide. Oxford is a registered trademark of
Oxford University Press in the UK and in certain other countries.

Published in India by
Oxford University Press
2/11 Ground Floor, Ansari Road, Daryaganj, New Delhi 110 002, India

ISBN-13 (print edition): 978-0-19-948760-8
ISBN-10 (print edition): 0-19-948760-X

ISBN-13 (eBook): 978-0-19-909715-9
ISBN-10 (eBook): 0-19-909715-1

Typeset in Trump Mediaeval LT Std 9.5/13
by The Graphics Solution, New Delhi 110 092
Printed in India by Nutech Print Services India

Contents

Acknowledgements

When I joined the Department of Political Science, University of Kashmir, as a student I would often hear from the distinguished Kashmir experts that the Kashmir issue was no doubt basically a political problem, but the patterns of rule, right from the demise of the princely order, had added fuel to the fire. At that stage I could not appreciate the import of this observation. But when my PhD supervisor, Professor Gull Mohammad Wani, suggested that I work on governance in Kashmir, the significance of the observation I used to hear during my novitiate days started unfolding with my reading of secondary works followed by a variety of primary sources. Thus the intellectual seeds of this book were sown during my PhD work. Immediately after the award of my degree in early 2014, I started to build upon my research in order to transform it into a book. I approached Oxford University Press for its publication. As per their set procedure, the publisher sent the manuscript to two anonymous reviewers who, while recommending it for publication, made some significant suggestions for its improvement. While making improvements in the light of their suggestions, the work has emerged as a substantial improvement over the original one, and I have no words to express my heartfelt thanks to the reviewers who unfortunately continue to remain anonymous to me.

This work is the outcome of the immense help I received from my teachers, friends, family, and well-wishers. It is my pleasure to thank them all. Professor Gull Mohammad Wani and his family always treated me as a family member besides having groomed me for a serious role in academics. I have no words to express my heartfelt gratitude to them.

I owe an immense debt to Professor Ayesha Jalal, Professor of History at Tufts University, USA, and Professor Niraja Gopal Jayal, Centre for the Study of Law and Governance, JNU, for their graciousness in accepting my request to opine on this book. I am greatly thankful to Professor Mark Bevir, Professor Rod Rhodes, Professor Haley Duschinski, Professor Amit Prakash, Alexander Evans, Professor K. C. Suri, Professor Ujjwal Kumar Singh, Professor Rekha Chowdhary, and Professor G. R. Malik for their help and guidance.

To my teachers and colleagues at the Department of Political Science, University of Kashmir, Professor Noor Ahmad Baba, Professor Asifa Jan, Professor Ravinderjit Kour, Professor Irshad Ahmad Shah, Dr Naseema, Dr Nazir Ahmad, Dr Anjum Ara, Dr Sanjeda Warsi, and Dr Javid, my debts are infinite.

I am also indebted to all those authors whom I quoted with their gracious permission.

This work has benefited immensely from discussions with my friend Javid ul Aziz. I thank him. My friends and well-wishers Ajaz ul Haq, Dr Mehmood ur Rashid, Bilal Ahmad Bhat, Mudasir Ahmad, Altaf Para, Inam-ul-Rehman, Tanveer Habib, Shahzad Wani, Neelofer, Feroz Ahmad Wani, Abdul Rouf, Tabzeer Yaseen Sajjad, Sakeena, Shahnawaz, Showket, Imran, Qayoom, Arshid, Tanveer, Imran, Muzamil, Irfan and Shazia rendered invaluable help. Words fail me in thanking them.

Farooq and Mehraj, my students as well as friends, deserve special thanks. I am sure without their unflinching support this work would not have been possible.

Special thanks to all my students who keep pushing me to work harder. I enjoy their company the most, and their inquisitiveness makes me an unending learner.

My combative debates and discussions with my friend, philosopher, guide, and teacher, my father Professor Mohammad Ashraf Wani, former Professor of History, University of Kashmir, gave a historian's touch to this work. I shall be failing in my duty if i do not thank Professor A. Q. Rafiqi, former Professor of History,

University of Kashmir, who not only initiated my father into the field of research, but also showed equal zeal and zest in my academic growth as well.

I am thankful to the non-teaching staff of my department, library staff of the Centre for Law and Governance, New Delhi, Indian Institute of Public Administration, New Delhi, Legislative Assembly, J&K, the Jammu and Kashmir Archives, Research and Reference Library, Department of Information J&K Government, Directorate of Economics and Statistics, Srinagar, and the Coalition of Civil Society, Srinagar, for their ungrudging cooperation. I am immensely grateful to my respondents in the field, who for various reasons wanted to remain anonymous, for their valuable feedback.

I would like to thank the editorial team of Oxford University Press, New Delhi, for their cooperation.

The love and support I received from Dr Aman Ashraf, Dr Shazia, Dr Javid Ahmad, Dr Fehmeeda, Dr Gowhar, Dr Sabina, Mukhtar Bhai, Rehana Baji, Dr Javid Ahmad Bhat, Arena Baji, Abu, Khali, Dr Shajer us Shafiq Jan, Dr Deeba, Rafiqa, Mr Ajaz, Mrs and Mr Gh. Rasool Bafanda, Mrs and Mr S. N. Jan, Mrs and Mr Gh. Hassan, Mrs and Mr Ab Rehman Shah, Mrs and Mr Ab. Karim Mohand, and Mrs and Mr Mohammad Abdullah is beyond words. Asifa, Aadil, Moonis, Tuba, Touyiba, Taha, Anas, Tazkiya, Basim, Zahra, Aayesha, Faliha, Nuha, Hayda, Usman, Hurrain, Adiyan, Izan, Kulsum, and Abida were a great source of joy throughout the period of study.

Dr Shaheena, my wife, and my two little children, Mohammad Mousa and Samra Noor, bore with exemplary patience my preoccupation with this work for months together when I should have been with them. They would feel to have been compensated for their purgatory should this book is seen to have slightly added to the existing body of knowledge on the subject.

This book is for Ami and Abajan, who made me what I am.

Abbreviations

BJP	Bharatiya Janata Party
CID	Counter Intelligence Department
CRPF	Central Reserve Police Force
IADP	Intensive Agricultural Development Programme
ICDP	Intensive Cattle Development Project
J&K	Jammu and Kashmir
JKLF	Jammu Kashmir Liberation Front
KCC	Kashmir Chamber of Commerce
KRBMS	Kashmir Raj Bodhi Maha Sabha
LBA	Ladakh Buddhist Association
MLA	member of the legislative assembly
MUF	Muslim United Front
PAK	Pakistan Administered Kashmir
RSS	Rashtriya Swayamsevak Sangh
YMBA	Young Men's Buddhist Association

Introduction

What Happened to Governance in Kashmir? documents the state of management of trouble-torn Kashmir by 'client governments'.[1] It is a story of engagement with political instability through the use of different instruments, strategies, and tactics to create 'order' in what Sanjib Baruah terms, in a different context, 'durable disorder'.[2] As the challenges are dynamic, and are expected to be the responses from an alert state, this book examines governance in the framework of the challenge-and-response continuum. In the process, the book unfolds the changes and continuities in the politics and governance of the state since 1948. It argues that since the crucial political decisions about the future of Kashmir were taken by the Indian and local leadership to the exclusion of the will of the people, opting to control the latter by sheer deception, coercion, corruption, and development, Kashmir became a smouldering volcano which finally exploded in 1989–90, with no end till date.

Jammu and Kashmir (J&K) is arguably the most ailing region, suffering as it does from political instability, financial crisis, and

---

[1] This term is used by Sumantra Bose to describe the post-1947 governments of Jammu and Kashmir imposed by New Delhi. Sumantra Bose, *Kashmir: Roots of Conflict, Paths to Peace* (New Delhi: Vistaar, 2003), 98, 190.

[2] See Sanjib Baruah, *Durable Disorder: Understanding the Politics of Northeast India* (New Delhi: Oxford University Press, 2005).

a deficit of the salient benchmarks of good governance. The ailment is not self-generated, it is ascribed; its sources lie beyond the borders of the state. A victim of colonialism and nationalism, the state still carries the disabilities imposed upon it by British colonialism, which, to serve its colonial interests, deprived the state of its locational advantages by snapping the commercial and cultural ties that connected it with Central Asia and Russia. In the 1950s, China closed the borders of eastern Turkistan and Tibet, resulting in the complete collapse of J&K's trade with Central Asia, Tibet, and China. The Indian state maintained the status quo with respect to closure of borders/trade routes, for it served its 'national interests'. Moreover, J&K became a bone of contention between the two dominion states, India and Pakistan, which emerged on its borders, resulting in the closure of the remaining age-old routes (Kashmir had been connected to the world through that part of the subcontinent which came to be called Pakistan) on the one hand, and, on the other, the creation of permanent instability in the state that adversely affected every sphere, including governance. In fact, for anyone interested in studying governance in Kashmir, it is a prerequisite to bear in mind this sovereign influence on the pattern of rule in the state. Choices regarding how the state is to be ruled; whether to treat it like other states or not; how selective the government should be in following its own constitution or the Indian constitution in their respective areas of jurisdiction; which policies should be adopted in their pure form and what should be modified or left out; whether to adopt the parameters of good governance or not; if it becomes necessary to respond, then what set of parameters should be followed in letter and in spirit, what should be regarded simply as a paper tiger, and what might be overlooked altogether—all these subjective choices are determined by the quest to maintain/restore order.

Being politically a crisis state, the priority of all the governments of J&K has been to win normalcy amidst troubled conditions by following the policy of coercion and consent. While coercion was and is being used against those who represent counter-voices challenging the dominant discourse, material development and socio-cultural policies have been deployed to create social consent and establish moral leadership, with the ultimate purpose of overcoming opposing political forces so as to establish hegemony. Coercive measures have included the use of armed forces, police, courts, and prisons, and the denial of democracy and civil liberties to liquidate or dominate antagonistic groups. Besides the use of coercive apparatuses, the state has also made use of what the French sociologist Pierre Bourdieu has called 'symbolic violence',[3] by marginalizing, discriminating against, and shaming 'outsider' groups both for reproducing its authority as well as for imposing silence, even generating consent.

Realizing that mere dependence on coercive and authoritarian means to enforce rule seriously compromises the credibility of the government besides producing a deceptive peace, it was thought by the policy pundits (the group associated with Kashmir policy, headed by Nehru himself during the critical one and a half decades, 1947–62) that both the centre and the state governments should in unison shore up rule in Kashmir through whatever ideological, economic, cultural, social, political, and legal resources they had at their disposal to resist/create social consent. State-led developmentalism was also favoured by the international environment of the time. 'Development' had become the major international concern since the emergence of the 'Third World'.[4] How to bring about speedy socio-economic changes to improve the quality of life of the people in the Third World assumed significance for the leaders of newly independent countries. Among these leaders, Nehru and Sukarno were at the forefront. International donor agencies and the US were also concerned about the development of the new nations, as they were apprehensive about the communist influence

---

[3] Symbolic violence takes the form of taste judgements, where outsiders are marginalized and shamed; of forms of physical behaviour and the 'way of living, where some feel confident and others feel awkward.... In these cases, a ruling power will see its authority reproduced, a subaltern group will aspire to the value and tastes of its superiors...'. Pierre Bourdieu, *Distinction: A Social Critique of the Judgment of Taste* (London: Routledge and Kegan Paul, 1984), quoted in Steve Jones, *Antonio Gramsci* (Routledge, 2006), 52.

[4] Bidyut Chakrabarty and Mohit Bhattacharya, eds, *The Governance Discourse: A Reader* (New Delhi: Oxford University Press, 2008), 45–6.

in poor countries. Thus, they mobilized resources and offered the services of their development pundits to advise 'underdeveloped' countries in their pursuit of planned change and development. The earliest phase after decolonization is, therefore, called 'developmentalism', connoting an emphasis on development as the essential means of modernizing and improving the condition of people.[5] Just as the development discourse of donor agencies and the US was informed by politics, India's development policy towards Kashmir is also informed by politics. Development became a tool to manage the state. Irrespective of the changes in the governments and ideologies that have ruled at the centre, there has been consistency in the political rhetoric of the Indian leadership that development will bring normalcy in Kashmir. Parallel to the flow of aid and the fanning out of development pundits from the US and donor agencies to Third World countries, the J&K government was provided liberal aid as well as planners, technicians, and subject experts by the central government to guide the state government in its pursuit of planned change and development. This policy received great impetus during the prime ministership of Bakhshi Ghulam Mohammad, to whom Nehru entrusted the job of restoring peace, parting ways with Sheikh Abdullah's development discourse within the framework of autonomous status for J&K.

Post-1947 Governance Model

Notwithstanding the federal character of the Indian state, the parameters of governance at both the national and the state level flow from the Indian constitution and from dynamic national policies which change with the times, under endogenous as well as exogenous pressures. It is binding upon state governments to follow the centre's imperatives at least in those spheres which are considered critical for the nation-state's needs. Although J&K has its own constitution, there is fundamental uniformity in the

---

[5] Bidyut Chakrabarty and Mohit Bhattacharya, eds, *The Governance Discourse: A Reader* (New Delhi: Oxford University Press, 2008), 45–6.

essentials of the two constitutions, besides the fact that most of the central laws now stand applied to J&K. Up to the beginning of the 1980s, the model of governance adopted by India under the domineering influence of Nehru was informed by Indian socialism, characterized by a mixed economy, the 'licence-permit-quota raj', a maximalist state, paternalistic behaviour of the government, inward-looking policy, planned development, and a high degree of centralization. The government in Kashmir also followed the same model even during autonomous rule (1948–53), evidently because the Naya Kashmir (New Kashmir) Programme drafted by National Conference during the freedom movement was also informed by a progressive ideology. Indeed, J&K took a lead in implementing its predominantly socialist-oriented Naya Kashmir Programme. The abolition of feudalism, nationalization of important industries, cooperative movement, the policy of license-permit-quota-raj (LPQRP), the promotion of public sector, and the policy of soft-subsidies, soft taxation, soft credit and soft administrative-prices, expanded service sector beyond requirements and other measures espoused by socialism and followed by the governments in Kashmir have to be understood as much in the context of the public policy guidelines of the central government and the financial support received from the latter for its implementations as in the context of J&K's own constitution and public policy.

The state-centric approach to governance was also necessitated by the compulsions of a fragmented nation, where national unity was not organic but state-facilitated and 'manufactured'.[6] Kashmir being a mini-India with regard to its sub-national diversity, the state-centric approach to governance became a compulsion here too. Moreover, both the Congress and the National Conference had identical ideological moorings, which, *inter alia*, favoured a centralized polity with the state as an overarching authority. 'The ideals of citizenship', says Srirupa Roy 'were also articulated in similar state-centered terms.'[7] 'Nehruvian India's most frequently

[6] Srirupa Roy, *Beyond Belief: India and the Politics of Postcolonial Nationalism* (Durham, NC: Duke University Press, 2007), 19.

[7] Srirupa Roy, *Beyond Belief: India and the Politics of Postcolonial Nationalism* (Durham, NC: Duke University Press, 2007), 20.

invoked figure was that of the "infantile" citizen and his need for state tutelage and protection in order to realize the potentials to citizenship,'[8] placing the state at the heart of individual and natural life. The Nehruvian policy remained vigorously in vogue until Indira Gandhi returned to power in 1980. She introduced some pro-liberalization and market-friendly economic reforms, alongside centralizing the party structure around a personalized and increasingly populist mode of leadership.[9]

The late 1970s and early 1980s saw a radical transformation in the role of government, propounded by neo-liberal theorists and promoted by capitalist democracies and their controlled global institutions. To overcome the fiscal crisis faced by advanced capitalist democracies, neoliberal theorists and management gurus conceived of the crisis and believed that the economic crisis would be overcome by changing the role of government from rowing to steering. Thus, calls for a minimal state, a stateless state, rolling back the state, reinventing the state, more governance and less government, network governance, and the like became the dominant notes for neoliberal theorists, leading to the public sector reforms of the 1980s and 1990s. With this, 'governance' came to be used as a specific term to describe the shift from a hierarchical bureaucracy towards a greater use of markets, quasi-markets, and networks,[10] especially in the delivery of public services. Experience has, however, shown that state and government

---

[8] Srirupa Roy, *Beyond Belief: India and the Politics of Postcolonial Nationalism* (Durham, NC: Duke University Press, 2007), 20.

[9] Atul Kohli, *Democracy and Discontent: India's Growing Crisis of Governability* (Cambridge: Cambridge University Press, 1990); Francine R. Frankel, *India's Political Economy, 1947–1977: The Gradual Revolution* (Princeton: Princeton University Press, 1979); Lloyd I. Rudolph and Susanne Hoeber Rudolph, *In Pursuit of Lakshmi* (New Delhi: Orient Longman, 1998); Stuart Corbridge and John Harriss, *Reinventing India: Liberalization, Hindu Nationalism and Popular Democracy* (London: Polity Press, 2000).

[10] See Mark Bevir, ed., *Public Governance*, vol. I (London: Sage, 2007), 2; Rod Rhodes, *Understanding Governance: Policy Networks, Governance, Reflexivity, and Accountability* (Buckingham: Open University Press, 1997), 15; Mark Bevir and Rod Rhodes, *Interpreting British Governance* (Basingstoke: Routledge, 2003), 55–6; Anne Mette Kjaer, *Governance* (Cambridge: Polity Press, 2004), 10–11.

continue to be central to the governance process.[11] Even in those countries where neoliberal reforms have been quite intensive, the state has 'scarcely rolled back at all'.[12] Emphasizing the centrality of state in governance, Jon Pierre and Guy Peters argue that, 'despite persistent rumours to the contrary, [the state] remains the key political actor in society and the predominant expression of collective interests'.[13] Mark Bevir, who favours the post-1980s definition of 'governance', is however not averse to using it as a blanket term for all patterns of rule. He says:

> More generally still governance can be used to refer all patterns of rule, including the kind of hierarchical state that is often said to have existed before the public sector reforms of the 1980s and 1990s.... However, if we are to use governance in this general way, perhaps we need to describe the changes in the state since the 1980s using an alternative phrase such as 'the new governance'.[14]

Apart from the arguments put forth by scholars supporting a state-centric, relational approach to governance,[15] I have special reason to agree with this thesis because, in Kashmir, the hierarchical governance structure continues to hold its ground, and the state has maintained its position as the pivotal player in framing and implementing policies and strategies to respond to challenges. In the difficult conditions of Kashmir, the state has assumed even more power, to the extent of forming part of the backdrop of everyday

---

[11] See Stephen Bell and Andrew Hindmoor, *Rethinking Governance: The Centrality of the State in Modern Society* (Cambridge: Cambridge University Press, 2009).

[12] Mark Bevir, 'Governance', in *Encyclopedia of Governance*, ed. Mark Bevir (London: Sage, 2007), 366; Mark Bevir, *Democratic Governance*, (Princeton: Princeton University Press, 2010), 31.

[13] Jon Pierre and Guy Peters, *Governance, Politics and the State* (Basingstoke: Macmillan, 2000), 25; also see Bell and Hindmoor, *Rethinking Governance*, 2–3.

[14] Bevir, 'Governance', 364–5.

[15] The 'state-centric relational approach' has been used by Bell and Hindmoor (Bevir 2007), by which they mean that far from receding, states are in fact enhancing their capacity to govern by extending top-down controls and developing closer ties with non-governmental sectors.

life. Governance through markets, associations, and community engagement is peripheral.

However, by a 'state-centric, relational approach' to governance, I do not mean the centrality of the state in governance process to the exclusion of the environment in which the state functions. Indeed, it would be naïve to take this position. Therefore, I have also taken note of the importance of situating the priorities and actions of governments within the context of their own respective environments, which in today's world are constituted by a mix of endogenous and exogenous pressures and opportunities. This approach is informed by the state-in-society discourse, which maintains that while states are responsive to social interests, they also retain some capacity to act independently against powerful social actors.[16] True, the will of the state and the priorities of the government do not always capture the interests of the society, not least in cases where, like Kashmir, there is a struggle for hegemony between the opposing narratives represented by the state and other powerful social actors. Yet, in such cases too, forging alliances with different sections of society by using power and patronage to disarticulate and disempower their opponents fits into the state's priorities. For example, in the difficult situation of Kashmir, where the support of the Muslims has always been crucial for the legitimacy of Indian rule, the central government employed two main strategies to win over the Muslims. One was to cultivate and make alliances with the ideologically favourable Muslim elite, and install some of them as heads of government to generate consent in the community, with the belief that it will somehow satisfy their psychological instinct of 'self-rule'. This policy was also aimed to draw mainstream discourse closer to the people by co-opting political society in the localities through the Muslim governing class.[17] Further, to impoverish its opponents

---

[16] See J. Migdal, 'The State in Society: An Approach to Struggles for Domination', in *State Power and Social Forces: Domination and Transformation in the Third World*, eds J. Migdal, Atul Kohli, and V. Shue (Cambridge: Cambridge University Press, 1994), 7–36.

[17] By 'political society' I mean those influential sections of the society who because of their comparatively better economic, social and cultural position have always been coopted by the state through patronage in order to establish its hegemony.

politically, the state, within the limits of its financial and other constraints, embarked on the policy of satisfying the 'acceptable' aspirations of the society to isolate the rebellious instinct of the people from the basic priorities of life, to preclude this 'instinct' from assuming violent expression. It is also observed that the state uses culture and ideology, by imposing its own meanings on them to reinforce the state narrative and weaken the counter-narrative, thus perpetuating its hegemony. The political institutions, tools, strategies, and tactics employed by governments have been crucial in setting constraints on human subjectivity and ensuring the implementation of 'official goals'; but at the same time, changing pressures from within and without have continuously modified (if not making structural changes in) the approach to governance. Indeed, governance in Kashmir is underlined by change in the broader framework of continuity (political and military status quo) in response to shifting endogenous and exogenous pressures.

Moreover, as articulated by Migdal, there is no human society where one incredibly coherent and complex organization exercises an extraordinary hegemony of thought and action over all the social formations intersecting that territory. The state, for its own survival, is always in constant interaction, negotiation, conflict, and conciliation with multiple groupings and power centres, making the state ultimately a site of contradictory practices and disparate alliances.[18] It is these multiple and contending structures, actors, and relationships that make up the modern state. The variety of encounters produces a state with different faces and 'languages of stateness',[19] a puzzling simultaneity—a state that is simultaneously all-encompassing and limited, omnipresent and non-totalizing, a 'ruling state' and a 'serving state', invincible and fragile,

---

[18] For details see Joel S. Migdal, *State in Society: Studying How State and Societies Transform and Constitute One Another* (Cambridge: Cambridge University Press, 2004).

[19] Thomas Hansen and Finn Stepputat, eds, *States of Imagination: Ethnographic Explorations of the Postcolonial State* (Durham, NC: Duke University Press, 2002).

timeless and new.[20] The 'ambiguities of domination',[21] rather than uniform application of state power, is also demonstrated by varied and disparate encounters across governance levels.[22] Encounters with the state have also varied significantly between 'peaceful' and 'disturbed' areas. Representations of state, nation, and citizen that emerge from 'disturbed' areas are quite different from those in 'peaceful' areas. The peaceful areas are the 'core' areas whose will is represented by the state, whereas 'disturbed' areas are the pockets on the margins of the state, in the sense that they constitute what Carolyn Nordstrom refers to as the 'shadows', [23] places existing as part of the formal state, but also excluded from it in terms of the violent realities of everyday life, the legal and extra-legal networks that support them, withdrawal of rights, and a series of invisible corrupting, messy, and profane strategies to silence dissent and perpetuate the dominant discourse.

From the early 1950s onwards, the situation in Kashmir has gone from bad to worse, to the extent that one is reminded of Philip Spratt's suggestion for the solution of the Kashmir problem in the early 1950s as one which should be 'tinged with morality, but more so with economy and prudence', and in which 'material interests should supersede ideological ones'.[24] In this book, I look for the reasons why the policies of the government have not yielded the desired results. Indeed, the present is the product of the past, and becomes intelligible only in the light of the latter. The different trajectories and processes under way between

[20] Roy, *Beyond Belief*, 21.

[21] See Lisa Wadeen, *Ambiguities of Domination: Politics, Rhetoric, and Symbols in Contemporary Syria* (Chicago: University of Chicago Press, 1999).

[22] J. Migdal, Atul Kohli, and V. Shue, eds, *State Power and Social Forces: Domination and Transformation in the Third World* (Cambridge: Cambridge University Press, 1994), 3.

[23] Carolyn Nordstrom, *Shadows of War: Violence, Power, and International Profiteering in the Twenty-First Century* (Berkeley: University of California Press, 2004), 34–9.

[24] Ramachandra Guha, *India after Gandhi: The History of World's Largest Democracy* (New Delhi: Picador, 2008), 259–60.

1947 and the 1980s have made Kashmir a smouldering volcano, which erupted in radical politics in the late 1980s. It was a period of engagement and negotiation with political instability through the development of different instruments, strategies, and tactics which created new challenges, but, except for cosmetic responses to these challenges, the governments held its ground. No doubt at the same time it performed important services for society, and it appeared to be a blessing. But ultimately it has been proved that these were only Kashmiri summers, masking autumn and presaging winter, for coercion and 'development' could not subsume the effects of idolizing outworn institutions, policies, and strategies.

As governance is a vast topic, I had to be necessarily selective in choosing only a few capital areas, issues, and measures which are important especially in the context of my subject bias. The work is based on primary sources, fieldwork, interviews, and the author's personal knowledge and experience gained as a 'participant observer' in contemporary Kashmir.

The first chapter, which contextualizes governance in Kashmir, considers the role of various factors that shaped and continue to shape the nature and character of governance in Kashmir. Among these, mention may be made of the impact of the Kashmir dispute, the identity politics of the state, the legacy of the authoritarian, feudal, and exclusivist princely order, the ideological orientation of the freedom movement, Kashmir's special position and its contestation, policy interventions from the centre, financial crisis, and the changing environment which characterize the period.

Being the main sponsor of the Naya Kashmir Programme, Sheikh Abdullah's brief reign of a little over five years (1948–53) is characterized by the process of speedy implementation of the programme. However, notwithstanding the revolutionary changes in agrarian relations and other aspects, the condition of the common people worsened, and the whole period became mired in controversies and conflicts leading to disillusionment among the people as well as the dismissal of the Sheikh. The second chapter revisits this period by examining the extent to which the Sheikh succeeded in replacing the old with the 'new' Kashmir.

The forces representing religion and region in the Hindu-dominated Jammu districts and among the Buddhists of Leh raised a banner of revolt against the Sheikh-led National Conference government, calling it 'pro-Kashmiri Muslim rule'. The nature of the revolt clashed sharply with the ideology of Abdullah that had prompted him to prefer India over Pakistan. Having become disillusioned with Indian secularism, on which he had pinned high expectations, and with India's constitutional promises of sovereignty, the Sheikh voiced his disappointment publicly and drifted towards a position in support of plebiscite, which led to his widely condemned dismissal. 'For maintaining hold over power', says Gramsci, 'the group in power has to be constantly alert to the volatile demands of the dominated and to the shifting contexts within which it exerts its authority.'[25] The deposition of Sheikh Abdullah and the imposition of Bakhshi Ghulam Mohammad in 1953 created a storm in Kashmir, followed by the formation of the Plebiscite Front under the patronage of Abdullah. At the same time, the central government felt the urgency to further integrate Kashmir with India, which the popular leader Abdullah had resisted. Thus emerged the need for Gramsci's 'expansive hegemony', to obtain the consent of the great mass of the people willingly and actively to the ruling establishment. The third chapter engages with the steps taken by Bakhshi under the patronage of the central government to change the tide in favour of the Indian nation-state.

Yet, some of the policy instruments, especially the curbing of civil liberties and promoting corruption and nepotism, did not go down well with the hegemony project, as goonda raj and the misuse of power evoked strong reaction both within and outside Kashmir. Most importantly, however, Bakhshi showed diffidence in cooperating with the further integrationist moves of the centre. Hence it was regarded as necessary to change the leadership in Kashmir and to install G. M. Sadiq, whose alternative views were clearly known. The fourth chapter examines the nature of and the changes in governance during the period of Sadiq and the extent of his success in a hostile environment, to which however

---

[25] A. Gramsci, *Selection from the Prison Notebooks* (London: International Publishers, 1971), 58.

he also added fuel, notwithstanding his celebration of 'liberalism' and clean government.

The failure of liberalization, state-led development, and clean government under Sadiq in neutralizing dissident voices further convinced the central leadership that peace in Kashmir was hanging by the eyelids. This realization, together with the pulls and pressures exerted by Sheikh Abdullah, led to the Indira–Abdullah Accord of 1975. With the coming back of Abdullah to power, Kashmir witnessed almost a decade of 'peace'. It seemed that the internal dimension of conflict, at least, had been buried for all time to come. However, these hopes were belied. Around the same time, a new voice was born out of the debris of the buried Plebiscite Front, culminating in armed resistance in 1989–90. The fifth chapter delineates the processes that had resulted in what was only a short-lived peace.

The Place

Kashmir—the term by which I have denominated the place of my study—is the popular name for that part of the erstwhile princely state of J&K which is administered by India. It comprises three distinct regions—the Kashmir Valley, Jammu, and Ladakh. Like the Lahore Darbar under the Sikhs in Punjab, the seat of power in J&K during the reign of the Dogras was known as the 'Kashmir Darbar', giving an international name—Kashmir—to the state, by which it continues to be known.

The princely state of Jammu and Kashmir came into existence in 1846, a creation of British imperialism.[26] The three main administrative entities within the princely state of J&K included the province of Jammu, the province of Kashmir, and the provinces of Ladakh and Baltistan. There were other distinct political entities which, as a result of their geographical location, had to formulate some type of political relationship with the princely state, like the Gilgit Agency, which the British attached to J&K for political

---

[26] Huttenback, A. *Kashmir and the British Raj* (New York: Oxford University Press, 2004).

convenience in 1889, and which the Dogra state leased back to them in 1935. Poonch was brought under the formal control of J&K in 1936.

The state was balkanized in 1947 following a war between the two newly created dominions of India and Pakistan, creating Indian Administered Kashmir or J&K, and Pakistan Administered Kashmir (PAK) or 'Azad Kashmir'. Later, Pakistan ceded some part of Kashmir to China in a boundary agreement. Both countries have been staking their claims to the erstwhile princely state, making J&K a disputed territory, and resulting in three wars between India and Pakistan besides the continued deficit of peace in Kashmir—the land which was once known as paradise on earth.

The Indian-administered state of J&K is not only a conglomerate of three distinct regions—Jammu, Kashmir, and Ladakh—but there are regions within each region marked off from one another by geography, culture, history, economy, and politics. According to the 2011 census, Islam is practised by 68.31 per cent of the total population, followed by Hinduism, practised by 28.43 per cent. The state is home to other minorities as well—Sikhs (1.87 per cent), Christians (0.28 per cent), Buddhists (0.89 per cent), and so on. Major ethnic groups of the state include Kashmiris (mostly living in the Kashmir Valley), Gujjars/Bakarwals, Paharis (spread both in the Kashmir and Jammu regions), Dogras (concentrated mostly in the Jammu region), and Ladakhis (living in the Ladakh region). In terms of religion, the Kashmir Valley is predominantly Muslim (around 97 per cent of the population). All the 10 districts of the Kashmir region are Muslim dominated. The population of the Jammu region is 60 per cent Hindu, 36 per cent Muslim, and 4 per cent Sikh. Hindus form the majority in 4 out of 10 districts, while in 6 districts Muslims are in the majority. As far as the Ladakh region is concerned, 50 per cent of its population is Muslim, 44 per cent Buddhist, and 6 per cent Hindu. Buddhists form the majority in Leh district and Muslims in the Kargil district of the region.

E. F. Knight, the late-nineteenth-century European traveller, titled his travelogue on Kashmir *Where Three Empires Meet*.[27]

[27] E. F. Knight, *Where Three Empires Meet* (London: Longmans, Green & Co., 1905).

Though by this he meant that the princely state was strategically a place of interest to the British, Chinese, and Russians during the period of intense colonial rivalries, he was also referring to another fact having enduring relevance: Kashmir is hemmed in by many countries, namely, China, Central Asian states, Afghanistan, Iran, India, and Pakistan.

1

Contextualizing Governance

Governance is both a cause and an effect. While it impacts every structure of the society, it is itself the product of its own specific context or, to be more accurate, contexts. Context matters to regional governance in two different ways. It is important in the sense that most of the explanatory factors that are causally related to governance are regional in origin, and their effectiveness is determined by the will and capacity of dominant and alternative discourses in the region to intersect, contradict, destabilize, cancel, or modify each other. Since governance involves multifaceted factors that are complex and interconnected, and as 'contextual diversity rules at the heart of the Indian nation—each region and sub-region having its own history, historical legacies, and political memories, political geography, social, cultural, economic, and ideological conditions, specific relations with the centre, institutional arrangements, elite initiatives, pressures generated from below, and the relative strength, diversity, and aspirations of stakeholders in regional politics—it is not difficult to see why different regions/states experience the problem of governance in different ways'. Even other changes which 'cascade down from the apex of the national political system or from beyond the national frontiers also vary from region to region'. This is why I have chosen to begin the discussion on governance in Kashmir by situating it in the contexts which have shaped its nature and character.

Political Instability

Perhaps no other factor has been as overarching in shaping governance in Kashmir as the political instability which constitutes the defining feature of the post-1947 history of the region. Instability produced a series of consequences, namely, a deficit of peace, intolerance of opposition, denial of democracy, installation of 'client governments',[1] and the central government's collaboration 'with the political practices, corrupt or otherwise, of their sponsored establishment faction'.[2] The deficit of peace and the resultant urgency of the maintenance of law and order appropriated the main attention of the central and state governments. The same reasons accounted for violations of human rights,[3] curbing of civil liberties, and restraints on the freedom of the media.[4] The state,

---

[1] For details, see Balraj Puri, *Jammu and Kashmir: Triumph and Tragedy of Indian Federalisation* (New Delhi: Sterling, 1981); Balraj Puri, *Kashmir: Insurgency and After* (New Delhi: Orient Longman, 2008); Bose, *Kashmir: Roots of Conflict, Paths to Peace*; Prem Nath Bazaz, *Democracy through Intimidation and Terror* (New Delhi: Heritage, 1978); Sumit Ganguly, *The Crisis in Kashmir: Portents of War, Hopes of Peace* (London: Cambridge University Press, 1997); G. N. Gauhar, *Elections in Jammu and Kashmir* (New Delhi: Manas, 2002).

[2] Ganguly, *Crisis in Kashmir*, 38–9.

[3] For a historical account see P. N. Bazaz, *The History of Struggle for Freedom in Kashmir* (Srinagar: Gulshan Books, 2009); also see Chapter 3, this volume. For a contemporary account of the issue, see Asia Watch, *Kashmir under Siege* (New York: Human Rights Watch, May 1991); Committee for Initiative on Kashmir, *Kashmir: A Land Ruled by Gun* (New Delhi, 1991); Asia Watch and Physicians for Human Rights, *Rape in Kashmir: A Crime of War* (New York: Human Rights Watch, June 1993); Asia Watch and Physicians for Human Rights, *The Human Rights Crisis in Kashmir: A Pattern of Impunity* (New York: Human Rights Watch, June 1993); Amnesty International, *India: Torture and Deaths in Custody in Jammu and Kashmir* (London: Amnesty International, January 1995); also see annual reports by the State Human Rights Commission and reports by the Coalition of Civil Society (a Srinagar-based human rights group).

[4] Bazaz, *History of Struggle for Freedom in Kashmir*, 399; Josef Korbel, *Danger in Kashmir* (Princeton: Princeton University Press, 1954), 10, 209,

both at the national and at the local levels, is more proactive here than elsewhere in imposing, promoting, and popularizing its own theology and world view through the use of power and patronage.[5] The disputed nature of the state is perceived as a major factor in making Kashmir a dependent economy. It is popularly believed that the Indian government is deliberately not mobilizing the resource endowments of Kashmir as it apprehends that economic prosperity would further intensify the separatist political aspirations of Kashmiris, and that the investment of huge resources in a state with an ambiguous future would be risky. Also, 'fostered dependence' is the best way to keep a government in control.[6]

214; Mir Qasim, *My Life and Times* (New Delhi: Allied, 1992), 83; Sanaullah Butt, *Kashmir in Flames* (Srinagar: Ali Mohammad & Sons, 1981), 42.

[5] Perhaps nowhere is the Indian state so passionate about imposing its own favourable version of Islam as in Kashmir. B. K. Nehru, *Nice Guys Finish Second* (New Delhi: Viking, 1997), 594–5; Also based on my study of the related publications of state managed J&K Cultural Academy; programmes broadcasted/telecasted through Radio Kashmir Srinagar and Doordarshan Srinagar; public addresses of the mainstream political leadership and governing class; state patronage to sufi shrines, their custodians and preachers, as well as the rituals performed there. This is also largely the impression of the educated section of Kashmir whom I interacted with.

[6] The perception of 'fostered dependence' is not only a popular one in Kashmir, it also rankles in the minds of the Kashmiri elite, even those with established pro-India credentials. For example, Haseeb Drabu, the former finance minister of J&K, in an interview to NDTV in 2011 said, 'Government of J&K is totally dependent on government of India. ... For a variety of reasons, if you take a larger political economy perspective, then there is a fostered dependence, which has been created in the system over the years. So the government was not obliged to raise resources to fund itself. This is the best way to keep the government in control.... . I have said it in many of my writings that these are precisely the chains that Nehru spoke of when he referred to the chains of gold. Nehru once famously said in Parliament that "I'll bind Kashmir in chains of gold."' Haseeb Drabu, 'We Must Try and Get Fiscal Autonomy', interview to NDTV, 18 September 2011, https://www.ndtv.com/india-news/we-must-first-try-and-get-fiscal-autonomy-haseeb-drabu-467944 (accessed 3 October 2017).

The deficit of peace has also made investors reticent to invest in Kashmir, besides acting as a serious bottleneck in the growth of the tourism industry.[7] What is more, the political dispute has appropriated the 'general will' of Kashmir, fogging the real governance issues in the absence of a vibrant civil society, except for the recently formed human rights organization, Coalition of Civil Society. In fact, until recently, 'civil society' was an unfamiliar word in Kashmir.

Prem Nath Bazaz, the contemporary historian and politician, has given an exhaustive and participatory description of the uncertainty caused by the Kashmir problem in the late 1940s and early 1950s, even though the popular leader of Kashmir, Sheikh Mohammad Abdullah, was still in power. We sense that the 'insecurity at the present and the uncertainty about the future due to the accession dispute'[8] that Bazaz recorded in 1952 have maintained continuity in the history of Kashmir up until our own times, despite more than 70 years having elapsed during which the international situation underwent a sea change, and revolutionary developments occurred in the economic, social, and cultural conditions of the state as well as in those countries which have been stakeholders in the Kashmir dispute.

The situation deteriorated further after 1952. The mass leader and the prime minister of Kashmir, Sheikh Abdullah, was dismissed and arrested on 9 August 1953, which eventually led to the launch of the second struggle, now against the Indian occupation of Kashmir. The otherwise pro-India National Conference

---

[7] In an interview with a Delhi-based journalist in 1968, Sheikh Abdullah said, 'Due to the strained relations between the two countries (India and Pakistan), there is a setback in the trade and commerce of Kashmir. This has an adverse effect on tourism which should bring a lot of revenue to Kashmir. No capitalist wants to make a large investment in Kashmir because of the uncertainties of the future. I have considered the situation seriously as have Bakshi Ghulam Mohammad and Ghulam Sadiq and they agree with me that tension between India and Pakistan must be removed at the earliest.' See Sheikh Abdullah, *The Testament of Sheikh Abdullah* (New Delhi: Palit and Palit, 1974), 84.

[8] Bazaz, *History of Struggle for Freedom in Kashmir*, 437–8.

was converted into the Plebiscite Front[9] under the patronage of the same 'Lion of Kashmir' under whose inspiring leadership Kashmir had struggled against the princely order, and with whose backing India not only claimed the legitimacy of Kashmir's accession but also established its foothold in the state against all conceivable odds.[10] The Plebiscite Front founded in 1955 demanded the final settlement of Kashmir through referendum. The struggle for self-determination continued for around 20 long years, during which, besides the intensification of political instability, the plebiscite sentiment became deeply embedded among the Muslims of the state.[11]

Although the successor of Sheikh Abdullah, Bakhshi Ghulam Mohammad, left no stone unturned to develop Kashmir and improve the condition of each section of society with the full financial support of the central government,[12] the *moe-e-muqaddas*

---

[9] For details about the Plebiscite Front, see *All Jammu and Kashmir Plebiscite Front*, vols I and II (Srinagar: Front Publication, 1964); *Constitution of All Jammu and Kashmir Plebiscite Front*, compiled by Ali Mohammad, General Secretary of the Plebiscite Front (Srinagar, 1965); Also see, All Jammu and Kashmir Plebiscite Front, 'The People's Voice in J&K State', *Proceedings of the Special Convention Held in July 1965 at Shah Masjid, Mujahid Manzil*, Srinagar, 1965. Sumantra Bose, *Transforming India: Challenges to the World's Largest Democracy* (Cambridge: Harvard University Press, 2013), 225–86.

[10] For the role Sheikh Abdullah played in the accession of Kashmir to India and subsequently in driving out the tribals and supporting India's claims both within and outside the country, see P. L. Lakhanpal, *Essential Documents and Notes on Kashmir Dispute* (New Delhi: Transnational Books, 1965); Bazaz, *History of Struggle for Freedom in Kashmir*, Chapters 13, 15, 16 and 23; Abdullah, The Blazing Chinar, Chapters, 38 to 43. S. L. Poplai, *Selected Documents on Asian Affairs: India 1947–50*, vol. I (Bombay: Oxford University Press, 1959); C. Thomas Raju, *Perspectives on Kashmir* (Boulder: Westview Press, 1992); Bose, *Kashmir: Roots of Conflict*, 36–7; Ian Copland, 'The Abdullah Factor: Kashmiri Muslims and the Crisis of 1947', in *The Political Inheritance*, ed. D. A. Low (New York: St. Martin's, 1991).

[11] See Chapter 5, this volume.

[12] For details, see Chapter 3, this volume.

agitation,[13] which completely paralysed life in Kashmir for more than a week (27 December 1963–4 January 1964) made it quite clear that the political mentality of the Muslims had largely remained stable, convincing Nehru that the solution of Kashmir lay in negotiating with Sheikh Abdullah.[14]

Meanwhile, Pakistan, and the people who had been forced to migrate to Pakistan after the large-scale massacres in Jammu, remained keen to take revenge against India. In 1965, Pakistan launched a plan known as Operation Gibraltar[15] to foment a rebellion in Kashmir. Pursuant to this plan, three to five thousand armed men infiltrated the region from Pakistan in mid-July 1965 and sneaked into Rajouri and Poonch. In local parlance, they were called *razakars* (volunteers) who were drawn from their own culture, spoke the same language, and knew the topography of the region. They spread across the rural areas of the modern districts of Poonch and Rajouri. The razakars received huge local support and established control over a vast area from Budhal in the north to Kalakote in the east. 'Hundreds of government employees left their jobs and joined ranks with *Razakars* who established their own "local governments" in the villages.'[16] This situation continued up to September, when the razakars left abruptly. Within a

---

[13] For details on the moe-e-muqaddas agitation, see B. N. Mullik, *My Years with Nehru-Kashmir* (New Delhi: Allied, 1971), 115–66; Qasim, *My Life and Times*, 94–6; Mohd Yousuf Saraf, *Kashmiris Fight for Freedom*, vol. II (Lahore: Ferozsons, 2005), 1235–43; Munshi Mohammad Ishaq, *Nida-i-Haq* (autobiography) (Srinagar: KBF Printers, 2014), 303–14; Butt, *Kashmir in Flames*, 85–98; also see Chapter 4, this volume.

[14] Mullik, *My Years with Nehru-Kashmir*, 172; Sheikh Mohammad Abdullah, *The Blazing Chinar* (English translation of Abdullah's autobiography *Ātash-i-Chinar*, by Mohammad Amin) (Srinagar: Gulshan Books, 2013), 486–7.

[15] For details see Saraf, *Kashmiris Fight for Freedom*, vol. II; Qasim, *My Life and Times*; A. M. Watali, *Kashmir Intifada: A Memoir* (Srinagar: Gulshan Books, 2016); Ishaq, *Nida-i-Haq*; Praveen Swami, *India, Pakistan and the Secret Jihad: The Covert War in Kashmir, 1947–2004* (London: Routledge, 2007); Butt, *Kashmir in Flames*; Zafar Choudhary, *Kashmir Conflict and Muslims of Jammu* (Srinagar: Gulshan Books, 2015).

[16] Choudhary, *Kashmir Conflict and Muslims of Jammu*, 156.

few days after they left, the Indian Army arrived in full force and launched Operation Clearance, which lasted from 17 September till 1 December 1965. The heart-rending scenes during this operation were related to Zafar Choudhary by contemporaries who were alive at the time.[17]

It was around the same time that a few hundred Pakistani guerrillas infiltrated into the valley too.[18] They were called *qabail* or *mujahid* in local parlance. Following their arrival, Kashmir reeled under violent disturbances, which were widely reported in the international media.[19] The rebellions were suppressed with a heavy hand,[20] and the government set ablaze Batamaloo, a quarter in Srinagar, which was considered the hideout of the Pakistani guerrillas.[21] Kashmir was converted into a garrison, the *Economist*, published from London, wrote on 28 October 1967: 'There are large military camps in and around all the main cities. Armoured troops perform regular duties in all urban areas and are invariably called in to deal with the civil disturbances. Kashmir looks like an occupied territory.'[22]

The political instability of Kashmir also found expression in the 'mysterious fires' that began in 1966 and continued up to 1969.[23] The Muslims held the Jan Sangh and the Rashtriya Swayamsevak Sangh (RSS) responsible for these fires, alleging that it was a

[17] Choudhary, *Kashmir Conflict and Muslims of Jammu*, 159–60; also see Chapter 4, this volume.

[18] Saraf, *Kashmiris Fight for Freedom*, vol. II, 1148–54; Qasim, *My Life and Times*, 108; Watali, *Kashmir Intifada*, 294–9; Ishaq, *Nida-i-Haq*, 329–30.

[19] The international media that covered the disturbances included *Frankfurter Allgemeine* (Germany), the *New York Times* (USA), the BBC (London), *Le Monde* (France), and Reuters. See Saraf, *Kashmiris Fight for Freedom*, vol. II, 1263–8.

[20] Saraf, *Kashmiris Fight for Freedom*, vol. II, 1265–8, 1271–3; Butt, *Kashmir in Flames*, 115–16; Ishaq, *Nida-i-Haq*, 330–33.

[21] Saraf, *Kashmiris Fight for Freedom*, vol. II, 1152; Butt, *Kashmir in Flames*, 107–9; Ishaq, *Nida-i-Haq*, 330. Also see David Devadas, *In Search of a Future: The Story of Kashmir* (New Delhi: Penguin/Viking, 2007), 107–8.

[22] *Economist*, 28 October 1967.

[23] Saraf, *Kashmiris Fight for Freedom*, vol. II, 1257–8.

gimmick to force the Muslims to migrate to Pakistan. The whole of Kashmir came under the sway of vigilantism.[24] The situation was so grim that the Indian National Congress had to send a Fire Enquiry Committee to Kashmir to submit a report on the mysterious fires. Also, Kashmiri leaders did some plain speaking in New Delhi, and many leaders visited Kashmir to take stock of the popular reaction.[25]

The unearthing of around 80 groups in the late 1960s that were indulging in espionage and subversive activities,[26] small incidents turning into big issues, like the riots that followed the rumour that British planes had bombed Aqaba to assist the Israelis, with Aqaba becoming confused with Kaaba in Mecca,[27] the marriage of a Hindu girl with a Muslim boy in July 1967 taking the form of Hindu–Muslim conflict,[28] the clashes that followed a football match between Kashmir University and the University of Punjab on 7 October 1967, and the violent incident in the Engineering College, Srinagar, in May 1968 between the students of two communities followed by a series of violent agitations in 1970[29]— these events clearly show that Kashmir was restive and that a serious situation was brewing underground. With the busting of underground organizations, especially Al-Fatah, in 1971, it was established that besides the above-ground secessionist movement, preparations for launching insurgency had also come to fruition.[30]

---

[24] Local informants.

[25] Saraf, *Kashmiris Fight for Freedom*, vol. II, 1257–8.

[26] Watali, *Kashmir Intifada* , 293–4; Swami, *India, Pakistan and the Secret Jihad*, 49–75; also see Chapter 4, this volume.

[27] John Ray, 'Kashmir 1962 to 1986: A Footnote to History', *Asian Affairs* 33, no. 2 (2010), 200.

[28] Qasim, *My Life and* Times, 117–18; Saraf, *Kashmiris Fight for Freedom*, vol. II, 1269–72. For details about this issue and the politics that played out on it, see Khalid Ahmad Bashir, *Kashmir: Exposing the Myth behind the Narratives* (New Delhi: Sage, 2017), 167–224.

[29] Saraf, *Kashmiris Fight for Freedom*, vol. II, 1271.

[30] Saraf, *Kashmiris Fight for Freedom*, vol. II, 1273–4; Qasim, *My Life and Times*, 128–9; also see Swami, *India, Pakistan and the Secret Jihad*, 76–103.

Sheikh Abdullah no doubt came to a compromise with New Delhi in 1975,[31] disbanded the Plebiscite Movement, revived the National Conference, and followed the politics of governance; but the stubborn political mentality created through a sustained movement for 22 long years refused to change. It expressed itself first in the formation of the Muslim United Front (MUF) in 1986 and the mass support it gained against the hegemonic power, the National Conference,[32] and ultimately in the armed struggle in 1989–90, which is now 28 years old.

Contestation on Special Status

The National Conference headed by Sheikh Abdullah favoured accession to India, in the belief that J&K would enjoy an autonomous position while remaining a part of the Indian Union.[33] And this is exactly what was laid down in the Instrument of Accession of 1947, which gave power to the centre to legislate with regard to three subjects only—defence, foreign affairs, and communication. This was a privilege the Sheikh could not expect in the alternative option of acceding to Pakistan. This is besides the fact that he had already allied himself with the Indian National Congress in preference to the Muslim League, and the former, especially Nehru, had been with him through thick and thin ever since 1939. Accordingly, in October 1949, India's Constituent Assembly inserted a special provision—Article 306A—in the Indian constitution which extended autonomy to J&K state. After India became a republic in January 1950, Article 306A became the basis of Article 370 of the Indian constitution, which asserts J&K's autonomy within the Indian Union.[34] Under the provisions of Article 370, India's federal

[31] For details, see G. R. Najar, *Kashmir Accord (1975): A Political Analysis* (Srinagar: Gulshan Books, 1988); Puri, *Jammu and Kashmir*, 184–7; *Statesman* (New Delhi), 25 February 1975.

[32] In the assembly election of 1987, the MUF got a 32 per cent vote share. Bose, *Kashmir: Roots of Conflict*, 49.

[33] Abdullah, *Testament of Sheikh Abdullah*, 40.

[34] For details on Article 370, see A. G. Noorani, *Article 370: A Constitutional History of Jammu and Kashmir* (New Delhi: Oxford University Press, 2011).

government can legislate on the three categories of subjects within its competence only in consultation with the government of J&K state, and on the other subjects in the union list only with the final concurrence of the J&K Assembly. The first Constitution Order of 1950 by the president of India, confirming the terms specified in the Instrument of Accession, also specified, in two different schedules, the powers of the union and the applicability of the constitution. It had already been declared by the Government of India that 'it was the people of the state of Jammu and Kashmir, who would finally determine the constitution of the state and the jurisdiction of the Union of India.'[35] Thus, J&K was the only state that could have its own constituent assembly and its own constitution.

Although the special status of J&K state was guaranteed by the Indian Constitution under Article 370, Nehru's real intention was to ultimately merge Kashmir with the Indian Union and to make J&K like any other state of the country.[36] The insertion of Article 370 as a temporary provision to grant internal autonomy during the transition period was dictated by political expediency. Here we should consider the international dimension of the Kashmir issue, and the compulsions of recruiting the support of the popular and autonomy-minded Sheikh Abdullah to legitimize the Indian claim on Kashmir and to sustain Indian control of the state at a critical phase when the practical support of Kashmiri Muslims was of crucial importance for the success of the Indian claim over Kashmir. However, for Sheikh Abdullah, autonomy for Kashmir was an inalienable

[35] Durga Das Basu, ed., *Sardar Patel's Correspondence 1945–50* (Ahmedabad: Navajivan, 1971), 275–309.

[36] During the Constituent Assembly debate on Article 370, responding to the question of Maulana Hasrat Mohani, who wanted to know why this discrimination was being made between J&K and other states, Shri N. Gopalaswami Ayyangar replied that 'discrimination is due to the special conditions of Kashmir. That particular state is not yet ripe for this kind of integration. It is the hope of everybody here that in due course even Jammu and Kashmir will become ripe for the same sort of integration as has taken place in the case of other states.' Ayyangar's detailed exposition of Article 370 in the Constituent Assembly on 17 October 1949, cited in Noorani, *Article 370*, 66. The subsequent policy followed by Congress governments from Nehru down to Indira Gandhi also clearly shows the undeclared intention of New Delhi vis-à-vis the future of J&K.

right considering the compromises he had made with the history, geography, economy, and culture of Kashmir in preferring India over Pakistan. The special position to Kashmir had been granted by India under the leadership of his long-time, trusted friend and the powerful prime minister of India, Jawaharlal Nehru. Notwithstanding his unflinching faith in the sincerity of Nehru's friendship, however, Sheikh Abdullah started to get suspicious of Nehru's hidden designs from the very birth of Article 370, as he expressed his disapproval of the use of the term 'temporary provision'.[37]

Abdullah's suspicion graduated into a firm belief when, under the pressure of the Praja Parishad's agitation—supported by Hindu nationalists with even some Congress stalwarts sympathizing with it[38]—he was coerced into surrendering more powers to the central government, leading to what is known as the Delhi Agreement in 1952.[39] The intensification of the Praja Parishad's agitation, even after the Delhi Agreement, for the complete abolition of the autonomous status of Kashmir and the veiled support it received from Delhi was something Abdullah could hardly believe, and more than he could swallow. Thus, the friend turned into a 'foe', providing an

---

[37] Showing his disapproval of Article 370 being described as a 'temporary provision', and of the fact that he and his colleagues representing Kashmir were not allowed to speak in the Constituent Assembly, Sheikh Abdullah wrote to Gopalaswami Ayyangar, the main architect of the article, 'As I am genuinely anxious that no unpleasant situation should arise, I would request you to see that if even now something could be done to rectify the position. In case I fail to hear from you within a reasonable time, I regret that no course is left open for us but to tender our resignation from the Constituent Assembly.' Sardar Vallabhbhai Patel, *Sardar Patel's Correspondence, 1945–1950*, vol. I (Ahmedabad: New Light on Kashmir, 1971), 306–10.

[38] For details see Bazaz, *History of Struggle for Freedom in Kashmir*, chapter 27, 'Clash with Dogra Hindus', 479–505, 563; Bhartiya Jana Sangh, *Kashmir Problem and Jammu Satyagraha* (Delhi, 1952); *Hindu* (Madras), 12 February 1952; Puri, *Jammu and Kashmir*, 98,103, 106; Balraj Madhok, *Kashmir, Centre of New Alignment* (New Delhi: Deepak Prakashan, 1963), 38; *Statesman*, 27 July 1952; Jyoti Bhusan Das, *Jammu and Kashmir* (The Hague: Martinus Nijhoff, 1968), 195.

[39] For details on the Delhi Agreement, see Noorani, *Article 370*, chapter 4, 'Negotiating the Delhi Agreement 1952', 123–72.

opportunity to those in Delhi for whom Abdullah's resistance to the complete merger of Kashmir with the Indian Union had become intolerable. After a split was engineered in his cabinet, Abdullah was dethroned, and his deputy Bakhshi Ghulam Mohammad replaced him on 9 August 1953. This event marks another watershed in the political and constitutional history of Kashmir with far-reaching ramifications for governance. The dismissal and arrest of Abdullah alienated the Muslims from India, and the National Conference launched the Plebiscite Movement demanding the settlement of the Kashmir dispute through referendum.

The Praja Parishad's agitation, which was also joined by the Ladakhi Buddhist leadership,[40] becomes meaningful when we consider that J&K state is, in the words of Dr Karan Singh, the living scion of the Dogra dynasty, 'a wholly artificial creation'.[41] It comprises three distinct geographical regions: Kashmir, Jammu, and Ladakh. It did not evolve out of the shared geography, history, ethnicity, language, or religion of the territory and people of the state. Historical memories are at variance with each other along communal lines. At no time in history did these regions and sub-regions constitute one political entity. Very rarely, during the period of mighty rulers, some of them were made tributaries of the Kashmir kingdom, but no sooner did weakness creep into the central authority than their having declared independence.

All in all, the historical, geographical, and cultural factors did not allow the state to move beyond an imposed political arrangement. The gulf was further reinforced by the sectarian and discriminatory

---

[40] Bazaz, *History of Struggle for Freedom in Kashmir*, 473, 500; Korbel, *Danger in Kashmir*, 226–31; Puri, *Jammu and Kashmir*, 96–8. Also see Martijn Van Beek, 'Dangerous Liaisons: Hindu Nationalism and Buddhist Radicalism in Ladakh', in *Religious Radicalism and Security in South Asia*, eds Satu P. Kimaye, Robert G. Wirsing, and Mohan Malik (Asia-Pacific Centre for Security Studies, 2004), 200, http://apcss.org/Publications/Edited%20Volumes/ReligiousRadicalism/ReligiousRadicalismandSecurityinSouthAsia.pdf (accessed 2 October 2017); Martijn Van Beek, 'Beyond Identity Fetishism: "Communal" Conflict in Ladakh and the Limits of Autonomy', *Cultural Anthropology* 15, no. 4 (2000): 534–5.

[41] Nehru, *Nice Guys Finish Second*, 589.

dynastic rule of the Hindu Dogras of Jammu, to whom the state was 'sold' in 1846. As per the British census of 1941, Muslims constituted 77 per cent of the total population of the state.[42] Yet they were treated as second-class citizens, oppressed and discriminated against, leading to their struggle for freedom.[43] Indeed, from 1846 to 1947, the Muslims constituted the alienated mass of the population aspiring to freedom from the exclusivist and oppressive Dogra raj. The turn of the tide came in 1947 when, after the demise of British paramountcy in India, the state of J&K came under the rule of the popular political party the National Conference, headed by the Kashmiri Muslim Sheikh Mohammad Abdullah. Also, as referred to earlier, the state was guaranteed an autonomous position under Article 370. The Dogras of Jammu in particular, and Hindu communal forces in general, could not swallow the demise of the Dogra raj. It was unacceptable to them to see the state coming under a party which was essentially Kashmir based, rooted in the support of the Muslims of the state, and, above all, which had fought against their rule. Thus, they were reconciled neither with the new political establishment and its programme, nor with the special position of Kashmir.[44] This is best captured by the contemporary historian Prem Nath Bazaz:

> In all the measures adopted and the policies pursued by the National
> Conference in power there was divergence in the views of National

---

[42] See https://aknandy.wordpress.com/2016/12/21/in-british-census-of-india-of-1941-jammu-kashmir-registered-as-a-muslim-majority-population-of-77 (accessed 21 February 2018).

[43] For details see Bazaz, *History of Struggle for Freedom in Kashmir*; Mohd Yousuf Saraf, *Kashmiris Fight for Freedom*, vol. I (Lahore, 1977); Saraf, *Kashmiris Fight for Freedom*, vol. II; Mirdu Rai, *Hindu Rulers Muslim Subjects* (New Delhi: Permanent Black, 2004); Walter Lawrence, *The Valley of Kashmir* (Srinagar: Gulshan Books, 2002 [1895]); Knight, *Where Three Empires Meet*; Robert Thorp, *Cashmere Misgovernment* (London: Longmans, Green & Co., 1870).

[44] Even as late as the 1970s, Karan Singh termed the post-1947 rule in J&K as the reverse domination of Kashmir over Jammu. Jawaid Alam, ed., *Kashmir and Beyond (1966–84): Selected Correspondence between Indira Gandhi and Karan Singh* (New Delhi: Penguin/Viking, 2011), 61.

Conference and those of Jammu Hindu leaders. Be it the question of the future of the royal family, centre-state relations, agrarian reforms, scaling down the debts, the national language, the national flag, the text books in the schools or 'anything under the sun'.[45]

Jammu had much earlier become a hotbed of communal forces—before the formation of the Praja Parishad under the banner of the RSS on 25 November 1947.[46] It was in the grip of such communal forces that, apart from the RSS headquarters in Nagpur, Jammu became the only town in India where the RSS openly revelled in the assassination of Gandhiji, distributing sweets among the people.[47] The city observed a shutdown on 5 February 1948 to protest the arrest of Veer Savarkar, the Hindu Mahasabha leader, on charges of conspiracy in Mahatma Gandhi's assassination case.[48] Leaving the details for the following chapter, it will suffice to say here that early in 1948, the Jammu Hindus (who were in a preponderant majority in the three districts of Kathua, Udhampur, and Jammu) rallied under the banner of the Praja Parishad and launched three aggressive agitations from 1949 to 1953 demanding the abolition of Article 370 and the complete merger of J&K with India, or at least the merger of Jammu and Ladakh with the Indian Union. Though the demand for abolition of the special position of Kashmir was largely fulfilled by the central government by hollowing out Article 370 with the active support of installed governments in Kashmir, the Hindu nationalist forces of Jammu continued their agitational politics against Article 370 and 'regional imbalances'. The struggle against 'Kashmiri domination' gained momentum in 1990 after the outbreak of militancy in the valley.

It may, however, be in place to mention that like any other society, Jammu society is not homogeneous. There are sharp ideological divisions among the Hindus of Jammu in general and the political elite in particular. It is a fact that the communal ideology

[45] Bazaz, *History of Struggle for Freedom in Kashmir*, 485.

[46] The Praja Parishad leadership comprised RSS workers, with Hari Wazir as president; Balraj Madhok as general secretary; and Prem Nath Dogra, Rup Lal Nanda, and others as office bearers.

[47] Abdullah, *The Blazing Chinar*, 319.

[48] Abdullah, *The Blazing Chinar*, 319.

was more pronounced in Jammu on the eve of independence. It is also true that the Praja Parishad mobilized the Jammu Hindus based on communal and anti-Kashmiri rhetoric. And the same was done by its successors—the state unit of the Jana Sangh from the mid-1960s, and the state unit of the Bharatiya Janata Party (BJP) from the 1980s. Also, the political space for secular-progressive parties was very peripheral in Jammu in 1947.[49] However, the assumption of power by secular parties continuously from 1947 to 2014 expanded their social base significantly and threw up a considerable secular elite affiliated with different national and regional parties. However, all the parties of Jammu, irrespective of their ideological differences, converged on the belief that the region did not have political parity with Kashmir, resulting in their demand for restructuring the power structure of the state.[50] While the secular Hindus demanded regional autonomy, communal forces led by the Praja Parishad in the 1950s and the BJP currently have demanded union territory status for Ladakh and statehood for Jammu.[51] Till recently, communal forces in Jammu complained vociferously of a discriminatory attitude against Jammu, a stance that has no appeal now in the face of the regional balance shifting in favour of Jammu.[52] Thus, the RSS and BJP have raised the pitch for the abrogation of Article 370 to stay relevant.

[49] Rekha Chowdhary, *Jammu and Kashmir: Politics of Identity and Separatism* (New Delhi: Routledge, 2016), 178.

[50] Rekha Chowdhary, *Jammu and Kashmir: Politics of Identity and Separatism* (New Delhi: Routledge, 2016), 159–96.

[51] For the regional autonomy argument, see Balraj Puri, *Jammu and Kashmir: Regional Autonomy (a Report)* (Jammu, 1999); and for the demand of Hindu right see, two articles in the *Statesman* by Hari Om: 'Jammu Region I: The Story of Neglect', and 'Jammu Region II: Not Too Late for Accession', 23 and 24 September 1996, respectively; Reeta Chowdhary Tremblay, 'Kashmir's Secessionist Movement Resurfaces: Ethnic Identity, Community Competition, and the State', *Asian Survey* 49, no. 6 (2009): 153–67. For a critical view on trifurcation ideas, see A. G. Noorani, 'In Pursuit of Trifurcation', *Frontline* 18, no. 8 (April 2001).

[52] As per the Finance Commission report, while the Kashmir region grew from 0.3481 (index value) in 1980–1 to 0.4349 in 2006–7, that is by 24.94 per cent (based on index value), the Jammu region demonstrated

Muslims constitute a significant population in Jammu province. Although reduced to a minority after the mass massacres in 1947 and forced migration to Pakistan, Muslims are still in a majority in 6 out of 10 districts of Jammu province.[53] Against the backdrop of communal politics cultivated in Jammu prior to and after independence, coupled with fears of majoritarianism, the Jammu Muslims followed a separate politics—separate not only from Hindu nationalist organizations, but also from secular Hindu discourse. Interestingly, when in April 1953 Sheikh Abdullah made an announcement granting autonomy to different cultural units in the state, the Muslim majority of Doda raised a hue and cry against the inclusion of the district in the Hindu-dominated Jammu province. 'In a convention of the Nationalist workers of the district', says P. N. Bazaz, 'a curious and significant development was witnessed. While the Hindu speakers supported the official move the Muslim leaders hotly contested and opposed it.'[54] A similar situation has continued unabated till date. While the Jammu-centric Hindu secular scholars framed an architecture of regional autonomy assigning a superordinate position to Jammu,[55] the secular Muslim elite divided Jammu into three equally powerful autonomous regions to allay the fears and suspicions of the Muslims of the region.[56]

---

growth from 0.3039 index value to 0.4333, that is, 42.58 per cent. As per the report during the period 1980–1—2006–7, there had been considerable improvement in the share of the Jammu region in development, contrarily the improvement in the share of Kashmir has declined substantially. Government of J&K, *Report of the J&K State Finance Commission*, November 2010, vol. I, 114.

[53] The six Muslim-majority districts of Jammu division as per the 2011 census are: Kishtiwar (57.75 per cent), Poonch (90.44 per cent), Rajouri (62.71 per cent), Ramban (70.68 per cent), Reasi (49.66 per cent), and Doda (60.71 per cent). https://www.census2011.co.in/data/religion/state/1-jammu-and-kashmir.html.

[54] Bazaz, *History of Struggle for Freedom in Kashmir*, 500.

[55] See Puri, *Jammu and Kashmir: Regional Autonomy*.

[56] See the National Conference's *Regional Autonomy Committee Report*, 13 April 1999.

Culturally, the Ladakh region is divided into two parts, Leh and Kargil—the former predominantly Buddhist and the latter over-whelmingly Muslim. Although both districts presented a dismal economic condition at the time of independence owing to rack renting under the Dogra rulers,[57] the rupture of commercial links with Central Asia,[58] educational backwardness, and overall neglect of the region vis-à-vis the Kashmir Valley and Jammu,[59] religion hijacked the enormous prospects of 'economic nationalism', espe-cially in Buddhist-dominated Leh where the post-1947 agrarian reforms did not have any effect because of the resistance of the lamas (the Buddhist clergy), who alone constituted the landlord class, besides being the only beneficiaries of the monasteries—the biggest land grantees of the time.[60] Given the prospects of fighting a struggle informed by economic and 'development'-related issues, the reasons why identity politics occupied almost the whole politi-cal space of Ladakh are both endogenous and exogenous. Internally the movement was, during its formative period, led by lamas and

[57] See J. N. Ganhar and P. N. Ganhar, *Buddhism in Kashmir and Ladakh* (New Delhi: Ganhar and Ganhar, 1956); Van Beek, 'Beyond Identity Fetishism', 534–5; Shridhar Kaul and H. N. Kaul, *Ladakh through the Ages: Towards New Identity*, 3rd edn (New Delhi: Indus, 2004), 107–17.

[58] For details see Janet Rizvi, *Ladakh: Crossroads of High Asia* (New Delhi: Oxford University Press, 1996), and Janet Rizvi, *Trans-Himalayan Caravans: Merchant Princes and Peasant Traders in Ladakh* (New Delhi: Oxford University Press, 1999).

[59] Van Beek, 'Beyond Identity Fetishism', 532.

[60] 'Isolated from the outside world, the Ladakhis lived a life of unrelieved toil, being dependent upon the Lamas. Land was the property of the Buddhist Church, rented to peasants under conditions of inhuman exploitation. It was the programme of land reforms of the Srinagar government which set the Lamas on the war-path. The Kashmir Government dared not touch the property of the monasteries. Land reforms made little progress in Ladakh. Though figures vary but in most estimates the land redistribution affected no more than 10,000 acres.' Korbel, *Danger in Kashmir*, 228–9; Kaul and Kaul, *Ladakh through the Ages*, 187–9; Ganhar and Ganhar, *Buddhism in Kashmir and Ladakh*, 196; Van Beek, 'Beyond Identity Fetishism', 533–4; Das, *Jammu and Kashmir*, 203–4.

religious fanatics with vested interests; and externally it was provoked by communal forces.

Indeed, the initial mobilization of the Buddhists was the handiwork of a small group of Kashmir-based neo-Buddhist Kashmiri Pandits, who formed the Kashmir Raj Bodhi Maha Sabha (KRBMS) in 1932. On the authority of Ladakh's foremost religious leader, Shushok Stagtsang Rapa of Hemis monastery, the KRBMS submitted a memorandum to the Glancy Commission.[61] Interestingly, the memorandum did not touch on the economic and infrastructural backwardness of the region or the plight of the Ladakhi masses owing to the oppressive policies of the state and the lamas; it only focused on educational development and the eradication of social evils.[62] The first local Buddhist organization, the Ladakh Buddhist Education Society, was formed in 1933; and this too was formed at the instance and with the active involvement of the Arya Samaji writer-activist Rahula Sankrtyayana, who had raised the bogey of 'the dangers of growing number of Muslims and the low birth rate among Buddhists due to monasticism and Polyandry' with prominent Buddhist leaders.[63] The Ladakh Buddhist Education Society was replaced by the Young Men's Buddhist Association (YMBA), 'which again had strong involvement of Kashmir Pandits, at least in the years until independence'.[64] Both the YMBA and KRBMS continued to raise the issue of Muslim progeny outnumbering the Buddhists in the region. 'From the start, then', says Van Beek, 'modern Ladakhi Buddhist activism was strongly informed by outsiders and their understanding of the Indian political system, as well as what they thought was good for Ladakhi Buddhists.'[65]

Already shaped on communal lines, the identity politics of Ladakhi Buddhists was further nourished by the Partition of India, the accession dispute, and the Praja Parishad agitation in Jammu. A memorandum submitted by Kalan Tsewang Rigzin, president of the Ladakh Buddhist Association (LBA) (as the YMBA had renamed

---

[61] Kaul and Kaul, *Ladakh through the Ages*, 277–86.

[62] Kaul and Kaul, *Ladakh through the Ages*, 277–86.

[63] Van Beek, 'Dangerous Liaisons', 200.

[64] Van Beek, 'Dangerous Liaisons', 201.

[65] Van Beek, 'Dangerous Liaisons', 201.

itself), in July 1949 to Prime Minister Nehru, asserted Ladakh's right to self-determination and the need for its secession from J&K state. It was argued that Ladakh was 'a separate nation' by all the tests—race, language, and culture—determining a nationality. It pleaded for direct central rule over Ladakh, or its amalgamation with the Hindu-majority parts of Jammu to form a separate province.[66] In the late 1960s, the demand for the introduction of central administration along the lines of the system applied to the North-East Frontier Agency was consistently made.[67] And in the 1969 agitation, the leadership played the communal card that eroded the possibilities of a united Ladakhi movement.[68] Following the policy of the Praja Parishad, the Jan Sangh and other Hindu nationalist forces consistently supported the demands of the Buddhists of Leh, including during the 1967, 1969, and 1981 agitations.[69] In August 1989, the LBA launched a violent agitation for union territory status and mobilized the Buddhists on the basis of the Hindu–Muslim divide.[70] The Buddhist agitators called for a boycott of Muslims and called on the local population to 'free Ladakh from Kashmir'. The social boycott of the Muslims continued for the next three years.[71]

In their bid to win support at the national level and bring pressure to bear upon the central government, the LBA revived with fanfare its liaison with its time-tested supporters—the Hindu nationalist forces—from 1989. In the words of Van Beek:

[I]t was not until the agitations of the late 1980's that the LBA was in more regular contact with national outfits such as the Vishwa

---

[66] 'Memorandum submitted by Shri Cheewang Rigzin, President Buddhist Association, Ladakh to Prime Minister of India on behalf of the people of Ladakh', cited in Navnita Chadha Behera, *State, Identity and Violence: Jammu, Kashmir and Ladakh* (New Delhi: Manohar, 2000), 311–14.

[67] Van Beek, 'Dangerous Liaisons', 202.

[68] Van Beek, 'Dangerous Liaisons', 203.

[69] Van Beek, 'Dangerous Liaisons', 203.

[70] Van Beek, 'Dangerous Liaisons',196.

[71] Navnita Chadha Behera, *Demystifying Kashmir* (New Delhi: Pearson/Longman, 2007), 115–16.

Hindu Parishad and Bharatiya Janata Party. This was partly the result of a conscious choice by a new generation of LBA leaders to play the communal card in the struggle for Union Territory for Ladakh. The new leadership had strong roots in the Congress Party and had become convinced that communalism was necessary to get the attention of the national government. Having watched agitation after agitation fail because of the cooptation of leaders and the splitting of the Ladakhi front, these younger leaders, most of whom had been educated at prestigious institutions in India, decided a more forceful approach was necessary to create a sustained movement. A more active relationship with nationalist forces was an element of this new approach.[72]

The Hindu nationalist forces obliged by actively supporting the demands of the LBA. This is clear from the repeated letters written by the BJP to the then prime minister Narasimha Rao, the questions asked by BJP members in Parliament, the national convention on Ladakh organized in March 1990, the participation of an LBA delegation in the BJP convention at Jammu in which Kashmiri Pandits also expressed their 'unstinted and unqualified support for the cause of Jammu and Ladakh'.[73] The then president of LBA was quoted as saying, 'for forty-three years the people of Jammu and Ladakh have been denied their constitutional rights. We have been struggling for justice, but separately. Let us unite, for our sufferings are common.'[74]

Given the strategically sensitive nature of Ladakh, the central government took advantage of the weak government in the state, and in October 1989 pacified the LBA by offering Autonomous Hill Council status, along the lines of the Darjeeling Gorkha Hill Council. The council was finally established in May 1995 during governor's rule after seeking a joint demand by all communities, though this was much to the dislike of the mainstream regional political party the National Conference.

The Muslims of Ladakh, who are mainly concentrated in Kargil district with a thin proportion in Leh, kept themselves aloof from

---

[72] Van Beek, 'Dangerous Liaisons', 204.

[73] *Daily Excelsior*, 2 September 1990.

[74] Van Beek, 'Dangerous Liaisons', 205.

the separatist politics of the LBA. The reasons are obvious. The separatist politics of Ladakhi Buddhists was essentially based on a distinct Buddhist culture, which, according to them, had greater affinity with Hinduism than Islam.[75] And the mobilization strategy they deployed involved the intensification of communal polarization. Just as the Buddhists considered themselves different from the Muslims, and as unsafe in a Muslim-dominated state, the Ladakh Muslims also refused to be a part of any political and administrative arrangement which would subordinate them to a Buddhist majority. In any case, in terms of the larger political cause they preferred to be a part of Kashmiri Muslim discourse without, however, radicalizing their politics; in matters of day-to-day life they affiliated themselves either with the Indian National Congress or the National Conference. When in the early 1950s Kushak Bakula demanded full autonomy for Ladakh, the Muslim leaders of Kargil did not support him. They preferred to be ruled by Srinagar and fiercely opposed the Buddhist monk.[76] Interestingly, unlike the Buddhists, 'these protestations of Ladakhi Muslims [were] never heard outside the State.'[77] Kargil Muslims continue to maintain this position.[78]

The Kashmir Valley is also far from being ideologically homogeneous. In 1947, there were three ideological groups—the secular nationalists represented by the National Conference, Muslim nationalists represented by the Muslim Conference, socialists represented by the Socialist Party and the Kisan Mazdoor Conference—and the status quoists who were mainly the Pandits and a few Muslim *jagirdars* and high officials. The National Conference represented the mass of the Muslim population who were greatly relieved by the struggle launched by Sheikh Abdullah

---

[75] See 'Memorandum submitted by Shri Cheewang Rigzin, President Buddhist Association, Ladakh to Prime Minister of India on behalf of the people of Ladakh', reproduced in Behera, *State, Identity and Violence*, 311–14.

[76] Bazaz, *History of Struggle for Freedom in Kashmir*, 474.

[77] Bazaz, *History of Struggle for Freedom in Kashmir*, 474.

[78] Bose, *Kashmir: Roots of Conflict*, 192.

and the reforms carried out by the National Conference government. So they switched sides along with their leader, at least as
long as there was mass illiteracy among them. From the 1980s,
the mass base of the National Conference was gradually eroded by
alternative discourses. The Muslim Conference consisted of followers of the two-nation theory, and certainly their number was
much lower than that of secular nationalists. They were largely
concentrated in the sensitive area of Srinagar. The Shias, as a community, were followers of Mohammad Ali Jinnah and therefore
supported Kashmir's merger with Pakistan.[79] The socialists and the
Kisan Mazdoor Conference consisted of some Hindu and Muslim
intellectuals who were dead against the National Conference,
the Indian National Congress, and accession to India. They were
ardent advocates of holding a plebiscite in Kashmir.[80] All these
parties which stood against the National Conference were ruthlessly suppressed by the Conference, and some of them were also
pushed to the other side of the border, and some left of their own
accord.[81] While some of these families were subsequently appropriated by the Congress and the National Conference, a section led
by the Mirwaiz family stuck to its stand. Though the pressures of
time diluted its rigidity, this section has been more receptive to
the anti-India creed.

The 'disloyal' voice was considerably augmented after the dismissal of Abdullah in 1953 and the formation of the Plebiscite
Front. The National Conference, interestingly enough, joined the
same plank which had been ruthlessly suppressed by it since the
heyday of the freedom struggle, and more so after 1947. While the
Muslims in general followed the reincarnated National Conference
into its new avatar, namely, the Plebiscite Movement, the chief
minister G. M. Sadiq—one of the leaders of the breakaway group
that deserted Sheikh Abdullah and occupied power in 1953—facilitated the establishment of the state unit of the Indian National
Congress in 1965. With power and patronage in the hands of the

---

[79] Ishaq, *Nida-i-Haq*, 195–8.
[80] Bazaz, *History of Struggle for Freedom in Kashmir*, 400.
[81] Bazaz, *History of Struggle for Freedom in Kashmir*, 401.

Congress for more than two decades, the party was able to win over a small section of the Valley to its fold and a sizeable number in the Jammu and Ladakh regions. Abdullah revised his stand again in 1975 and revived the erstwhile National Conference. However, the environment created by his inspiring leadership and cadre-based party proved more powerful than the 'Lion of Kashmir' himself, especially in the changing scenario—the onset of a new generation with a modern exposure and outlook, facing new issues and entertaining new aspirations. The religious groups Jami'at-i-Tabligul Islam and Ahli Hadis were supporters of the National Conference. The only religious party which took part in electoral politics and posed a challenge to mainstream parties including the National Conference was the Jamat-i-Islami. Jamat's participation in electoral politics ironically became the cause of the direct and 'symbolic coercion' to which they were subjected by the ruling parties,[82] which also accounted for their decision to join the armed struggle later.

The Kashmiri Pandits constitute a small percentage of the Kashmiri population. However, given their overwhelming influence and role in post-1947 Kashmir,[83] and considering that the

---

[82] For details, see Saif-ud-Din Qari, *Vadi Ya Purkhar* (Srinagar: Chinar, n.d.).

[83] As a classic example of the patronage the Kashmiri Pandits enjoyed in the corridors of power in New Delhi, the principal of Tyndale Biscoe school, John Ray, states, 'In the school we had come through a serious strike in 1973.... one of the main instigators of difficulty was a particular Pandit teacher. Things settled down and then in 1975 Mrs Gandhi declared the "Emergency". The teacher disappeared, wanted by the police, and was finally dismissed as being absent without leave. After the Emergency I was asked to take him back by the State Education Department. Then came a more ominous enquiry from the Union Home Ministry. I was getting concerned when one evening the phone rang. "Sheikh here. Mr. Ray, you have dismissed Mr. Madan. Can you tell me why?" I told him. He thanked me and I heard no more.' Ray, 'Kashmir 1962 to 1986', 202–3. Another example could be what Sheikh Abdullah himself said about his dismissal in 1953. In an interview he remarked, 'Our plan affected both Hindu and Muslim capitalists and *zamindars* (landlords) equally, but the Hindus had

level of their satisfaction has always been an important variable in measuring governance in Kashmir, a special mention of the issues and aspirations of the community is in order. The Pandit community of Kashmir has always been a literate community with stakes in government service. Given their expertise in manning the revenue administration, the change of governments had little effect on the position of the community. They held influential positions even during the reign of the Afghans,[84] though the latter are portrayed as brutes in the legendary histories written or remembered by the Pandits.[85] However, with the establishment of Sikh rule, the community came to be privileged over the Muslims, as is clear from the overwhelming number of land grants appropriated by Pandit community members.[86] This privileged position was further augmented by Dogra rulers who conferred *jagirs*, *chaks*, and *maufi* grants mainly upon the Pandits as per the canons of medieval polity.[87] True to their tradition, the Pandits were the first to learn English to meet the demands of congruence and compatibility to maintain their hold over the administration in the colonial period. And it was also for the same purpose that they launched the Kashmir for Kashmiris movement,[88] resulting in the introduction of the provision of 'hereditary state subject' on 31 January 1927. According to this provision, 'all persons born and residing in the state before the commencement of the reign of Maharaja Gulab Singh Bahadur and also persons who settled therein before the commencement of Samvat 1942 (1885 AD) and have since been permanently residing in the country are hereditary subjects

---

direct lines to Delhi.' Y. D. Gundevia, 'On Sheikh Abdullah', in Sheikh Abdullah, *The Testament of Sheikh Abdullah* (New Delhi: Palit and Palit, 1974), 117.

[84] R. K. Parumu, *A History of Muslim Rule in Kashmir* (New Delhi: People's Publishing House, 1969), 352.

[85] Lawrence, *Valley of Kashmir*, 190–8.

[86] See *Dastur-ul-Amali Kashmir* (MS), composed during the Sikh period (1819–46).

[87] For details see Rai, *Hindu Rulers Muslim Subjects*.

[88] Bazaz, *History of Struggle for Freedom in Kashmir*, 122–3.

of the state.'[89] It was also for reasons of maintaining their monopoly over the service sector, and thereby over land revenue grants and assignments, that they launched the Roti agitation in 1932 against the Glancy Commission Report,[90] which recommended special treatment to Muslims to enable them to enter the service sector.[91]

Indeed, the Pandit community was a beneficiary of the feudal and sectarian rule of the princely order. It is, therefore, no wonder that they constituted its dedicated support structure, and decried, even worked against, the freedom struggle along with the Hindu Dogras and communal organizations of the subcontinent.[92] For the same reason they, along with the Hindu Dogras, supported the last maharaja in his plans to opt for an independent Kashmir rather than joining India or Pakistan. However, when subsequent developments put the plan in jeopardy and the maharaja considered acceding to India a better option, the Kashmiri Pandits, like the Hindu Dogras, also followed suit.[93]

The post-1947 history of the Kashmiri Pandits also resembles that of the Hindu Dogras. Barring a handful who, like a similar number of Dogra youngsters, saw the writing on the wall, the Kashmiri Pandits were unhappy with the National Conference, as were the Hindu Dogras. After all, the National Conference overthrew their benefactor, and, more than that, gave their land to the tillers who were mostly Muslims, scaled down debts, affecting the Hindu *sahukar*s, and appointed Muslims in large numbers and gave them prize posts even if they were less qualified.[94] Following the

---

[89] For details see Ghulam Shah and G. N. Reshi, *State Subjectship in Jammu and Kashmir* (Srinagar: Jupitor, 1988); Bazaz, *History of Struggle for Freedom in Kashmir*, 123.

[90] Saraf, *Kashmiris Fight for Freedom* (Lahore: Ferozsons), vol. I, 441.

[91] For details see Government of J&K, *Report of the Kashmir Constitutional Reforms Conference* (henceforth Glancy Commission Report) (Jammu: Ranbir Government Press, 1932).

[92] Bazaz, *History of Struggle for Freedom in Kashmir*, 151; Abdullah, *The Blazing Chinar*, 179.

[93] Bazaz, *History of Struggle for Freedom in Kashmir*, 484–5.

[94] For details, see Chapter 2, this volume.

regime change and the uncertain political conditions in the state, a large number of Kashmiri Pandits left and settled in the plains and worked against the National Conference.[95] Although the Kashmiri Pandits were happy with the overthrow of Abdullah, and they extended their fullest support to Bakhshi for being the saviour of Indian interests in Kashmir,[96] they turned against him too for being favourably inclined to 'appoint Muslims in proportionate to their population ratio'.[97] The Kashmiri Pandits were more assertive against Bakhshi's successor Sadiq who, true to his socialist ideology, substantially increased the reservation quota for backward classes. The agitation which the community launched against the marriage of a Pandit girl with a Muslim boy was as much an expression of grievance against Sadiq's policy of pushing the subalterns onto the path of progress at the cost of the hereditary official class, as was it a demonstration of the overriding concern of the community to save their identity, which they felt was threatened in the face of the growing prosperity of the dominant majority—Muslims.[98]

Significantly, at the social and interpersonal levels, the relations between the two communities were proverbially cordial, and this was not shaken even during the most critical and testing times, like 1947, 1953, the two long decades of the Plebiscite Movement, or during the widely publicized communal clashes in India. Indeed, the two communities were closely wedded to each other thanks to the dense network of interdependence,[99] which acted as a bulwark

[95] Bazaz, *History of Struggle for Freedom in Kashmir*, 475–6.

[96] Saraf, *Kashmiris Fight for Freedom*, vol. II, 1223.

[97] Saraf, *Kashmiris Fight for Freedom*, vol. II, 1223; Butt, *Kashmir in Flames*, 77; Koul, M.L., *Kashmir, Past and Present: Unraveling the Mystique*, Accessed from http://www.koausa.org/pastpresent/chapter11.html , Accessed on 9 November, 2017.

[98] Alam, *Kashmir and Beyond*, 61; Qasim, *My Life and Times*, 117; Bashir, *Kashmir*, 212–13.

[99] For details, see Ashiq Hussain Dar, 'Inter-community Relations in Kashmir (Sixteenth to Twelfth Century)', PhD thesis, Department of History, University of Kashmir, 2015. Also see T. N. Madan, 'Ritual and Religion among Muslims in India', in *Religious Ideology and Social Structure: The Muslims and Hindus of Kashmir*, ed. Imtiyaz Ahmad (New Delhi: Manohar, 1981).

against the tremors coming from within and without. The secular politics championed by the National Conference also contributed to the tolerant ethos of Kashmir.[100] However, at the level of ideology the Pandit community refused to change with the changing circumstances. It did not become a part of Kashmir's collective consciousness. And thus it broke with the context. The prominent Kashmiri Pandit polymath and freedom fighter, Prem Nath Bazaz, who was a rebel in this regard, is regretful of this attitude of his community. He not only frankly captures the mentality of his fellow members, but also warns them of the bitter consequences of the collective decision of the community not to rise above their parochial interests. His warnings in the early 1950s proved prophetic:

> But it must be frankly recognized that the majority of the Pandits have kept themselves aloof and refused to follow the democratic path.... It is time that they realize the stern reality that if the dispute is not settled amicably and peacefully between India and Pakistan the internal conditions in the State can in no way improve; indeed, they will deteriorate and someday unawares something might happen which will jeopardize the life of the community. It is, therefore, wise and sagacious to take the time by the forelock and prepare the community psychologically and otherwise, for the inevitable.[101]

For obvious reasons, the stability of any regime in Kashmir was in no small part contingent on the support of the Pandit community. After all, the Pandits' presence symbolized the presence of India in Kashmir; it gave Kashmir a touch of plurality and thus contributed to the legitimacy of Indian control over Kashmir. The presence of a Hindu minority in Muslim-dominated Kashmir was a factor contributing to the much-needed social harmony in India. Moreover, the community constituted a support structure of unflinching loyalists of the centre in a troubled state; they represented Kashmir's erstwhile position as a centre of Hinduism and *Saradadesa* (land of the goddess of learning); and, above all, they had considerable

---

[100] Even during the stormy period of the Plebiscite Movement, one of the main slogans of the Front was 'Hindu, Muslim, Sikh Itihad—Zindabad' (Long Live Hindu-Muslim-Sikh Unity).

[101] Bazaz, *History of Struggle for Freedom in Kashmir*, 469.

influence in the corridors of power at the centre and easy access to media desks.[102]

Before concluding the discussion on this sub-theme, it is necessary to mention that, alongside these major fault lines which gave birth to different religious identities, none of these communities has been an undifferentiated category of people. Indeed, there are fragments within the fragment based on ethnicity, race, language, belief, class, status, and ideology. This social reality is important to bear in mind, as it has a tremendous functional value in realpolitik as well as in measuring governance. The Hindus of Jammu consist of the Dogras (the largest ethnic group in Jammu at 67 per cent population), Rajputs, Punjabis, Patwaris, Harijans, and so on. The Muslims of Jammu province comprise Paharis, Gujjars, and Bakarwals. In some pockets of Jammu, there are both Sunnis and Shias. The Muslims of Kashmir comprise two sects, the Sunnis and the Shias, the former constituting the majority and the latter with a population of around 13 per cent. Ethnically speaking, the Kashmir Muslims are divided among Kashmiris, Gujjars, Bakarwals, and Paharis. Leh Buddhists (77.30 per cent) consist of Bodh, Gara, Mon, Beda, Changpa, Brokpa, and other ethnic groups. Kargil has a predominantly Shia population and a small minority of Sunnis (around 7 per cent) and Buddhists (15 per cent). However, all these fragments share one common feature: they are class-ridden and ideologically divided, though the latter kind of division is more predominantly found among the comparatively large and modernized communities. Yet, when larger community interests are threatened, the otherwise divided members often join hands forgetting their differences. This is best captured by a Kashmiri maxim: *shal shal beun beun, tongi wizi quney* (Though jackals live separately from one another, they howl together).

Financial Crisis

It goes without saying that good governance is *inter alia* dependent on the sound financial conditions of the people and the state.

---

[102] Tavleen Singh, 'Most national newspapers tended to employ Hindu rather than Muslim stringers to cover the valley', *Kashmir: A Tragedy of Errors* (New Delhi: Penguin, 1996), 31.

The question of proper management and judicious use of resources arises only after the resources are made available. Therefore, for measuring governance and making an appraisal of the performance of post-1947 governments in Kashmir, it is important to know the economic condition of the state in 1947, the financial constraints under which successive governments had to function, and the forces and factors that created the resource crunch. Consideration of the level of backwardness of Kashmir on the eve of independence would make the benevolent nature of the state and the resultant state-led development policy intelligible. The 'first-aid' solution of meeting economic crisis via grants, loans, and packages has no doubt helped the pain subside, but without curing the disease, which has become further aggravated with each passing day, and with the ever-increasing gap between supply and demand

It is surprising to hear a socialist intellectual, historian, and freedom fighter such as Prem Nath Bazaz rueing the 'great misfortune of the Kashmiris that the British ... did not take the Valley under their own control' and instead handed it over to the medieval-minded Dogras who 'brought nothing but misery'.[103] This bemoaning is not, however, absurd. To be sure, the Dogras did not want to modernize Kashmir, and if some improvements took place it was because of British intervention from the 1880s.[104] The princely order was mainly interested in draining wealth out of the outlying parts of the state to transfer the same to the maharaja's hometown—Jammu—and sharing a part of it with co-religionists and collaborators. The pauperized and tyrannized people were ruled by the maharajas through a class of jagirdars, *chakdar*s, m'afidars, intermediaries, corrupt officials, and notorious police officers all connected with one another to form, in the words of Walter Lawrence, 'a powerful ring of iron, inside which the village

---

[103] Bazaz, *History of Struggle for Freedom in Kashmir*, 108.

[104] See Bashir Ahmad Sheikh, 'Kashmir's Response to European Technology', MPhil dissertation, Department of History, University of Kashmir, 1984; Shirin Bakshi, 'Social Change in Kashmir with Special Reference to European Impact (1846–1947)', PhD thesis, Department of History, University of Kashmir, 1992; Rai, *Hindu Rulers Muslim Subjects*, 136–44.

tax payer lay fascinated, and if he were wise, silent'.[105] Since the maharaja and his collaborators were interested only in extortions, infrastructural development was the lowest priority of the government, save establishing a few departments, hospitals, schools, and cart roads at the behest of the British colonial power, and subsequently under pressure of the freedom movement.[106] The small class of landlords-cum-bureaucrats accumulated their wealth in gold and silver, or wasted it on luxurious lifestyles. They were too 'innocent' to invest their wealth in any productive sector. About the abysmal economic condition of Kashmir on the eve of 1947, the Godbole Report states:

> In 1947 Jammu and Kashmir was one of the least developed states, which was reflected in the abysmal mass poverty, deprivation, hunger, disease and ignorance. In 1950 the state had a per capita income of Rs. 208 (at 1960–61 prices). The rate of literacy was 5%. Agriculture which was the dominant sector was stagnant. Industrial development was almost negligible. Infrastructure bottlenecks crippled the state economy.[107]

What is worst, colonialism did enduring and irreparable damage to Kashmir by snapping one of the lifelines of the state's economy. Until the beginning of the twentieth century, J&K was one of the major entrepôts of trade between India and Central Asia, and thence with China and Russia.[108] Kashmiri traders were seen playing a

---

[105] Lawrence, *Valley of Kashmir*, 401.

[106] Sheikh, Bashir, *Kashmir's Response to European Technology*, Bakshi, Shirin. *Social Change in Kashmir with Special Reference to European Impact* (1846–1947), Rai, *Hindu Rulers Muslim Subjects*.

[107] Government of J&K, *Report of the Committee on Economic Reforms for Jammu and Kashmir* (henceforth Godbole Report) (Srinagar/ Jammu: General Administration Department, August 1998), 3.

[108] William Moorcroft and George Trebeck, *Travels in Ladakh and Kashmir*, vol. I (Calcutta: Asiatic Society, 1841), 207, 306, 319–20, 323–6, 329–30, 385, 388–92; Hasan Shah Khohami, *Tarikh-i-Hasan*, vol. I, trans. Sharif Hussain Qasim (Srinagar: Ali Mohammad and Sons, 2013), 487–90; Knight, *Where Three Empires Meet*; K. Warikoo, *Central Asia and Kashmir: A Study in the Context of Anglo-Russian Rivalry*

prominent role in this trade, with their business houses located as far as in Russia.[109] Srinagar and Leh were important meeting grounds for traders from different empires of the time.[110] This thriving business, however, came to an abrupt end at the beginning of the twentieth century following the Russian advancement towards Central Asia, and subsequently with Chinese occupation of eastern Turkistan and Tibet.[111] The borders continue to be closed, as free India also preferred to maintain the colonial policy for its 'national interest', but this has been to the great economic disadvantage of J&K state. Kashmir continues to be deprived of the commercial and cultural relations with its neighbourhood that had been built laboriously over centuries.

The Partition of India, and the Indo-Pak dispute over Kashmir, dealt the final blow to the ramshackle economic structure of the state. In 1947, J&K's trade with the Indian subcontinent was conducted through three highways and four waterways.[112] All these highways and waterways connected J&K with that part of Indian subcontinent which now forms part of Pakistan. The value of trade and other business conducted through these routes amounted to crores of rupees every year (at 1952–3 prices),[113] and all classes of people living in different regions of the state were beneficiaries of this trade as the highways and waterways were spread over all the

(New Delhi: Gyan Publishing House, 1989); K. Warikoo, 'Trade Relations between Central Asia and Kashmir Himalayas during the Dogra Period (1846–1947)', *Cahiers d'Asie centrale (En ligne)*, 1 February 1996; Rizvi, *Ladakh: Crossroads of High Asia*; Rizvi, *Trans-Himalayan Caravans*.

[109] Moorcroft and George, *Travels in Ladakh and Kashmir*, 383–91.

[110] Rizvi, *Trans-Himalayan Caravans*.

[111] Rizvi, *Trans-Himalayan Caravans*; Rizvi, *Ladakh: Crossroads of High Asia*; see also Fida Mohammad Hassnain, *British Policy towards Kashmir (1846–1921)* (New Delhi: Sterling, 1974).

[112] The four waterways were Sind, Vitasta (Jhelum), Chenab, and Ravi, and the three highways were the Jhelum Valley Road from Srinagar to Kohala via Baramulla and Dome; Banihal Road from Srinagar to Sialkot via Banihal and Jammu; and Abbotabad Road, from Domel to Abbotabad via Ramkot.

[113] Bazaz, *History of Struggle for Freedom in Kashmir*, 416.

important areas of the state. With the Partition of India and the subsequent dispute between India and Pakistan over Kashmir, all these commercial routes were closed, and with this the state lost a vast market and traditional trade centres extending from the north-west frontier through western Punjab to Sind. The roads still exist, but, except for a recent half-hearted initiative, the traffic on them across the border has come to a standstill. The closure of borders also affected seasonal migration. Before Partition, the favourite destinations of Kashmiri labourers were invariably Rawalpindi, Gujrat (Punjab), Gujranwala, Lahore, Sialkot, Layalpur, Amritsar, and Ludhiana.[114] Except the last two, the markets were now closed to Kashmiri seasonal labourers, with obvious consequences.

The political instability that has been an ever-present thorn in the post-1947 history of Kashmir has produced a devastating effect on Kashmir's economy. Since the dispute has maintained continuity, its economic consequences have remained unchanged. It is, therefore, not surprising that what we read in this regard in the sources of the late 1940s and early 1950s, and the statements and public speeches of today's political leaders—both mainstream and separatist—echo the same concern. In a public speech on 8 May 1952, Sheikh Mohammad Abdullah observed:

> Kashmir is today standing at crossroads. The fate of this country has not been decided yet. This question has been before the U.N. for many years. Attempts are being made to solve it, but no one knows what the solution may be. There is a tension in the atmosphere owing to this. It is difficult to complete any work in these circumstances. For instance people do not invest any capital in business and when people do not invest it is difficult to progress. After all government cannot do everything. Many things are to be achieved by the people themselves. As Kashmir is a place where visitors come and go. For their comfort hotels must be opened. But many friends have told me that due to uncertain conditions they cannot open any hotels. This is only an instance. I can give other instances of the same kind to show how, due to the present conditions, the progress of our country has stopped.[115]

---

[114] Bazaz, *History of Struggle for Freedom in Kashmir*, 433.
[115] *Khidmat*, 10 May 1952.

Similar comments were reiterated by Sheikh on 25 July 1953:

> The sword of uncertainty is hanging over the heads of state people
> which has ruined them. All the works of progress and prosperity
> have come to stop. No construction work can be accomplished. No
> one seems to be prepared to invest any capital and no one comes
> forth to start a big industry, business or trade with composure of
> mind. Granted that nature has given us forests, mines and similar
> other sources of wealth. But to tap them capital is needed and that
> can be available only when uncertain conditions are ended.[116]

Recent newspapers published from J&K are replete with similar
statements by chief ministers emphasizing that there is a positive
relationship between *amun* (peace) and *khushali* (prosperity), as
no one wants to invest in the trouble-torn state. It is, therefore, not
surprising that the Godbole Report finds Kashmir unmoved by the
current environment of foreign investments.[117]

Similarly, when one reads about the damning effect of political
uncertainty on Kashmir tourism during 1948–52 in contemporary
sources, it appears as if these sources capture the present scenario
of Kashmir, except that the term 'Nationalists' used for the then
ruling party has to be overlooked:

> A great part of income of Kashmir people and the state government
> is derived from influx of visitors. Experts in the tourist trade are
> of the opinion that by proper publicity and by affording reasonable
> amenities and facilities to the visitors, at the modest estimation
> more than a lakh people would like to visit the valley every year.
> Before 1947 nearly forty thousand outsiders came annually. It gave
> an impetus to the local business and brought substantial revenue
> into the State coffers. A well-to-do class of people, the houseboat
> workers called Manjis, lived entirely on the visitors; they did not
> have any other source of income. But since the Nationalists were
> installed in power tourist business has flagged.
>
> Every year at the beginning of the season the Nationalist
> Government assures the Kashmiris that the coming year will bring
> business beyond expectations and more visitors than had ever come

---

[116] *Khidmat*, 27 July 1953.

[117] Godbole Report, 1.

to Kashmir will be pouring into the valley but at every close of the season it is known to have been worse than the last. One unfortunate communal incident somewhere in the subcontinent, some controversy between India and Pakistan or an un-statesmanlike speech by some leader, mars the bright prospects.[118]

Again, for 'national interests', the state sacrificed potential power generation of about 15,000 MW owing to the restrictions imposed by the Indus Water Treaty.[119] Moreover, all the major power projects are centrally controlled.[120] Considering that J&K has suffered huge economic losses for the 'national interest', the state deserves special treatment for overcoming its infrastructural bottlenecks. This is why committees constituted at the state level and central initiatives to suggest economic reforms have advocated compensating Kashmir for the loss suffered owing to the transfer of the power projects to central control.[121] Unfortunately, the central

[118] Bazaz, *History of Struggle for Freedom in Kashmir*, 420–1.

[119] The Indus Water Treaty was signed on 19 September 1960 in Karachi by Prime Minister Nehru and President Ayub Khan. According to Article 2 of the treaty, all the waters of the three eastern rivers—the Sutlej, Beas, and Ravi—would be available to India, and Pakistan would receive 'for unrestricted use all those waters of the western rivers—the Indus, Jhelum, and Chenab'. A report prepared by the International Water Management Institute in collaboration with Sir Ratan Tata Trust, Mumbai, in March 2005 states that the Indus Water Treaty signed by India and Pakistan in 1960 put J&K behind by an estimated Rs 6,500 crore annually. The treaty has also badly hit power generation (the report estimated a loss of around 20,000 MW) and agriculture potential in the state. See Mitul Thakkar, 'Indus Water Treaty Hit J&K Growth: IWMI', *Business Standard*, 6 February 2013.

[120] The National Hydroelectric Power Corporation (NHPC) controls all the major power projects in J&K like the 690 MW Salal, 390 MW Dulhasti, 1,000 MW Pakal Dul, 1,020 MW Bursar, 600 MW Kiru, 690 MW Ratla, 520 MW Kawar, 330 MW Kishanganga, 240 MW Uri (II), and 48-MW Uri (I).

[121] Godbole Report, 194–5. Also, it was in this regard that the report of the Working Group on Economic Development of Jammu and Kashmir constituted by Prime Minister Manmohan Singh in May 2006 suggested immediate handover of major projects like the 390 MW Dulhasti and 1,020 MW Bursar from NHPC to J&K state, as well as enhancing states' share in free power in Central Hydel Projects. Government of India, *Economic*

government has not taken any worthwhile interest in making Kashmir economically self-sufficient. Significantly, J&K (and Assam) did not become 'special category' states in 1969 when the scheme was introduced; according to this scheme, a 'special category' state is entitled to receive 90 per cent of its funds as grants and 10 per cent as loans. It was only in 1990, after the popular uprising, that Kashmir was included among the 'special category' states.[122]

Kashmir is one of the states that depends most heavily on central assistance. The dependence of the state on loans and packages has generated huge debt liabilities similar 'to the debt crisis facing Africa where resources required for productive investment are being diverted to debt repayments'.[123] Jammu and Kashmir is a land-scarce state. Only 30 per cent of its total area is cultivable, the remaining being under mountains, hillsides, and water bodies. Hence, the size of landholdings is extremely low—90 per cent of the arable land is constituted by marginal and sub-marginal holdings.[124] The arable land in J&K was 0.14 hectares per person in 1981 and has further shrunk since then.[125] Not surprisingly, therefore, the state has been facing a considerable deficit in food and other essential items.[126] The industrial sector has also been

---

Development of Jammu and Kashmir, Report of Working Group no. III, Home Department, New Delhi, March 2007, 7–8.

[122] Gautam Navlakha, 'Kashmir: Political Economy of Fiscal Autonomy', *Economic and Political Weekly* 38, no. 40 (2003), 4213.

[123] Siddhartha Prakash, 'The Political Economy of Kashmir since 1947', *Economic and Political Weekly* 35, no. 24 (June 2000), 2053.

[124] Nisar Ali, 'Structural Changes in Jammu and Kashmir Economy and Conflict', in *Conflict and Politics of Jammu and Kashmir: Internal Dynamics*, eds Avineet Prashar and Puvan Vivek (Jammu: Saksham Books, 2007), 62.

[125] Mukeet Akmali, 'In J&K, Agricultural Land Conversion Continues under Government Nose', *Greater Kashmir*, 27 October 2015, http://www.greaterkashmir.com/news/kashmir/in-jk-agricultural-land-conversion-continues-under-government-nose/200091.html (accessed 1 March 2018).

[126] See Mulk Raj Saraf, ed., *Jammu & Kashmir Trade Guide (J&K Guide)* (Delhi: Universal Publications, 1969).

very peripheral, and most of the units run by the state consistently suffered huge losses. Those units which were congruent with the conditions of Kashmir also failed to sustain themselves or take off because of huge recruitments and corruption.[127] In the absence of any significant changes in this situation, the state government has been the only employer, making the service sector oversaturated and encumbered by an inflated salary bill. Therefore, infrastructural development suffered and continues to suffer from lack of investment. All these factors have resulted in a classic 'backwardness trap' of low economic activity, low employment, and low income generation.[128]

The problem was further compounded by the prevalence of rampant corruption—both political and bureaucratic—ever since the *awami hukumat* (people's government), as the post-1947 government was called, took over. Exceptions apart, it is an inherent human weakness to misuse power for amassing wealth, creating assets, and living a luxurious life. This innate instinct gets translated into practice in the absence of institutionalization and technologization of the governance system. It is also promoted by the absence of the policy of reward and punishment, and want of any consideration as to the moral credentials of candidates for nomination, appointment, and promotion. These best practices have been missing in the history of governance in Kashmir, and even today they exist only on paper and not in practice. Hence the prevalence of corruption is quite understandable. But what is peculiar about Kashmir is that corruption has been used as an instrument of managing the state.[129] In the hostile environment of Kashmir, it has been and is being used to create a support structure. And with this it has received unwritten legitimacy by the state. No wonder, then, that all those who became rich and powerful by looting public wealth earned respectability in

---

[127] *Report of the Committee on Economic Reforms for Jammu and Kashmir* (1998: 208–35).

[128] Government of India, *Economic Development of Jammu and Kashmir* (Report), 2.

[129] Riyaz Punjabi, 'Corruption: A Factor in Kashmiri Alienation', *Mainstream Weekly*, 29 (March 1991), 22–3; also see Chapter 3, this volume.

society, and those very few who stuck to being honest and therefore lived from hand to mouth, came down in the public estimation.[130] That J&K figures among the most corrupt states is thus understandable. Corruption became a drain on the public wealth. It hampered development and consequently hamstrung the prospects of creating jobs; accounted for individual and collective injustice; eroded the credibility of institutions; and factored in the price rise and the widening gulf between rich and poor. All these factors ultimately conspired to create a deep social discontent, particularly among the young, educated sections of the people.

Naya Kashmir Programme

The nature and character of post-1947 governance in Kashmir has in some respects been shaped by the Naya Kashmir Manifesto issued by the National Conference—the front-ranking, mass-based political movement which fought for inaugurating a new era in J&K. Drafted by the communists, who were then associated with the National Conference, the Naya Kashmir Manifesto was adopted by the annual session of the National Conference held in Srinagar during 29 and 30 September 1944. The Naya Kashmir Manifesto was a draft constitution and included economic planning for the future Kashmir. It is often referred to as the Magna Carta of the Kashmiri people and it became the Bible of Kashmir's economic life.[131] In the introduction, Sheikh Abdullah declares that the National Conference fights

> for the poor, against those who exploit them; for the toiling people of our beautiful homeland against the heartless ranks of the socially

[130] Based on my personal knowledge as a keen observer of Kashmir society, and also on the basis of interactions with people.

[131] The Naya Kashmir Manifesto (New Kashmir Programme) was first published in 1944 by the National Conference (Kashmir Bureau of Information, New Delhi). First published as a booklet, its cover was decorated with the red flag and white plough that later became the national flag of Kashmir. It was subsequently republished by the National Conference.

privileged.... The history of freedom movements ... had only one lesson to teach—that freedom from all forms of economic exploitation is the only true guarantee of political democracy.... In our times, Soviet Russia has demonstrated before our eyes, not merely theoretically but in her actual day-to-day life and development that real freedom takes birth only from economic emancipation.

The Naya Kashmir Programme promised a democratic and decentralized polity besides a socialist mode of production. It made breathtaking promises to take all conceivable steps to eradicate poverty, oppression, discrimination, disease, illiteracy, and ignorance.

When the National Conference came to power, the Naya Kashmir Manifesto became its theory of governance and an unwritten constitution. Indeed, Naya Kashmir became the guiding principle of all governments which succeeded the Sheikh, regardless of their political affiliations. Not surprisingly, when the constitution of J&K was adopted on 17 November 1956 during the prime ministership of Bakhshi, the Naya Kashmir Manifesto became the basis of the Directive Principles of State Policy.[132] G. M Sadiq (1963–71) who succeeded Bakhshi was a known leftist, and so was Mir Qasim (1972–3). Thus, that their priority was to implement Naya Kashmir was inevitable. To the National Conference, Naya Kashmir was its brainchild, a gift to Kashmiris, a dream to be realized, and an effective instrument of mobilization. This is the reason that the National Conference leadership has continued to harp on the programme. It should, however, be mentioned that Naya Kashmir was only selectively followed; democratic deficit and misuse of power underlined the functioning of all governments in J&K, including the National Conference.

Shifting Contexts

Sheikh had two biggest advantages of the time. One, illiteracy was common, and two, technology was not known. Today he couldn't be

---

[132] For details see A. S. Anand, *The Constitution of Jammu and Kashmir: Its Development and Comments* (New Delhi: Universal Law Publishing, 1980).

pro-Indian in Delhi, anti-Indian in Kashmir. Today even the smart-
est liars are trapped in their own lie. You can't disown your statements as seamlessly as you could do then. Not morality, technology
makes it impossible.

Aijaz-ul-Haque, *Greater Kashmir*, 24 December 2017[133]

I will add: today is not the early 1970s when the then chief minister, Mir Qasim, said proudly, 'All graduates who had obtained their
degree till 1972 were employed without any condition.'[134]

The post-1947 history of Kashmir is the history of change, of
development. The changes in different spheres occurred gradually,
dividing the post-colonial history of Kashmir broadly into two
phases: 1947–75, and 1975 to the present. The first phase saw great
changes in the agrarian sector, much to the relief of the peasantry.
Besides institutional and technological changes in the agrarian sector, the debt of the masses was written off. The peasantry became
the owners of the land and of the produce. Efforts were made to
promote crafts by extending adequate state patronage. People in
urban areas were provided cheap food and fuel. A network of canals,
roads, and hospitals was constructed. Education and employment
received a great fillip. No less a relief for the people was the abolition of state-sponsored religious discrimination. To be sure, if we
compare the Kashmir of these decades with the conditions obtaining in 1947, the region really appears to have entered into an era of
revolution in the history of subalterns.[135] This is the reason that at
the mass level, Kashmir presented a charming picture.[136] With the
masses still emotionally attached to their hereditary occupations,
the number of matriculates among them was not more than 2–3
per cent; the percentage with undergraduate degrees was around
0.25; and those with postgraduate degrees were even fewer.[137]

---

[133] Aijaz-ul Haque, 'Who Divided Whom?', *Greater Kashmir*, 24
December 2017.

[134] Qasim, *My Life and Times*, 145.

[135] For details, see Chapters 2, 3, and 4, this volume.

[136] See Ray, 'Kashmir 1962 to 1986', 195–205.

[137] G. Rasool and Minakshi Chopra, *Education in Jammu and Kashmir:
Issues and Documents* (Jammu: Jay Kay Book House, 1986), 16. This is
also substantiated by my case study of my home tehsil, Tral.

According to my father, for instance, in 1975–6, he was the lone candidate from tehsil Tral who sought admission to a postgraduate course at Kashmir University. Since the number of educated youth was still marginal, Mir Qasim could afford to give employment to all unemployed graduates in 1973;[138] and a little more than a decade before Qasim, Bakhshi could proudly say that 'those who would not be benefitted during his period, they would never be able to improve their position.'[139] He was referring not only to his own munificence, but to the inevitability of every educated person getting a government job during his time.

Up to the early 1970s, unlike the present, a government job was not a matter of life or death in the villages and among the common urban folk. During the interviews with different people that I conducted for this book, not only was I told about the absence of temptation for government jobs in the villages in those times, I also met with some semi-educated people who had left a government job immediately after getting it.[140] The reasons are not far to seek. The villages of Kashmir were still 'medieval' at the beginning of the 1970s, in that urban culture had not made any serious inroads into villages. The aspirations of village folk were still modest: a simple rice meal (*batta*) two times a day with the domestically cultivated green vegetable, *hak* (knol khol), was considered a sign of a normal lifestyle; salt tea with milk (people at large still could not afford tea with milk, called *duddār chai*); a pair of shirts and trousers, a *pheron* (traditional Kashmiri outfit) (which lasted two to three years); a modest house (pucca houses with tin roofs were still a distant dream); and a little money to celebrate the marriages of children. If a household could bear the educational expenses of its children, it symbolized a good economic position. Household property comprised utensils of clay, aluminium, some copper and

---

[138] Qasim, *My Life and Times*, 145.

[139] This is a commonplace account in Kashmir.

[140] In my native village Nowdal (district Pulwama), three persons, namely, Gh. Hassan Bhat (age 75), Abdul Salaam Shah (age 65) and Gh. Mohi-ud-din Shah (age 62) coming from modest backgrounds had procured government jobs; all of them gave up their jobs after having served for a brief period.

china wares; agricultural tools; domestically woven grass matting called *patej*, and bedding which among the poor was partly made of worn-out woollen clothes.[141]

The villages were almost self-sufficient. All the work including the construction of houses and farming was done by labour-sharing arrangements among households. All the basic needs were met internally by households, including *pattu* (woollen cloth) for pheron. Eggs, walnuts, almonds, cereals, and oil seeds were the popular mediums of exchange for purchasing certain items, namely, tea, salt, and clothes, for which people were dependent on the market. Vegetables were domestically available for six months; for the remaining six months (October to March), they were dried (called *hukh seun*) and preserved. Sheep, cattle (called *māl* meaning money), and trees for firewood and construction (called *tā'mir*, literally meaning construction, but technically meaning assets) were sources of wealth for meeting major demands. Sheep, cattle, and trees were called *muhimuk yar*, meaning the friend in need.[142] Indeed, the peasants had an emotional attachment with their land. Besides being an issue of mentality, which changes slowly, the peasants had no cause to be disappointed with their profession. After the land reforms in the 1950s and the introduction of high-yielding crops and modern fertilizers in the late 1960s, a revolutionary change occurred in their conditions, especially when the size of holdings was reasonably good and there was also scope for further expansion, with considerable fallow land still available.

The urban poor were also not yet completely dissatisfied with their lot. They were mainly artisans, craftsmen, petty merchants, drivers, conductors, *shikarawalas* (boatmen), vegetable growers, and petty officials working in the government offices and recently established industries. By the beginning of the 1970s, in the words of Ashraf Wani, 'The urban artisan had emotional attachment to his hereditary craft, and unlike his modern counterpart the other lucrative professions, which were still peripheral to Kashmiri

---

[141] Based on my extensive interactions with my late grandmother, who was around 80 at the time of the interview in 2012.

[142] See note 141.

Society, had not taken his heart out of his occupation.'[143] The demands of the urban poor were also modest: state patronage for the promotion of crafts and reasonable pricing of their products, cheap food, and fuel. A little later, the education of children too began to catch the imagination of the urban poor. These aspirations were met by the state, which followed these benevolent policies with conviction.

Clearly, the radical economic and social measures buoyed the world of the masses for about two and a half decades. And it was due to the perception of mass contentment and innocence that misgovernance did not become a burning issue during the first phase of governance in Kashmir. It was also for the same reason that the affinity of the masses with their leader, Sheikh Abdullah, could not be converted into the movement he had expected; nor did religious and political sentiment become radicalized during the two Indo-Pak wars of 1965 and 1972, except among a section of the urban educated youth. A. M. Watali, then superintendent of police, Anantnag, one day asked Afzal Beg, the point man of Abdullah, the reasons for rolling back the Plebiscite Front in the wake of the Indira–Abdullah Accord of 1975. The reply given by Beg is interesting:

> While Beg-Parthasarthy talks were in progress, he would often come to Anantnag, where I would also call on him and spend some time with him. Once in Khanabal Dak Bungalow, when only two of us were there, I ventured to ask him under what circumstances the PF [Plebiscite Front] leadership has climbed down from their stated position of securing right to self-determination and implementation of UN resolutions. What has made them to start a dialogue with the Centre? He said that this was a valid question which could be in the minds of every Kashmiri. He explained that the demand for Plebiscite could not be converted into a mass movement, particularly in rural areas. It had remained confined to the city of Srinagar and a few towns of Kashmir Valley and Jammu. It was therefore very easy for the administration to tackle the problem by arresting the

---

[143] Mohammad Ashraf Wani, 'Religion, Economy and Political Crisis in Kashmir', in *Identity Politics in Jammu and Kashmir*, ed. Rekha Chowdhary (New Delhi: Vitasta, 2010), 185.

leaders and prominent workers across the Valley and by imposing curfew at a few places. There would be no reaction in rural areas and the situation would be controlled by the authorities within a few days. Everything would be forgotten thereafter. People, particularly peasantry, would be more interested in economic development, government employment and green revolution. That is what forced them to negotiate an acceptable solution with New Delhi.[144]

The situation, however, started changing drastically by the middle of the 1970s; it branched off in the mid-1980s, and reached a stage of serious crisis by the mid-1990s. Education and development, which had received a great fillip in the 1950s and 1960s, started producing visible results—an unprecedented increase in the numbers of the lower middle classes, the emergence of a new rich class associated with new economic activities and administrative machinery, the proliferation of the rich class as compared to the small upper class of the pre-1950s period, wage incomes, dependence on government initiatives, and the intrusion of the market in the villages.[145] After 1975, the process of development and the expansion of education received a great push. In the 1980s, the number of persons who had passed matriculation had grown by two and a half times, about 50 per cent growth among those who passed graduation and about 18 per cent growth among those who passed post graduation, and about 18 per cent increase in the number of postgraduates[146]

Investment in infrastructural development, the establishment of new departments and institutions, and the expansion of existing ones nearly doubled from 1975–6 to 1989–90. The result was that the employment sector expanded considerably, and those who were not employed in the service sector or were not directly connected

[144] Watali, *Kashmir Intifada*, 337–8.

[145] In comparison to a few thousand government employees in the early 1960s, the number increased to 168,000 in 1980 and around 300,000 in 1997–8. Godbole Report, 257–9; Government of J&K, *Jammu and Kashmir 50 Years*, Department of Information, Srinagar/Jammu, 1998, 350. For detailed statistical information about different economic activities, traditional as well as modern, see Saraf, *Jammu & Kashmir Trade Guide*.

[146] Godbole Report, 257.

with infrastructure development also benefited from the circulation of money. Consequent upon the emergence of new sectors, agriculture lost its primacy as the mainstay of the economy. Now its place was taken by business, industry, and the service sector.[147] In comparison to modern sources of income, agriculture became a losing occupation, as it does not yield more than what is invested in it. Crafts and other old surviving petty occupations were also no longer attractive in the face of government jobs and other professions linked with modernization. With the great influx of money and the growth of wage labour, labour-sharing arrangements in the villages and towns broke down;[148] and with these developments also came to an end the traditional self-sufficiency of the villages and their simple living.

Having observed the vast difference between the living standards of government employees and those of workers (even rich peasants and urban craftsmen), together with the relative prestige enjoyed by the former in society, government jobs became the most sought-after aspiration among the young generation— whether educated or uneducated. True, state service was not a new phenomenon in Kashmir, but what was new was the impressive proliferation of the service class and the mass demand for government jobs. While previously, government employees in villages were too scarce and too dispersed over a large area to be noticed, in the new scenario the growth of the service class was so huge that there was hardly any village/mohalla which did not have a few government servants. This brought the difference in the quality of lifestyles of the two categories of people to the doorsteps of the unprivileged, causing frustration among them. The government job did not only carry material attractions and a badge of social respectability, it was also physically and mentally much less exacting.

It is also germane to mention that since ancient times, state service had been the calling of the Pandit community, and they continued to be dominant in this sector even after the change of policy

---

[147] Godbole Report, 3–5.

[148] This is a pervasive phenomenon in the whole of J&K state. For Ladakh, see Van Beek, 'Beyond Identity Fetishism', 538.

after 1947. Clearly, until the 1960s, when the Muslim community in the villages had yet to produce a good number of matriculates and graduates, the monopoly of the service sector by the Pandits was taken for granted. The situation, however, changed with the increasing number of Muslims in the government sector towards the end of the 1970s. It intensified by mid-1980s and reached to a stage of serious crisis by mid-1990s. Now for the first time, in imitation of their neighbours, friends, and acquaintances, there grew among the rural Muslim masses an increasing appetite for government service.

With the passing of the generation which had experienced both pre- and post-1947 conditions, there emerged new challenges requiring new responses. Having been born and brought up in relieved conditions, the new generation spent their youth in educational institutions and got exposure to modern life-ways through different mediums. Unlike their parents/grandparents, they did not compare their own times with pre-1947 conditions; instead, they saw themselves in the mirror of the fast-moving modern world. And to them it was not following a mirage, especially when they saw people in their own neighbourhoods and surroundings enjoying the modern facilities. These privileged people were not peasants or artisans; they comprised different rungs of the modern bureaucracy as well as the political and economic classes whose number had increased manifold, making their presence, with their high standard of life, felt everywhere. Comparing the condition of a peasant or artisan with a government servant of even a lower rung is bizarre; but in the conditions of Kashmir, it was even more absurd because by the mid-1970s, the size of landholdings had begun decreasing considerably; in the 1980s, it had reached a dismal point, not more than 6 *kanals*.[149] In terms of input–output, the income was almost zero, except providing a few working days to the adult members of the household. Moreover, Kashmir has a mostly one-crop-a-year economy; orchards, the only cash crop worth the name, were restricted to a few pockets, and with the steeply shrinking size of landholdings and exorbitant input costs, apple growing was also increasingly becoming not an attractive

---

[149] 1 kanal is equal to 0.0505857 hectares.

economic activity. Apart from this unattractive scenario of farming, the young generations, having spent their youth in educational
institutions, found it mentally as well as physically repulsive to
adopt the professions of their parents. They were trained and pro-
grammed to do a government job only.

Development without safeguards was accompanied by two further inimical consequences. First, it created vast disparities in the
society; and second, it broke down the traditional self-sufficiency
of the village without replacing it by an alternative income-generating policy. Development threw up a sizeable class of politicians,
businessmen, industrialists, engineers, contractors, forest lessees,
hoteliers, transporters, journalists, bureaucracy of the higher and
lower rungs, and so on. As money was pumped into Kashmir to
create a strong local support structure through political manipulation, patronage politics, subsidies, packages, corruption, loot,
and nepotism, a new rich class comprising the segments just mentioned emerged in Kashmir from the 1950s. These sections circulated the privileges and opportunities among members of their
own group, giving birth to tensions in the society.[150] This situation created a class war that was both open and passive.

It is true that corruption was not a new thing in Kashmir. It had
also been rampant in pre-1947 Kashmir.[151] But what was really

---

[150] Riyaz Punjabi, 'Corruption: A Factor in Kashmiri Alienation'.

[151] Munshi Mohammad Ishaq relates an interesting episode in this
regard. He says that during his posting as an emergency officer after the
tribal raid, one villager applicant approached him and said that since he
did not have one rupee to attach with his application as *nathi*, he would
come the next day. 'I asked him what does *nathi* mean?, he replied that
after returning from your office one police man took me to a *thanedar*
(officer of the jail) and he asked me to give one rupee for doing *nathi* to the
application. After hearing this I called the *thanedar* to my office and asked
him what does "getting *nathi* done" mean? He bent his head in shame and
said that the villagers do not get satisfied with simply writing an order on
application unless they pay some money for it. Since your office does not
take any money from them, so for their satisfaction we take this money
from them. In this way our job is also done because with little salary we
are not able to maintain our families.' Ishaq, *Nida-i-Haq*, 206.

new was that the volume of corruption and the number of its beneficiaries had increased beyond any precedent. The patronage money pumped in by the state for a vast array of its local allies, and the opening of a large number of new departments, expansion of existing ones, and the multifarious new economic and socio-cultural activities introduced after 1947, provided a variety of opportunities for political and bureaucratic corruption in alliance with local economic classes and crooks, exacerbating the gulf between the rich and the poor and creating heartburn among the people, especially among young, educated, unemployed youth, who have been expressing their dissatisfaction through different means since the mid-1960s.

Besides amassing wealth through dubious means, the rich class became what the sociologists call a 'reference group'—the group whose lifestyle becomes a standard and norm for people wanting to obtain a respectable place in society. As it was not possible for the mass of the population to obtain the standards set by the 'reference culture'—characterized by dynamism and prohibitive cost—it created mass frustration and an insatiable lust for money and resources. Further, as the 'respectable' class got wealthy by dubious means, corruption and other varieties of the misuse of power became a part of the reference culture, instead of being considered a vice. The rich class did not only grab national wealth and create a cost-prohibitive reference culture, it also determined the market rates which were exorbitantly high, so high that it is not possible for the poor to purchase a few yards of land and construct a house.[152]

Development also accounted for the extinction of traditional sources, the breaking down of the self-sufficient village economy and labour-sharing arrangements consequent upon the influx of machine-made goods, penetration of the market, and growth of wage labour and the money economy. The result was that the villages of Kashmir became completely dependent for every item— even eggs, meat, milk, rice, flour, and vegetables—on the market, necessitating a good economic condition in the villages. This however is a dream in an environment of increasing unemployment,

[152] Based on my field study.

abysmally small size of landholdings, absence of a private sector, withdrawal of subsidies, withdrawal of youth from the hereditary professions of their families, dependence on imported labour, absence of a sustainable income-generating policy, and, last but not least, a deficit of peace. The ultimate brunt of this crisis was faced by the state. To substantiate this fact, Kashmir since the 1980s has been a laboratory for social scientists.[153]

There is a positive relationship between media landscapes and governance outcomes. Until recently, neither were these media landscapes diverse, nor were they available to the whole populace: a liberalized media system has been absent as the control of ideas

---

[153] Since the inception of militancy many works have been produced explaining the possible reasons for insurgency in Kashmir. While the political factor is predominant in most of the writings, however, many scholars have been researching the link between economic factor and the crisis. Although more serious study is required to examine in depth the role of economic factor in mass based militancy, a few studies carried so far have tried to look for relation between economic factor and militancy/uprisings. Cf Also, Ganguly, Sumit (1996), 'Explaining the Kashmir Insurgency: Political Mobilization and Institutional Decay', *International Security*, Vol. 21, No. 2, pp. 76–107; Prakash (2000); Bose (2003) p. 149; Wani, Mohammad Ashraf, 'Religion, Economy and Political Crisis in Kashmir' in Rekha Chowdhary (ed.), *Identity Politics in Jammu and Kashmir*, Vitasta, New Delhi, 2010; Dar, Zubair Ahmad, *Behind the Numbers: Profiling those Killed in Kashmir's 2010 Unrest A study on socio-economic conditions of families who lost a member in 2010 protests and the processes of justice that followed the death of these young men and women*, Centre for Dialogue and Reconcilaition, http://www.cdrindia.org/pdf's/Kashmir's%20unrest%202010,.pdf. (Accessed on 15 September, 2018); Rasool, Alson, *In Search of the Causes of Militancy: A Case Study of the Militants of Bijbehara*, M.Phil Dissertation (unpublished), Department of History, University of Kashmir, Srinagar, 2009; Umar, Adil, *Social Background of Militancy in Kashmir: A Case Study of Anantnag Tehsil*, M.Phil Dissertation (unpublished), Department of History, University of Kashmir, Srinagar, 2009; Ahmad, Wajahat, 'Hurriyat and the Economic Questions of the Struggle', *Kashmir Ink*, 11 January 2017, http://kashmirink.in/news/perspective/hurriyat-and-economic-questions-of-the-struggle/228.html Accessed on 3 April 2018.

was considered fundamental to the state-building project. We have now arrived at a situation that has never existed before. One important development is the diverse and liberalized media system, and a second is the access and choices available to people. We have access not only to satellite and domestic media, but humanity as a whole is soon to become almost ubiquitously connected, with almost everyone on the planet having some kind of access to a mobile phone.[154]

Though the print media began in Kashmir in the nineteenth century,[155] a free press was systematically undermined by successive governments by misuse of power. The media was composed, therefore, of either the state-owned press or the patronage press, and thus purveyed more propaganda than news.[156] It was during the period of Sadiq that the press got some freedom,[157] which received momentum from 1977 onwards. Proprietors installed offset printing presses which helped in expanding their readership, which was then limited to the upper strata of society (the ruling elite, bureaucrats, and the rich).[158] Up to the late 1970s, no villager remembers having had access to newspapers. Generally, people relied on government-controlled radio, and on *afwah* (rumours), *sena gazettee* (concoctions), and *tar* (lies), or on the occasional visitor to Srinagar or some *shahar bash* (city dweller) who visited the

---

[154] James Deane, 'Media and Communication in Governance: It's Time for a Rethink', in *A Governance Practitioner's Notebook: Alternative Ideas and Approaches*, eds Alan Whaites, Eduardo Fyson, and Graham Teskey (OCED, 2015), 268.

[155] 'Chronology of the Print Media in J&K', https://www.slideshare.net/Mehraj9906/history-of-press-in-kashmir (accessed 22 February 2018).

[156] Masood Hussain, 'Mass Media and Kashmir in 1900s', Kashmir Portal, 16 June 2009, https://kashmirian.wordpress.com/2009/06/16/mass-media-kashmir-in-1900s (accessed 22 February 2018).

[157] Masood Hussain, 'Mass Media and Kashmir in 1900s', Kashmir Portal, 16 June 2009, https://kashmirian.wordpress.com/2009/06/16/mass-media-kashmir-in-1900s (accessed 22 February 2018).

[158] Masood Hussain, 'Mass Media and Kashmir in 1900s', Kashmir Portal, 16 June 2009, https://kashmirian.wordpress.com/2009/06/16/mass-media-kashmir-in-1900s (accessed 22 February 2018).

village to meet his villager friend or to do some business. However, from the late 1970s, many newspapers began to be issued and circulated in good number in district and tehsil headquarters.[159]

The radio was introduced in 1948. But according to Josef Korbel, in the mid-1950s, 'with the exception of a few wealthier families and high state officials, there are no privately owned radio sets.'[160] Even in the late 1960s, a radio set was the privilege of only rich families in rural areas.[161] The only radio available to the common people was what was popularly called 'panchayat radio', the radio installed in the panchayat buildings for the purposes of serving the interests of the state in the face of opposition both within and beyond the ceasefire line. Radio, however, became ubiquitous from the beginning of the 1970s. A television centre was established in Kashmir in 1972. By 1984, the posts and telegraph department had issued 20,846 licences.[162] In 1992, Kashmir had 118,000 television sets.[163] Significantly, the revolutionary developments in neighbouring Central Asia and Eastern Europe and their live telecast provided a great stimulus to Kashmiris to follow suit.[164]

Before concluding this discussion, it is important to remember that in the context of J&K, governance has been and is being largely scripted by New Delhi within the broader framework of the policy of coercion and consent to meet the challenges emanating from the state's disputed nature. It is also common knowledge that the local government has no control over the most critical wings of the state—the army and the paramilitary forces whose

[159] Masood Hussain, 'Mass Media and Kashmir in 1900s', Kashmir Portal, 16 June 2009, https://kashmirian.wordpress.com/2009/06/16/mass-media-kashmir-in-1900s (accessed 22 February 2018).

[160] Korbel, *Danger in Kashmir*, 208.

[161] As per the statistical data, 2,125, 6,839, 18,827, 20,287, 32,149, 38,262 and 45,358 radio licences were issued in J&K during 1950–51, 1955–6, 1960–61, 1961–2, 1962–3, 1963–4, and 1964–5, respectively, by the Superintendent of Post and Telegraph. Cited in Saraf, *Jammu & Kashmir Trade Guide*, 265.

[162] Ganguly, *Crisis in Kashmir*, 35–6.

[163] Ganguly, *Crisis in Kashmir*, 35–6.

[164] Watali, *Kashmir Intifada*, 60.

role in managing the conflict in Kashmir is crucial. Even the local police are also practically under the Ministry of Home Affairs.[165] Yet, the personality and the priorities of individual rulers and the specific conditions in which each of them worked also influenced the tone and tenor of governance.

---

[165] Paul Stainland, 'Kashmir since 2003: Counterinsurgency and the Paradox of Normalcy', *Asian Survey* 53, no. 5 (2013), 953.

2

How *New* Was the New Kashmir (1948–53)?

The popular political party of Kashmir, the National Conference, which fought the struggle for freedom against despotism, feudalism, and sectarianism under the leadership of Sheikh Abdullah, was ideologically informed by socialist thought. It was thus passionate about changing the complexion of the state and society by inaugurating a new era of people's government underlined by democracy, justice, equity, and the uplift of subalterns by freeing them from all kinds of exploitation and discrimination. Indeed, Sheikh Abdullah's entry into politics was prompted by the pathetic condition of the masses owing to the oppressive and discriminatory rule of the princely order.[1] And it was because of his representation of the fundamental economic problems of the people with sincerity and boldness that the Sheikh became the heartbeat of the masses and was almost deified by them. In fact, right from the days when Sheikh Abdullah was leading the Muslim Conference, he and his comrades had always talked in terms of the exploiter and the exploited rather than Muslims and non-Muslims.[2] Seeing

[1] For his description of the pain which Sheikh Abdullah felt on seeing the pathetic conditions of the working class, see Abdullah, *The Blazing Chinar*, 339–40; Abdullah, *Testament of Sheikh Abdullah*, 65–6.

[2] For details, see *The Annual Resolutions of Muslim Conference*, cited in *Kashmiri Mussalmanun ki Siyasi Jadujahad, 1931–1939: Muntakhib Dastawaizat* (Srinagar, 1991).

in his own surroundings both Muslims and non-Muslims fragmented into classes, and all communities divided into haves and
have-nots, exploiters and exploited, good and bad, just and unjust,
honest and dishonest, Abdullah came to be inspired by socialist
ideology rather than the two-nation theory:

> Our experience had convinced us that the basic conflict among the
> various sections of people was not that of religious but material
> interests. It was essentially a conflict between the exploiters and
> the exploited, between the oppressors and the oppressed. We had
> realized that we were confronted by a despotic regime, not by any
> one person. It was a quarrel between us and a feudal set-up, not with
> the person of the feudal lord. To put it differently, we hated the sick-
> ness, not the sick man.[3]

Even though, barring a very few, the Hindus did not respond to
the National Conference's clarion call for fighting a joint struggle,[4] the Sheikh read this cold shouldering from the perspective
of the realities of life rather than from a communal angle, as the
Kashmiri Pandits were beneficiaries of the status quo. After all,
those Muslims who were the beneficiaries of the maharaja's rule
also did not join the movement, and instead opposed it as vehemently as did the co-religionists and favourites of the maharaja.[5]
Sheikh Abdullah's own convictions were further reinforced by the
socialists who joined the Sheikh in his struggle. The progressive
ideology of the National Conference was, however, a synthesis
of communism and democracy.[6] It was because of the socialist ideology of the National Conference that Jawaharlal Nehru
could successfully persuade the Sheikh to change the name of
the Muslim Conference to the National Conference.[7] Given the

---

[3] Abdullah, *The Blazing Chinar*, 217. Also see *Testament of Sheikh Abdullah*, 58.

[4] Abdullah, *The Blazing Chinar*, 74–5; Bazaz, *History of Struggle for Freedom in Kashmir*, 151.

[5] Abdullah, *The Blazing Chinar*, 44, 52, 85.

[6] 'Indeed, we too favoured combining the communist ideology with democracy and liberal humanism.' Abdullah, *The Blazing Chinar*, 218.

[7] Abdullah, *Testament of Sheikh Abdullah*, 72–3.

secular-progressive nature of the National Conference, the Indian communists found a patron in the Sheikh.[8] The local sections of youth belonging to the Indian Communist Party[9] did not find any need to launch a parallel organization, as they saw a ready-made platform in the National Conference not only to propagate their ideology, but also to mould it perfectly in their crucible of political activism. They succeeded in their mission, as it was the Indian communists, namely, B. L. Bedi, Mohammad Din Taseer, K. M. Ashraf, Daniel Latifi, Ehsan Danish, and Farida Bedi who, in 1944, at the instance of the National Conference leadership, drafted the Naya Kashmir Manifesto[10]—the political and constitutional document for building a new Kashmir after the lapse of British paramountcy, which did not seem a distant circumstance in 1944. And as expected, the Naya Kashmir Plan was framed along socialist lines. Significantly, this was the first ever endeavour of this kind in the subcontinent.

The Naya Kashmir Manifesto contained a proposal for a constitution, a national economic plan, and a women's charter. It guaranteed equal rights and status to all the people of the state without any discrimination. It also guaranteed freedom of faith and expression as well as freedom of speech, freedom of press, freedom of assembly, and freedom of association. It recognized individual freedom and privacy and guaranteed that, except through due process of law, private and personal correspondence would not be interfered with. The manifesto regarded disloyalty to the country and conspiracy with enemies of the nation as heinous crimes. It guaranteed the right to work and employment to every individual. In case the state failed to provide employment to any person, he/she would have the right to employment insurance. Every person was to have the right to education. In the new Kashmir, it was the responsibility of the government to finance the education of the poor. Primary education would be imparted in the mother tongue, and efforts would be made to develop the regional languages of the state. The official language

---

[8] Bazaz, *History of Struggle for Freedom in Kashmir*, 357–61.

[9] Among them, mention may be made of G. M. Sadiq, D. P. Dhar, Girdhari Lal Dogra, Mir Qasim, and G. R. Renzu.

[10] Abdullah, *The Blazing Chinar*, 217–18.

of the state would be Urdu. The manifesto ensured legal protection and equality before law, promised an independent judiciary, mobile courts and legal aid, and the right to property and inheritance. It made it binding on all citizens to abide by the law and promote the basic objectives of secularism, socialism, and democracy.

The country's highest organ, according to the New Kashmir Manifesto, was the national assembly, with its usual legislative powers as well as with the jurisdiction to represent the state in external relations, to organize the defence of the state, and to prepare the national economic plan. The New Kashmir Manifesto visualized a cabinet government and promised a democratic and decentralized polity. Political power 'shall be equitably distributed among the regions of the state and would be further decentralized at lower levels like Districts, Blocks and Panchayats'. The section on the economy buzzes with a socialist tone and tenor. The national economic plan would direct the economic life of the state. As per this plan, cooperative enterprise would be stressed, as against 'destructive competition'. Marketing should not be spontaneous but controlled and organized. The new Kashmir would be free from landlordism. The manifesto called for the abolition of parasitic landlordism without compensation, transfer of land to the tillers, and the establishment of cooperative associations of peasants. The peasant's charter guaranteed freedom from debt, and a number of material, social, and health facilities besides the right to work.

There would be people's control of forests, organized cultivation, self-reliance, and development of agriculture, bee keeping, and fish farms. A national agriculture council was envisaged to execute and supervise the national agricultural plan.

While emphasizing the paramount role of industrial advancement for progressive standards of living, the New Kashmir Programme kept all the key industries in the hands of the people's government, abolishing large private property and monopoly; only private small-scale enterprise was allowed in strict conformity with the national plan. It also included a workers' charter which guaranteed the right to work, a standard living, the right to associate with trade unions, and other facilities. The New Kashmir Manifesto included a national health charter, national education council, and national housing council to take the care of the health of citizens, foster education, and develop housing, respectively. Banking, currency, and financial matters were nationalized under the control of the national economic council. Assuring that the transport would receive a high priority to connect far-off areas with cities and towns by using surface and river transport, the New Kashmir Plan envisaged the elimination of 'vested interests' and monopolies in the road transport system. Bus and truck transport on major roads would be gradually nationalized.

The New Kashmir Manifesto concludes with the women's charter, assuring Kashmiri women their just and rightful place in society. 'In every field of life, including economic, social and cultural as also in government employment, women shall have equal rights with men,' reads the women's charter. Indeed, the New Kashmir Programme was an ambitious plan covering all spheres of life and promising a paradise which had been lost in the medieval rule of the princely order.[11]

The National Conference government, called variously *quami haqumat* (national government), *awami hukumat*, and 'popular government' in the official documents[12] of the period, assumed

[11] The contemporary poet, Mehjoor versified the rosy future of Kashmir:

> The dreary wastes shall no more lie desolate,
> A new world shall settle therein
> The thorny poisonous bush [*arak hal*] shall get a graft of the pine;
> The willow shall get the durability of the Sandal Wood
> The hills shall give birth to precious stones
> Divers shall detect rubies in Dal.
> The ranges of the mountains shall yield gold
> Pearls shall emerge out of the Wular lake.
> Mahjoor, quoted in Bazaz, *History of Struggle for Freedom in Kashmir*, 253.

[12] See *Jammu and Kashmir 1947–50: Achievements of the Three Years of Sheikh Mohammad Abdullah's Government* (Jammu: Government Ranbir Press, 1951), Archives Reference Library, Srinagar, Serial no. 293, Accession no. 149/GACC), 10. The official sources of the period (1948–53) also called J&K as *muluk*/country. Even Prem Nath Bazaz has frequently used the term 'country' for J&K.

office on 5 March 1948, with Sheikh Mohammad Abdullah as prime minister. Having inherited a very difficult legacy, it was not a bed of roses for the new government. Kashmir was politically unstable, socially fragmented, and economically impoverished. Kashmir had become an international issue, with Pakistan claiming Kashmir as its legitimate part, and internally there was a vocal segment of the population supporting the claim of Pakistan.[13] There was a lack of coherence between the individual regions of the state. The sectarian Jammu and Ladakh factions were interested in nothing short of overthrowing the autonomous regime and settling scores with Sheikh Abdullah and his National Conference.[14] Financially, Kashmir was reeling under a serious crisis owing to the closure of the highways following the Partition of India and the hostilities between India and Pakistan. Its exports and imports had come to a standstill. There was acute scarcity of basic necessities.[15] The dislocation of trade and tourism together with large expenditures on relief and rehabilitation of refugees continued to plague the economy of Kashmir in the succeeding five years.[16] The devastating flood of autumn 1950 made matters worse. And the violent protests in Jammu during 1952 and 1953 leading to the frequent closure of the Pathankot–Srinagar route sounded the death knell to the already fragile economy of Kashmir.[17]

Harmonious social relations were badly vitiated by the massacre of Muslims in Jammu[18] and the plundering raids of tribals and their savagery in north Kashmir.[19] What is more, in 1947, J&K was one of the least developed states in India, as reflected in the abysmal mass poverty, deprivation, hunger, disease, and ignorance.

[13] Bazaz, *History of Struggle for Freedom in Kashmir*, 207–32, 287–93.

[14] See Chapter 1, this volume, and the section titled 'Major Crises' in the present chapter.

[15] See the section titled 'Major Crises' later in this chapter. The *nun dreag* (salt famine) of late 1947 and 1948 is still alive in public memory.

[16] Michael Brecher, *The Struggle for Kashmir* (Toronto: Ryerson Press, 1953), 153.

[17] Saraf, *Kashmiris Fight for Freedom*, vol. II, 1207.

[18] Saraf, *Kashmiris Fight for Freedom*, vol. II, chapter 22, 808–34; Choudhary, *Kashmir Conflict and Muslims of Jammu*, 95–6.

[19] Saraf, *Kashmiris Fight for Freedom*, vol. II, 906–8.

Even in 1950, the state had a per capita income of Rs 208 (at 1960–1 prices).[20] The rate of literacy was 5 per cent.[21] Agriculture, which was the dominant sector, was stagnant. Industrial development was almost negligible. Infrastructural bottlenecks crippled the state economy and accentuated the poverty syndrome. Power was concentrated in the hands of the ruler and his councillors, the legislature being still a nominal institution. It was a raj of jagirdars, chakdars, and maufidars who constituted the bureaucracy and the religious class;[22] the dominant majority of these privileged rights holders, even at the lower rungs of the bureaucracy, were monopolized by the Rajput Dogras and Kashmiri Pandits—the privileged people of the princely order.[23] The Muslims in the valley and lower castes in Jammu were mainly peasants, and a large section comprised the tenants of absentee upper-caste Hindu jagirdars, chakdars, and maufidars. Corruption and nepotism were so rampant that the poor masses could hardly dream of improving their wretched position. The common people lived in such appalling poverty that it was not possible for them to manage the few annas which primary and secondary education entailed.[24] State terrorism was so wanton that the sight of a policeman would send shivers down one's spine, especially in rural areas.[25] The slaughter of cows/buffaloes/oxen by famine-stricken people would bring merciless reprisals from the corrupt and communal police.[26] The

[20] Godbole Report, 3.

[21] Government of J&K, *Administrative Report of Jammu and Kashmir State*, General Administration Department (Srinagar, 1951).

[22] For details see Mirza Mohammad Afzal Beg[0], *On the Way to Golden Harvest: Agricultural Reforms in Kashmir* (Government of J&K, 1951)

[23] Riots Enquiry Committee Report, Witness of Pirzada Ghulam Rasool, Headmaster, Islamia High School, Srinagar, July 1931, Witness no. 87; Glancy Commission Report, cited in *Kashmiri Mussalmanun ki Siyasi Jadujahad*, 110–11; *Inqilab*, Lahore, 8 July 1930, 4 January 1931, 15 February 1931, and 26 June 1931.

[24] Based on interviews conducted with people who were contemporary with the times.

[25] Based on interviews.

[26] Based on interviews.

highest functionaries of the princely order, not mincing words, called Jammu and Kashmir a Hindu state, though Muslims constituted 77 per cent of the total population of the state.[27] Sir Benegal Narsing, the prime minister, who according to Sheikh Abdullah was 'a noble soul', said in a press conference at the time of his taking over as prime minister that 'Jammu and Kashmir is a Hindu state. But I do wish that non-Hindu population also makes progress.'[28]

Keeping with its commitment to building a new Kashmir, the National Conference government initiated the process of implementing its manifesto immediately after it assumed power, resulting in some remarkable structural changes, although the party stayed in power for less than six years. Critics of Sheikh accuse him of having carried out the reforms, especially agrarian reforms, in a 'breathless hurry' on account of 'political exigencies'.[29] Indeed, Sheikh had a political agenda in being hasty. He wanted to sell to the world, which was debating Kashmir, the idea that the people of Kashmir, especially Kashmiri Muslims, were happy with their present political position. He was also facing the serious challenge of sustaining his mass following in the circumstances of an impending plebiscite. This is what he unhesitatingly underlined in his speech to the Constituent Assembly in 1951.[30] Yet, we cannot ignore the Sheikh's deeply felt desire to end the exploitation of the masses and inaugurate a new era as soon as possible, without being sabotaged by forces inimical to the reforms. He thus began

[27] The British census of 1941.

[28] Abdullah, *The Blazing Chinar*, 219.

[29] 'Sheikh Abdullah's Downfall', *Amrita Bazar Patrika* (Calcutta), 17 March 1955, quoted in *Kashmir after August 9, 1953* (Srinagar: Lalla Rookh Publications, 1955), Archives Reference Library, Srinagar, Accession no. 554/GACC, 13. Legislative Assembly Jammu and Kashmir, Assembly Debate Part I First Volume (1951-1955), 252, http://jklegislativeassembly. nic.in/debates/debate%20Part%20I.pdf

[30] 'We have already heard that news of our Land Reforms has travelled to the peasants of the enemy-occupied area of our state, who vainly desire like status, and like benefits.' Inaugural speech made by Sheikh Mohammad Abdullah in the Constituent Assembly of Jammu and Kashmir, 1951. Also see Qasim, *My Life and Times*, 44.

the implementation of the Naya Kashmir Programme with promises which, without entailing any serious financial implications, were of a revolutionary character and in the interest of the general masses. Implemented in haste, these measures suffered from serious infirmities. Yet, a good beginning to ameliorate the condition of the toiling masses was made, and that too with a big bang. The most revolutionary steps taken during the period are described in what follows.

Agrarian Reforms

As late as 1947, the peasantry of Kashmir was still labouring under medieval conditions as the princely order refused to dismantle the feudal agrarian relations.[31] The state of J&K was a landlord-dominated society when British paramountcy over the princely state lapsed. There were about 13,000 landlords in J&K in 1947–8. Out of them, 396 were the biggest landlords, called jagirdars, to whom were alienated the land and other revenues of a vast area yielding an annual revenue of Rs 566,313.[32] The area assigned to them was called 'jagir'. The jagirdars were not owners of the land; they were, however, owners of the revenue of the areas assigned to them. Along with the alienation of land revenue, the state also transferred the people of the jagir to the control of the jagirdar. The remaining 9,000 landlords comprised those landed magnates who, unlike the jagirdars, were owners of the land. They were called chakdars, and the land owned by them was known as chak. Each chakdar on an average owned around 73 acres of *ābi* (paddy) land,[33]

---

[31] For an account of the abject poverty of the peasantry, see Government of J&K, *4 Years*, Archives Reference Library, Accession no. 501/GACC, 11.

[32] The jagir of the raja of Chinani was 95 square miles in area with a population of 12,000. His jagir was revoked and in return he was given a monthly allowance of Rs 300. Beg[0], *On the Way to Golden Harvest*, 209; Mirza Mohammad Afzal Beg, 'Land Reforms in J&K', *Mainstream* 15 (1976), 27.

[33] This is calculated on the basis of the following facts:

1. Total number of chakdars = 9,000
2. Maximum ābi land a chakdar was permitted to retain = 22¾ acres

excluding orchards and other lands classified under fuel and fodder resources. Besides the jagirdars and chakdars, there was also a class of cash grantees called *mukarares*, numbering 2,347. The number of landless peasants (called *kashtkars*) was around 300,000.[34] Apart from this, 250,000 peasants were part-tenants-part-owners;[35] that is, part of the land cultivated by them was actually under the possession/ownership of the jagirdars/maufidars/chakdars. From 1948 to 1950, the government took a series of steps to liquidate feudalism and relieve the peasantry from the burden of over-taxation and fear of arbitrary eviction.

Abolition of Jagirs

In April 1948, the government abolished the jagirdari and mukarare systems. Accordingly, the jagirs of 396 jagirdars and the cash grants of 2,347 mukarare holders were revoked. Further, 4,250 acres of land were transferred to landless tillers.[36]

Amendment to the Tenancy Act of 1924

In October 1948, the government amended the State Tenancy Act 1924, reducing the rental payment by a tenant. He was now liable to pay not more than three-fourths of the produce in the case of irrigated land, and not more than two-thirds of the produce in the

---

3. Total surplus ābi land expropriated = 4.5 lakh acres, which on an average amounted to 50 acres per chakdar.

Since these 50 acres were in excess of the 22¾ acres a chakdar was allowed to retain, it means that on an average the chakdar owned 73 acres of ābi land.

[34] Wolf Ladejinksy, 'Land Reforms: Observations in Kashmir', in *Agrarian Reforms as Unfinished Business*, ed. L. J. Walinsky (Oxford: Oxford University Press, 1952), 179.

[35] Wolf Ladejinksy, 'Land Reforms: Observations in Kashmir'.

[36] Government of J&K, *Land Reforms in Jammu and Kashmir*, Department of Information, Archives Reference Library, Srinagar, Accession no. 666/GACC, 1.

case of dry land, provided the tenancy holding exceeded 12½ acres. In case the size of holdings was 12½ acres or less, the landlord was entitled to one half of the produce.[37] Also, by amending the law, fixity of tenure was granted to tenants in respect of tenancy holdings not exceeding 2⅛ acres of irrigated land or 4⅛ acres of dry land in the Kashmir province, and about double the size in Jammu province.[38] Moreover, through this amendment, tenants were protected from arbitrary eviction without court procedures.

Big Landed Estates Abolition (*Kahtima-i-Chakdari*) Act 1950

On 13 July 1950, the government announced, in the words of Brecher, 'the most sweeping agrarian reform undertaken in the Indo-Pakistan subcontinent since the partition'.[39] The announcement became a reality with the enactment of what is known as the Big Landed Estates Abolition Act on 17 October 1950. This legislation set a maximum limit of 22¾ acres on the holdings of landowners excluding orchards, fuel, and fodder resources and uncultivable wastelands.[40] Ownership rights of land in excess of this amount was transferred to the tiller. Importantly, the tiller did not have to pay any compensation to the original owner. Land belonging to the religious shrines of different communities, however, remained untouched. Also, to appease the politically powerful lamas (the Buddhist clergy), these reforms were not introduced in Ladakh. 'Extremely interesting', says Joseph Korbel, 'was the provision for the confiscation of the property of the "enemy agents", these agents being largely defined as persons who had expressed a desire to join Pakistan.'[41]

---

[37] State Tenancy (Amendment) Act VII of Samvat 2005.

[38] Government of J&K, *Land Reforms in Jammu and Kashmir*, 1.

[39] Brecher, *Struggle for Kashmir*, 159.

[40] These 22¾ acres of land consisted of different categories—20 acres of agricultural land, 1 acre of land for residential use or vegetable gardening, ½ acre as residential site, and 1.25 acres of orchards. Korbel, *Danger in Kashmir*, 211; Beg, *On the Way to Golden Harvest*, 28; Beg, *Land Reforms in J&K*, 28.

[41] Korbel, *Danger in Kashmir*, 211.

As a result of this drastic reform, by the end of March 1953, 188,775 acres of land had been transferred to 153,399 tillers, indicating that each peasant received thereof an average of 1.23 acres of land under this programme.[42] The process of transfer of land continued. By the end of April 1953, 192,652 acres of land had been attested in favour of 160,939 tillers. In addition to this, more than 93,500 acres had come to be vested in the state.[43] According to an official document under the title *Land Reforms in Jammu and Kashmir*, issued by the Department of Information, J&K Government, in the late 1950s, in all 9,000 landowners were expropriated of their surplus land, amounting to about 450,000 acres, out of which ownership rights of around 230,000 acres were transferred to the tillers.[44] The government also established collective farms on lands which were not distributed among the peasants. Collective farms were established at Gopalpora, Shalteng, and Harwan. By April 1953, 87,500 acres of land were government owned.[45]

Corruption in the administrative machinery and some loopholes in the scheme, such as the exemption of orchards from the ceiling limit and no distinction being made between the 22.75 acres of dry and irrigated/fertile land, and other such limitations, vitiated the effectiveness of the reforms in practice. Mir Qasim, who was actively associated with preparing the law, observed:

> There were some lacunas in the land reform. We had fixed 182 kanals for both Kashmir and Jammu; although in Jammu the land fertility was low. Secondly, the landlords had been given the right to choose the area they wanted to retain. This gave a landlord the tool to extort money from his tenant on the threat that he would choose to keep his tenant's portion of land with him. Thirdly, there were complaints that the implementation of the land reforms had been

[42] Korbel, *Danger in Kashmir*, 212; Government of J&K, *4 Years*, 4; H. D. Malaviya, *Land Reforms in India* (New Delhi: AICC, 1954), 422; Sisir Gupta, *Kashmir: A Study in India Pakistan Relations* (New Delhi: Asia Publishing House, 1966), 396.

[43] *Hindustan Times*, 17 July 1953.

[44] Government of J&K, *Land Reforms in Jammu and Kashmir*, 3–4.

[45] Korbel, *Danger in Kashmir*, 212.

left to the whims of the corrupt bureaucracy. It was a revolutionary program which had fallen a prey to large-scale corruption.[46]

Many other contemporaries concurred with Mir Qasim's frank admission in saying that land reforms were executed by the local corrupt bureaucracy in favour of the rural aristocracy. In the words of Joseph Korbel, 'many landless peasants received considerably less than the average (acres of land), because many local officials were given more and better land, sometimes even above the maximum.'[47]

The *Statesman* reported in its 28 August 1953 issue:

> Experience has shown that what intended to be a free conferment of land on the tiller has usually involved him in great expense. . . . the distributing agencies as well as some of the landlords have become richer, as proprietors always selected the best land for their own use, tillers had to pay the cost of land to secure its exclusion from the proprietor's unit.[48]

Speaking in the Constituent Assembly in March 1952, a prominent National Conference member, Rampiara Saraf, complained that 'since the land reforms in the state were being implemented by a bureaucratic officialdom tillers have to pay bribes which in no case are less than the actual compensation.'[49]

Daniel Thorner, the agrarian historian and economist, who visited the valley in 1953, made a balanced critique of the land reforms:

> Land reform in Kashmir has clearly done away with the jagirs, and has weakened the position of all the great landlords. It has distinctly benefited those individuals who, at the village level, were already the more important and substantial people. It has done the least for the petty tenants and landless labourers. These two categories being the largest in the countryside.[50]

[46] Qasim, *My Life and Times*, 45.

[47] Korbel, *Danger in Kashmir*, 2012.

[48] 'Many Anomalies in Working of Land Reforms', *Statesman*, 28 August 1953.

[49] Bazaz, *History of Struggle for Freedom in Kashmir*, 430.

[50] Daniel Thorner, 'The Kashmir Land Reforms: Some Personal Impressions', *Economic and Political Weekly*, no. 37 (1953[0]), 999–1002.

Clearly, as observed by P. K. Bardhan in the case of land reforms in India as a whole, 'laws were frequently enacted with deliberate loopholes and tell-tale exemptions designed to induce fictitious transfers of land to close and distant relations and to keep the permissible retentions high.'[51] Taking advantage of the deliberate limitations in the law, the landlords devised their own ways to defeat the purpose of the law. For example, in order to evade resumptions, they showed their families as having broken up into separate households, which entitled each adult male to the limit of 22.75 acres.[52] Likewise, since the act exempted orchards from appropriations, it paved the way for big landlords to escape the ceiling by converting cereal acreage into orchards. It is, therefore, no wonder that in the neighbourhood of old Srinagar, for example, we find thousands of kanals of land being owned by the descendants of the Dogra bureaucratic class.[53] Indeed, a large number of peasants still continued to be landless tillers working on the lands of these landlords who, besides having 182 kanals of ābi land, now possessed huge orchard lands.

Despite these limitations, the act was a progressive measure. It signalled a new era of peasant emancipation and marked the beginning of the ultimate liquidation of the feudal set-up in the state. Thousands of peasants previously living in virtual slavery became landholders. Moreover, as Wolf Ladejinsky observed,

> whereas virtually all land reforms in India lay stress on elimination of the Zamindari [large estates] system with compensation, or rent reduction and security of tenure [for tillers], the Kashmir reforms call for distribution of land among tenants without compensation to the erstwhile proprietors ... [and] whereas land reforms enforcement

[51] P. K. Bardhan, *The Political Economy of Development in India* (New Delhi: Oxford University Press, 1984), quoted by Prakash, 'Political Economy of Kashmir since 1947', 2054.

[52] Robert I. Crane, ed., *Area Handbook on Jammu and Kashmir State* (Chicago: University of Chicago Press, 1956), 361.

[53] A huge number of orchards commonly called *bhags* (gardens), for example, Dhar Bagh, Sikh Bhag, Bata Bhag, Mirza Bhag, Miskeen Bagh, Chai Bagh, Pandit Bagh, and Sahani Bagh, on the periphery of old Srinagar were owned by the erstwhile Dogra bureaucracy.

in most of India is not so effective, in Kashmir enforcement is unmistakably rigorous.[54]

However, in parts of the Jammu region and among the Kashmiri Pandits, who were the main beneficiaries of feudalism, the land reform 'catalysed a tenacious movement of social and political reaction'.[55] The land reform was dubbed as anti-Hindu and pro-Muslim. This is rubbished by Mir Qasim, the secular-progressive former chief minister of Kashmir. As deputy minister in Sheikh Abdullah's government, Mir Qasim was associated with the programme of land reforms:

> After the law was enacted, problems came to the surface and criticisms were made—some genuine, some motivated. The gravest charge, though not well-founded, was that the land reforms were designed to pass on the land of Hindu landlords to Muslims. The critics conveniently ignored the fact that this law equally applied to Muslim landlords and that a large number of Hindu tillers, especially Harijans, in Jammu benefited from it.[56]

Sheikh Abdullah was also distressed to see his reforms being given a communal colour rather than being appreciated for their progressive nature:

> The exploiters of this state and their supporters in the Centre did not like our land reforms. Sardar Patel especially opposed these reforms. The main reason for this opposition was that the Hindu Jagirdars of state had made him to believe that these reforms were carried out to satisfy the sentiment of our religious fanaticism as most of the landlords affected by these reforms were non-Muslims. With the help of the statistics I tried to satisfy Sardar Patel that there is no question of religious fanaticism involved in introducing these reforms as the jagirdars as well as the tenants belonged to both the communities. But Sardar Patel could not be motivated and he continued opposing these reforms. However, Jawaharlal Nehru and Maulana Abul Kalam Azad were in favour of these reforms.[57]

---

[54] Ladejinksy, 'Land Reforms', 179–80.

[55] Bose, *Kashmir: Roots of Conflict, Paths to Peace*, 2003, 28.

[56] Qasim, *My Life and Times*, 44.

[57] Abdullah, *Atash-i-Chinar*, 493.

Although Sheikh Abdullah tried to convince his opponents that the agrarian reforms, far from being driven by any communal agenda, were motivated by the desire to legitimize his political preference (of supporting accession) by economic logic,[58] they could not be convinced. According to Gundevia, the foreign secretary during Nehru's government, Sheikh Abdullah's dismissal was a conspiracy hatched by the 'reactionary elements' in the Home Ministry to see him out of power before the Kashmir constitution sanctioned the 'no compensation' part of the Big Estates Abolition Act.[59] Mir Qasim also corroborates Gundevia's account, saying, *'in my opinion these land reforms were the beginning of the mistrust between New Delhi and Sheikh Abdullah'.*[60]

Distressed Debtors Relief Act 1949

In 1946, H. M. Brailsford, a famous British writer who had studied debt and poverty in the villages of India, visited Kashmir and observed: 'The peasants are sunk in unimaginable poverty. Their mud huts contain hardly a trace of visible property, save a few pots and water jars. When I put my questions in a typical village, every household was in debt, and the usual rate of interest was 48 per cent.'[61] Another contemporary observer of Kashmir also provides revealing information about the rural indebtedness in the state:

> Though not as important as the land problem the staggering rural indebtedness is by no means ignorable. Poor, illiterate and

[58] Sheikh Mohammad Abdullah, Inaugural Speech in the Constituent Assembly of Jammu and Kashmir, 1951.

[59] 'From Sheikh's speeches in the Kashmir Constituent Assembly that followed, it was clear that the "no compensation" part of the Big Estates take-over was going to be sanctified in the Kashmir Constitution, which, in turn was next on the anvil. The only hope, the only way out, if a lot was to be averted, was to throw Sheikh Abdullah out before the two Committees appointed by him to go into these various matters could submit their recommendations to the Kashmir Constituent Assembly.' Gundevia, 'On Sheikh Abdullah', 110–11.

[60] Qasim, *My Life and Times*, 44 (emphasis mine).

[61] Government of J&K, *4 Years*, 11.

simple-minded working classes in the state have been literally enslaved by money lenders, a small amount once lent to any working man has often made him a debtor for ever. He is forced to go on paying in the shape of interest for years and decades. Not infrequently a debt incurred by father is inherited by the son. Freedom has no meaning for such a worker if he is not liberated from the clutches of the shylock.[62]

In such a situation, the great service that could be rendered to the poverty-stricken people was to free them from indebtedness. Official reports estimated rural indebtedness in Kashmir at about Rs 31,000,000 and urban indebtedness around Rs 5,600,000.[63] To give immediate relief to the peasants and other poor sections of the population, the government passed Ordinance no. XXI of Samvat 2004, which stayed for the period of one year all suits or proceedings for the realization of the debt against agriculturists and others.[64] To further alleviate the distress of the poverty-stricken people of the state, especially agriculturists, the J&K Distressed Debtors Relief Act no. XVI, Samvat 2006 (1949) was enacted. Following the passage of this act, five debt conciliation boards were appointed in the districts of Anantnag, Baramullah, Kathua, Jammu, and Udhampur to bring about voluntary conciliation between debtors and creditors.[65] The boards, which included officials and people's representatives, were empowered to liquidate a debt if a borrower had already paid interest equal to the principal sum borrowed. Debt claims of about Rs 17.5 million were conciliated by them and scaled down to Rs 8.5 million. Moreover, mortgaged debts of the value of Rs. 1.459 million were liquidated.[66] This relieved the agriculturists, artisans, and village menials of the burden of their accumulated debts. Moreover, to enable mortgagors to recover their property in a summary manner and to allow them the benefit of

[62] Bazaz, *History of Struggle for Freedom in Kashmir*, 428.

[63] Das, *Jammu and Kashmir*, 189–90.

[64] Government of J&K, *In Ninety Days: A Brief Account of Agrarian Reforms Launched by Sheikh Mohammad Abdulla's Government in Kashmir* (Jammu: Land Reforms Officer, 1948), 18–19.

[65] *Jammu and Kashmir 1947–50*, 74.

[66] Government of J&K, *Land Reforms in Jammu and Kashmir*, 3.

income received by the mortgagee during the period of mortgage, a new act known as the 'Mortgaged Properties Act' was enacted. Under the provisions of this act, if the court found that the value of the benefits enjoyed by the mortgagee equalled or exceeded the cost of improvements, if any, effected by the mortgagee, plus 1½ times the amount of the principal money or the pecuniary value of goods actually advanced, the mortgage would be extinguished. In no case should the principal sum plus the interest at a rate not higher than 6 per cent exceed 1½ times the principal sum or the value of goods actually advanced.[67] Consequent upon this measure, immoveable property worth Rs 3.7 million, which had been mortgaged by 34,000 persons, was resituated.[68] According to Michal Brecher, by June 1953 debts amounting to Rs 11,122,054 had been reduced by approximately 80 per cent through the instrument of the debt conciliation boards.[69] This is also substantiated by Korbel, who says that the debt conciliation boards disposed of 48,000 applications in less than three years, and scaled down the debts by about 80 per cent from 11.1 million to 2.43 million rupees.[70]

However, a new problem arose with the promulgation of this law, as it left a grave vacuum in the credit system in rular areas that the regime, for the want of finnaces, was not able to fill. This had the result that the new owners had no money to purchase the inputs. 'The capitalists or usurers no longer help him because they feel aggrieved on account of the new law,' reported *Khidmat*.[71] The Jammu National Conference adopted a resolution revealing that 'needy peasants and workers are made to sign documents holding them responsible for double the amounts than are actually lent to them, and this is done in presence of magistrates so that it may be proved, if need be, that the transactions are authentic.'[72]

---

[67] Javid ul Aziz, 'Economic History of Modern Kashmir with Special Reference to Agriculture (1947–1989)', PhD thesis, Department of History, University of Kashmir, Srinagar, 2010, 63.

[68] Government of J&K, *Land Reforms in Jammu and Kashmir*, 1.

[69] Brecher, *Struggle for Kashmir*, 159.

[70] Korbel, *Danger in Kashmir*, 214.

[71] Quoted in Bazaz, *History of Struggle for Freedom in Kashmir*, 430.

[72] Bazaz, *History of Struggle for Freedom in Kashmir*, 434.

The *Hindustan Times* also pointed out the problems faced by rural people in the absence of alternative credit facilities after the passing of the act:

> Ironically enough, the beneficiaries [of the land reform] themselves have yet to reap the full advantage out of their ownership which is currently operating more as a liability than as asset; for the cancellation of agricultural indebtedness, with the stroke of a pen, has left a serious void in the rural credit system which the regime has been unable to fill, with the result that more often than not, the new owner has no money to buy a bullock or agricultural implements with.[73]

The Grow More Food Policy

Apart from introducing structural changes in production relations, which itself largely accounted for an increase in agricultural production, the Sheikh government was passionate about making Kashmir self-reliant in food so as to realize twin objectives: economic development, and safeguarding the political autonomy of the state which was threatened in the face of overbearing economic dependence on the centre.[74] The government thus followed the policy of 'Grow More Food' to achieve self-reliance. The Grow More Food scheme was launched in 1948. Under this scheme, the government allowed peasants to bring all that cultivable land under the plough over which the state exercised ownership rights. The only condition was that such land should not belong to the category of *gas charai* (grazing land), and the peasant had to pay half of its produce to the state.[75] In 1948–9 alone, 185,583 kanals of cultivable wasteland were allotted to landless peasants.[76] As a

[73] *Hindustan Times* (Delhi), 23 May 1953.

[74] Abdullah, *Ātash-i-Chinar*, 497, 500–1.

[75] Government of J&K, Order no. 48-Cof, 17 April 1948; *In Ninety Days*, 20–21; *Jammu and Kashmir 1947–50*, 4; Government of J&K, *On the Road to New Kashmir*, Ministry of Information and Broadcasting (Jammu: Ranbir Government Press), Archives Reference Library, Srinagar, Accession no. 1241, 11–12.

[76] *Jammu and Kashmir 1947–50*, 4–6.

result of this measure, the increase in agricultural produce was estimated at about 200,000 maunds.[77]

Besides this, the development of irrigation was made a top priority, and efforts were made to introduce better seeds and manure. To give further impetus to agriculture, an independent department of irrigation was established, and the development of irrigation received the highest priority in the First Five Year Plan as 60.5 per cent of the total plan allocation was earmarked for this sector.[78] Ten old and new canals were either restored or established. One of these canals, the Awantipora canal which was completed at the cost of about Rs 8 million, irrigated an area of 4,000 acres of land.[79]

Education

Promoting education and the reorganization of the educational system to meet new demands figured among the two most important priorities of the Abdullah government. In the Sheikh's opinion, 'land to the tiller and education for everyone were the two basic needs, if the people of the state . . . were to emerge into the wider lands of plenty and enlightenment, the "golden threshold" to a fuller life.'[80] The taking over of the education portfolio by the prime minister himself signalled the recognition of the importance he attached to education in inaugurating a new era in Kashmir. As against the imperial policy of Macaulay, Abdullah wanted men and women with creative and constructive minds rather than mere clerks. He also wanted that the obsession with a purely scholastic outlook should end, and so must the present aimless drift of boys with a vague idea of getting into the state service at the end.[81]

[77] *Jammu and Kashmir 1947–50*, 4–6.

'Maund' is the term for a traditional unit of weight used in British India. One maund is equal to 37.32 kilograms.

[78] Government of J&K, *First Five Year Plan Document*, 1951 (Srinagar: Department of Planning and Development).

[79] Korbel, *Danger in Kashmir*, 212.

[80] Government of J&K, *Report of the Educational Reorganisation Committee* (Jammu: Department of Education, 1950), i.

[81] Government of J&K, *Report of the Educational Reorganisation Committee*, 8.

The purpose of the government was to reorganize and restructure the system in such a way that, as enunciated in the New Kashmir Programme, students who wished to make matriculation the final stage of their studies would be fully equipped to enter society as responsible earning members without having to proceed further. And a student of 'scholastic bent of mind should find it intellectually a satisfying stage preparing him for further studies'.[82]

Accordingly, immediately after Abdullah took over the government, an eminent educationist Mr M. A. Kazmi was appointed director of education, and under his chairmanship an Education Reorganization Committee was set up to suggest improvements in the existing educational system. The important recommendations of the committee were: arts and crafts would form an integral part of the syllabus at the primary stages, and crafts would be selected by individual schools from among those which were popular in their locality and which possessed both educational and economic value; and a four-year course would be instituted for secondary education, where, besides the ordinary subjects, special care should be given to impart scientific knowledge, with an agricultural, technical, and industrial bias to meet the needs of the state and society.[83] A beginning was made in putting these recommendations into practice by setting up two urban and one rural multipurpose schools.[84]

The second measure taken by the new government to meet the demands of the New Kashmir Programme was to constitute a textbook advisory board in 1948, with the prime minister as its chairman, to prepare new textbooks to be written in the spirit of the New Kashmir Manifesto.[85] According to government sources, 2,790,500 volumes of textbooks were published up to 1950. Refresher courses were also organized to 'indoctrinate new teachers' in the spirit of the New Kashmir Manifesto.[86] Another important innovation of the time was what was called 'social education'

---

[82] Government of J&K, *Report of the Educational Reorganisation Committee*, 8.

[83] See *Jammu and Kashmir 1947–50*.

[84] Government of J&K, *4 Years*, 7.

[85] Government of J&K, *4 Years*, 5.

[86] Government of J&K, *4 Years*, 5; Korbel, *Danger in Kashmir*, 209.

for adults, the purpose of which, in the words of Sheikh Abdullah, was 'to educate public opinion in the ideology of New Kashmir'.[87] A network of social education centres spread over the entire state (60 each in Jammu and Kashmir provinces).

It was envisaged in the New Kashmir Plan that education in all primary schools would be imparted in the mother tongue. To implement this 'sound psychological' principle, two committees were set up, namely, a language committee and a script committee.[88] The script committee adopted the Persian-Arabic script (Naskh style), introduced the necessary letter symbols to denote the sounds peculiar to Kashmiri phonetics, and in a short time produced the first primer on the language. It was thus possible to introduce Kashmiri as the medium of instruction in the first and second primary classes from April 1949, and in the third primary class, from the session of 1950.[89] It was expected that by 1952 Kashmiri would have become the medium of instruction at the primary stage in all subjects throughout the Kashmiri-speaking areas. A script training centre was set up where teachers were instructed in the new Kashmiri script.[90]

That the government attached great importance to education is also evidenced by the fact that 35 per cent of the annual budget was spent in the expansion of education and the development of educational infrastructure.[91] As a result, according to Korbel, 'much has been done in this field.'[92] The Jammu and Kashmir University was established in the year 1948.[93] New schools were opened, some 60 of them for the age group 3–5. Two intermediate colleges were opened in Anantnag and Sopore, and one such college was started exclusively for girls in Srinagar.

---

[87] *Jammu and Kashmir 1947–50*, 5, quoted in Korbel, *Danger in Kashmir*, 209.

[88] Government of J&K, *4 Years*, 5.

[89] Government of J&K, *4 Years*, 5.

[90] Government of J&K, *4 Years*, 5.

[91] Abdullah, *Ātash-i-Chinar*, 499.

[92] Korbel, *Danger in Kashmir*, 209.

[93] Korbel, *Danger in Kashmir*, 209; Government of J&K, *4 Years*, 4.

The Policy of Nationalization

With the firm belief that a backward society necessarily requires state intervention for improving the condition of the people in particular and the state in general, the authors of the New Kashmir Programme insisted on cooperative enterprise 'as opposite to destructive competition', and a controlled and organized market instead of spontaneous marketing and trade. They also underlined the need for keeping all industries (except for small-scale enterprise) in the hands of the 'people's government', and the abolition of large private capitalist enterprise and private monopoly. Accordingly, the development of industry and trade with full state support formed an important thrust area of the economic policy of the National Conference government.

The turmoil of 1947 and the Partition of India had dealt a big blow to the timber trade in the state, which depended on waterways passing through the newly created dominion of Pakistan. The government restored it quickly by constructing the Jammu–Pathankot road, and selling the wood at subsidized prices. Motor transport was made available at cheap rates so that the timber did not become cost-prohibitive in the Indian markets. Similarly, considerable government support was given to the silk industry; silkworm eggs were imported from foreign countries and distributed among rearers. The already existing government-owned silk-weaving factories continued to be patronized by the government.[94] Other industries which were supported by the state included the manufacture of wool, sports goods, drugs, matches, and carpets. The available sources of information do not indicate which of these industries were government owned, but they all depended on governmentally distributed supplies and marketing.[95] Among the major industries which were established by the government during the period were a joinery mill and a ply board factory at Pampore, a cement factory at Wuyan, and the Drug Research Laboratory in Jammu.[96] The government also owned two printing houses, Ranbir and Partab Government Press.

[94] Korbel, *Danger in Kashmir*, 215.

[95] Korbel, *Danger in Kashmir*.

[96] Korbel, *Danger in Kashmir*. Also see *Jammu and Kashmir 1947–50*.

The government organized and subsidized cottage industries and helped new entrepreneurs to start small-scale manufacturing units.[97] Special attention was given to handloom weaving, hosiery, furniture, ceramics, paper manufacturing, embroidery, and papier mâché. Artisans of these industries were organized into industrial societies.

The government also laid great emphasis on the cooperative movement, as it was felt that the problems of the rural population would not be solved unless the entire village life was brought within the fold of the cooperative movement. To achieve this purpose, the cooperatives were tasked with scaling down the debts of the members to the extent of their repaying capacity. The cooperatives had also to spread the reduced debt over a number of years, lease the expropriated land to the members, finance crops, supply daily commodities, and facilitate marketing.[98]

To supply the necessaries of life, the government transferred the work of distribution of basic goods to the cooperative societies. For this purpose, it was decided to organize a network of multipurpose cooperative societies—one each in Patwar Halqa.[99] By the end of April 1950, there were 306 multipurpose cooperative societies with a membership of 41,221 and a paid-up share capital of Rs 178,993 in Kashmir province, and 30 societies with a membership of 3,190 and paid-up share capital of Rs 25,225 in Jammu province, increasing steadily thereafter.[100] In 1950, there were 1,731 agricultural cooperatives, 386 purchase and sale cooperatives, and 378 non-agricultural credit cooperatives.[101]

The multipurpose cooperative societies drew their supplies from the cooperative stores, which in turn were fed by the central institution, the Kashmir People's Cooperative Society Limited. Essential commodities like salt, sugar, kerosene, cloth, and cotton

[97] P. N. Bazaz, *Kashmir in Crucible* (Srinagar: Gulshan Books, 2005 [1967]), 47–8.

[98] Government of J&K, *4 Years*, 14–15; also see *Jammu and Kashmir 1947–50*.

[99] *On the Road to New Kashmir*, 4.

[100] Government of J&K, *4 Years*, 15.

[101] Korbel, *Danger in Kashmir*, 215–16.

yarn were supplied through these cooperatives at fair prices. The Supplies Department, which imported these commodities, fixed the quantity per consumer as well as selling price in such a way as to make the price level uniform in all the rural areas. The deficit, if any, was met from the departmental premium. The Supplies Department also fixed the ceiling price of general consumer goods.

Though the cooperative movement could have been of great help to poor peasants in particular, and others in general whose association in cooperatives would help relieve their financial misery, in practice it turned into an instrument of the National Conference's party politics.[102] Moreover, as the government itself had to admit in the summer of 1953, the cooperatives collapsed completely because of corruption and the malpractices of government officials.[103] 'Only relatives and friends of the Nationalists in the country side get these goods at the controlled rates,'[104] says Prem Nath Bazaz. As these cooperatives were run by National Conference workers, it was risky for the government to displease them. Thus, the cooperative movement became a scourge for the common people as it served only the influential sections of the society.[105]

All foreign trade was conducted through the office of the Trade Commissioner in Delhi. It was through this department that art emporiums were established at different places in India, namely, New Delhi, Bombay, Shimla, Lucknow, Madras, Calcutta, and Amritsar. It was also through this office that goods for different departments were purchased. This agency was also entrusted with the task of promoting tourism. Nationalized bazaars were organized for trade within the country.[106]

---

[102] *Noonas gauos National Waanus / Thoupham gode ral Hindustanus Seeth, / Zoojan wandha Hindustanus, / Dil chum Pakistanus Seeth.* (I went to a shop, run by a National Conference worker, to purchase salt. The shopkeeper told me that without pledging support to India, I cannot get salt; I replied, I do not hate India, but my heart is for Pakistan.) Butt, *Kashmir in Flames*, 48. Also see Korbel, *Danger in Kashmir*, 216.

[103] See Gupta, *Kashmir: A Study*, 396–7.

[104] Bazaz, *History of Struggle for Freedom in Kashmir*, 431.

[105] Bazaz, *History of Struggle for Freedom in Kashmir*, 431–2.

[106] *Jammu wa Kashmir 1947–50: Sher-i-Kashmir Sheikh Mohammad Abdullah ki hukumat kay teen salah kar kardagi* (Urdu) (Jammu:

The transport was managed by the government Transport Department, which owned some 400 vehicles, most of which operated between Srinagar and Jammu.[107] About 1,000 people were employed in the government Transport Department.[108]

The government's monopoly over trade and business did not go without a reaction from the Kashmir Chamber of Commerce (KCC), which was still dominated by Hindu members. In the spring of 1953, the KCC submitted a memorandum to the government requesting the abolition of state trading; decontrolling of all commodities; introduction of free and healthy competition in purchase, distribution, and sale; removal of restrictions on private transport; and cancellation of existing monopolistic licences in favour of a few individuals and firms.[109]

Local Governance

Korbel, a sober critic of Sheikh Abdullah, was impressed by the system of dense local organizations—panchayats and cooperatives—to govern at the grassroots level, but, like other observers of the period, he also found the otherwise attractive system being destroyed by the hegemonizing role of local National Conference organizations:

> In truth the National Conference is the only effective political party in Kashmir, with local organizations in almost every village. Although on the surface considerable progress has been achieved in local self-government, in reality it is the local National Conference organization which decides everything: who is going to be elected to what office, who will get a job, who will receive the supplies which it alone distributes.[110]

  [107] Government of J&K, *4 Years*, 7.
  [108] Government of J&K, *4 Years*, 9.
  [109] Korbel, *Danger in Kashmir*, 216.
  [110] Korbel, *Danger in Kashmir*, 208.

Not surprisingly, therefore, the rule of the National Conference came to be contemptuously called 'Halqa President Raj'.[111]

Empowering the Disempowered

The Sheikh government, true to its socialist policy, took steps to empower the Muslims who, despite being the largest segment of the population of the state, were abysmally under-represented in the state services. The government initiated the policy of appointing Muslims as per their population ratio, and this policy was maintained by Abdullah's immediate successors. As a result, the Muslim representation in gazetted positions rose from 30 per cent in 1947 to 50 per cent in 1953.[112] For the first time, the state had a Muslim inspector general of police, a Muslim accountant general, and a sizeable number of local Muslims as deputy commissioners and high police officials. And with the establishment of the National Militia, thousands of Kashmiri-speaking Muslims received, for the first time after 100 years of deprivation, the opportunity to serve in the military.[113] The Sheikh's policy of giving Muslims a share in the service sector as per their population ratio exacerbated social tensions, because until 1947–8 this sector had been monopolized by the favourites of the maharaja, who were mainly Kashmiri Pandits and Dogra Rajputs. Like the policy of land to the tiller, the new recruitment policy also invoked New Delhi's sympathies with the affected community of Kashmir. To quote Sheikh Abdullah:

---

[111] Butt, *Kashmir in Flames*, 48. Being a cadre based party, the National Conference had a well-knit organizational structure operating from village/mohalla to state level. At lower level the most important unit operated at the halqa level. Halqa was a conglomerate of a few villages headed by politically important/powerful person of the unit. He was called as *Halqa President*. Since during the NC rule the Halqa President was the real power to reckon with at the Halqa level, that is why the NC government commonly came to be called as *Halqa President Raj*.

[112] Saraf, *Kashmiris Fight for Freedom*, vol. II, 1217.

[113] Saraf, *Kashmiris Fight for Freedom*, vol. II, 1217.

when the country became free and we took over, we thought that justice and equality demanded that we take measures to uplift the backward sections of the Muslim community. Members of the Hindu community, who dominated the administration, misconstrued it as an assault on their monopoly. They knew that their position was weak. So, in order to confuse the minds they raised a hue and cry, and gave a distorted picture of the consequences of the land reforms. Patel on many occasions made queries in this regard. When we presented him with statistics concerning the number of employees working in government departments with a ratio between Muslims and non-Muslims, he was taken aback. The complaints should have come from the Muslims, and not from the Hindus, said he. In response, I said, 'Probably the Hindus believed that the centre is there for the protection of their interests alone and that it should be indifferent to the Muslims.' This brought a smile to his face.[114]

Resettlement of Refugees

One of the graver problems that the National Conference government had to face was settling the refugees who had either immigrated from the newly created dominion of Pakistan, or had to forcibly migrate to Pakistan after the communal frenzy in Jammu. Around 11,086 refugees came to the Kashmir Valley. Besides providing immediate relief in the relief camps in Kashmir Valley, the refugees were settled in Srinagar, Baramulla, Ganderbal, and Sopore.[115] They were provided free rations, and schools were opened for their children. In Jammu, a large number of buildings were vacated for their accommodation. Also, the Department of Dharmath Trust established camps for their temporary residence. The government sanctioned loans and provided timber to the refugees for the construction of houses. And they were settled in Udhampur, Ram Nagar, Jammu, Ranbirsinghpura, Akhnoor, Reasi, and Rajouri.[116] Most of them were settled on the lands of those persons who had migrated to Pakistan. However, for a proper

---

[114] Abdullah, *The Blazing Chinar*, 342.

[115] *Jammu wa Kashmir 1947–50*, 54.

[116] *Jammu wa Kashmir 1947–50*, 54.

resettlement of refugees, a joint committee of the Government of India and the Government of Kashmir was constituted under the chairmanship of Major General Tara Singh. The Government of India had earmarked Rs 4.2 million for the permanent settlement of these refugees.[117]

Following the communal frenzy in Jammu, a huge number of Muslims were driven out of their homes in 1947–8. They took shelter in Pakistan or in Pakistan Administered Kashmir. Although a large number of them were resettled on evacuee properties, thousands were still living in tents at Mansar and Kala. They belonged mainly to Rajouri, Reasi, and Mehndar. Towards the middle of 1949, a movement for the return of refugees started on a small scale, gaining momentum by the end of 1950 under the patronage of the state. A fair estimate of the returnees enumerates them at about a hundred thousand. Sheikh Abdullah's government resettled them on their abandoned properties, advanced *taqqavi* loans, and appointed special staff to address their problems.[118] The influx alarmed New Delhi, and Sheikh Abdullah was asked to cooperate in stopping their entry.[119]

Major Crises

The National Conference government suffered some serious crises—some created by circumstances and some self-created—which came in the way of the implementation of its agenda of New Kashmir, impaired the position of Sheikh Abdullah as a leader of the masses, and also led to his tragic dismissal with enduring consequences. The major crises were: (a) the financial crisis; (b) corruption; (c) intolerance of opposition; and (d) political crisis.

---

[117] *Jammu wa Kashmir 1947–50*, 54.

[118] The loan provided by the state to the peasant/artisans to tide over the problems caused by a natural or man-made calamity was called as *taqqavi* loan. It was to be returned in easy instalments once the condition of the affected peasant/artisan improved.

[119] Saraf, *Kashmiris Fight for Freedom*, vol. II, 1217.

Financial Crisis

The most serious problem faced by the government was the acute scarcity of financial resources on account of the snapping of traditional routes and waterways. The problem was exacerbated by political instability. Prior to 1947, Kashmir was connected with the outside world through many highways and waterways which passed through what came to be the newly created dominion of Pakistan. The business conducted through these routes and waterways had amounted to crores of rupees every year.[120] Since relations between India and Pakistan were strained over the Kashmir dispute, and the Kashmir government had declared Pakistan an enemy country, traffic on these routes and waterways came to a standstill. The Indian government no doubt constructed the Jammu–Pathankot road to connect Kashmir with India, but in no way was it an adequate substitute for the traditional connectivity of Kashmir with the Indian subcontinent. In the words of a contemporary observer, 'it is . . . not very difficult to see the irreparable loss that the state people have suffered by the new arrangement and will continue to suffer until the closed waterways and highways and the railway are reopened and traffic on them restored under normal and peaceful conditions.'[121]

Apart from the fact that Kashmir lost the neighbourhood with which it had had close commercial and cultural relations from ancient times, with the closure of existing routes and the construction of the Srinagar–Pathankot road, the distance between Kashmir and the neighbouring markets increased manifold. For instance, it now took around 48 hours for a goods lorry to cover the distance of 255 miles between Srinagar and Pathankot. Before 1947, the main *mandi* (market) of the Kashmir fruit industry was in Rawalpindi, which is only around 150 miles away and the distance was covered in about 12 hours. Considering the longer distance between Kashmir and the new markets, the freight charges became costlier than before, which hit the fruit industry hard, at least while the Sheikh was in power. The higher freight charges

---

[120] Bazaz, *History of Struggle for Freedom in Kashmir*, 416.
[121] Bazaz, *History of Struggle for Freedom in Kashmir*, 417.

also increased the prices of essential commodities, namely, salt, tea, kerosene, sugar, soap, and clothes, which were imported from outside.[122]

The timber trade with west Punjab through riverways was a significant source of income for the state prior to Partition. The trade suffered a serious jolt after the closure of waterways in late 1947. The National Conference government tried to revive the timber trade by building a government transport service. The Congress government in Punjab also helped the Kashmir government in finding a market. However, the cost of Kashmir timber was prohibitive. To overcome the problem, the government came up with a strange solution. It subsidized the timber trade and hiked passenger fares to make up for a part of the loss. Yet, despite these measures, the timber trade could not be revived to pre-Partition levels.[123]

After the closure of the historic routes, Kashmir remained cut off from the outside world for four to five months during winter, as the Banihal pass was closed after the first snowfall. The misery, especially from the scarcity of essential goods and price rise, faced by the marooned people during these months is easy to imagine.[124] Before 1947, the closure of the Banihal pass did not make any difference as the traffic was diverted to other routes which generally remained open. As a result of the snapping of ties with the west and north-west neighbourhood of Kashmir, the Kashmir government lost a substantial source of income, as business worth crores of rupees had been transacted through the routes which were closed after Partition. It took much time to compensate for this loss. Statistical information on revenue figures, as shown in Tables 2.1 and 2.2, shows that even until 1953–4 the revenue receipts had not reached their pre-1947 levels.

Another important source of income for the state during the pre-1947 period was the tourist trade. Before Partition, 30,000 to

---

[122] Bazaz, *History of Struggle for Freedom in Kashmir*, 416–17.

[123] Korbel, *Danger in Kashmir*, 216.

[124] When Sheikh Abdullah's relations with the Government of India became strained in mid-1953, he began to point out these problems. For his remarks at a gathering of peasants in Ganderbal on 3 July 1953, see Bazaz, *History of Struggle for Freedom in Kashmir*, 419.

Table 2.1   Revenue of the State (Lakhs of Rupees)

Year	Revenue
1945–6	557.00
1946–7	538.00
1947–8	274.45
1948–9	307.95
1949–50	434.11
1950–51	444.72
1951–2 (estimates)	457.56
1952–3 (estimates)	466.03
1953–4 (estimates)	479.53

Source: Brecher, *Struggle for Kashmir*, 153.

Table 2.2   State Budget, 1950–53 (Lakhs of Rupees)

	1950–51	1951–2	1952–3
Revenue	444.72	457.56 (estimates)	466.03 (estimates)
Expenditure	500.28	562.48 (estimates)	607.78 (estimates)
Deficit	55.56	104.92 (estimates)	141.75 (estimates)

Source: Brecher, *Struggle for Kashmir*, 153.

40,000 tourists had visited Kashmir every year,[125] providing liveli-hoods to many sections of the people associated with the industry, besides significantly contributing to the state coffers. According to Korbel, tourism provided an income for about 200,000 people.[126] However, because of the post-Partition developments and insta-bility in Kashmir, the tourism business flagged considerably. In 1949, only 3,700 tourists visited Kashmir, of whom only 426 were foreign tourists. In 1950, the figure rose to 5,355.[127] The decline in the tourist trade devastated all those people who had been directly or indirectly associated with the business, besides inflicting a severe loss of revenue on the state:

[125] Korbel, *Danger in Kashmir*, 217; Bazaz, *History of Struggle for Freedom in Kashmir*, 420.

[126] Korbel, *Danger in Kashmir*, 217.

[127] Korbel, *Danger in Kashmir*, 420.

They (those associated with tourist trade) have gradually exhausted their ancestral properties and incurred huge debts. . . . The condition of the Manji class and those shopkeepers who depend solely on the visitors has become desperate. It is reported on good authority that a large number of house boats in Srinagar and its suburbs are now devoted to immoral purposes.[128]

The Partition, Indo-Pak hostility, tribal raids, and the unstable political conditions of Kashmir had an adverse impact on its economy in many other ways too. Not only was no one prepared to invest money in Kashmir, even local or non-local businessmen who had well-established businesses in Kashmir felt threatened in the environment of uncertainty about the future, and they too closed their concerns and left the state.[129]

The recurrent Praja Parishad agitations in Jammu strangulated an important source of income which had to some extent compensated the loss Kashmir had suffered after the closure of its historic routes. Income from the Custom House at Pathankot and earnings from government transport plying on the Srinagar–Pathankot route were among the principal sources of revenue for the state.[130] But the agitations led almost to a suspension of traffic, not only to save life and property from the arsonists, but also because of the necessity of sealing all routes connecting Punjab with the state to check the arrival in Jammu of Jan Sangh volunteers from different states of India in order to intensify the agitation.[131]

The government was in such a weak economic position that it could not fulfil even the modest promises it had made; it widened the tax base; imposed taxes even on education,[132] expanded the existing tax structure, nationalized trade, and charged exorbitant

---

[128] Bazaz, *History of Struggle for Freedom in Kashmir*, 421.

[129] Among those who closed their business in Kashmir, the overwhelming majority were non-Muslims. This information is based on oral history. Also see Bazaz, *History of Struggle for Freedom in Kashmir*, 439.

[130] Saraf, *Kashmiris Fight for Freedom*, vol. II, 1207.

[131] Saraf, *Kashmiris Fight for Freedom*, vol. II; Mullik, *My Years with Nehru-Kashmir*, 34.

[132] Government of J&K, Department of Education, File 1651 Ed-726-US/1948, Archives Reference Library, Srinagar.

prices to earn money at the cost of the poor masses. And what was worse, as we will see in the following pages, it forcibly extracted grains from the poor, and that too at cheaper rates to overcome the drain of resources on the import of foodgrains.[133] For the same reasons, people were told to subsist on substitute foods to satisfy their empty stomachs.

The New Kashmir Programme had promised a better quality of life to all workers and underpaid officials. It was in this context that soon after Sheikh Abdullah assumed office, he made a heart-warming declaration that the minimum salary of the government employees would be fixed at Rs 100 per month. But the promise was never fulfilled. The low-paid employees continued to get Rs 30–35, the amount they had been paid prior to 1947, even though the prices of all essential commodities had substantially risen since then.[134] For example, while the maximum price of rice per maund was Rs 20 on the eve of Partition, it was sold at Rs 30 a maund during the National Conference regime. When Sheikh Abdullah was reminded of his promise in the legislature on 19 May 1952, he replied, 'I have been reminded of my promise to fix the minimum pay of Rs. One hundred. Until and unless the economic conditions of the country improves, it is not possible to redeem the pledge.'[135] When Abdullah was asked why the government did not make alternative arrangements to supply credit to the peasantry after the liquidation of debt and the centuries-old moneylending system, the reply was that the government did not have the requisite finances.[136] The National Conference government closed many hospitals on the plea that these dispensaries existed only in name; they were without doctors and even without buildings. This did not, however, justify the closure of these dispensaries because, according to contemporary sources,

---

[133] See sub-section titled 'Revenue or Plunder?', this chapter.

[134] Government of J&K, 'Aid from India', text of the speech by Bakhshi Ghulam Mohammad, Prime Minister, J&K, in the State Assembly, 17 March 1955, Ministry of Information, *Kashmir Today* series II, Archives Reference Library, Srinagar, Accession no. 552/GACC, 5.

[135] Bazaz, *History of Struggle for Freedom in Kashmir*, 422.

[136] Bazaz, *History of Struggle for Freedom in Kashmir*, 435.

no dispensaries could be found within 30 to 40 miles. 'It was disheartening', complained *Khidmat*, 'that at places both Allopathic as well as Unani and Ayurvedic dispensaries were closed down.'[137] Instead of making these dispensaries functional, they were closed because the government was facing an acute financial crisis.

Revenue or Plunder?

In his quest to maintain the 'political independence' of Kashmir, Sheikh Abdullah followed the policy of 'self-reliance' to save Kashmir from getting buried beneath the overbearing burden of debt from the Indian government. In a fierce speech at Ranbirsinghpura in 1952, Sheikh said, *inter alia*, 'It would be better to die than submit to the taunt that India was our bread-giver.'[138] This was, however, a romantic idea in the absence of resources and any worthwhile business or industry. The Kashmir government did not have the power to throw its borders open and revive the pre-colonial scenario in which Kashmir had trade relations with the whole of the neighbouring world. Nor was the Kashmir government empowered to raise international finances. Trade and tourism had declined; the industrial sector was nominal; and the overwhelming number of agriculturalists did not produce enough to fulfil their own food needs. Kashmir had a one-crop-a-year economy; commercialization of agriculture was still in its infancy; and per unit productivity was very low in the absence of modern fertilizers and high-yielding crops.

There was hardly any sector/section of the society, especially while Kashmir was still reeling under the adverse impact of Partition and war, that could bear heavy taxation. Notwithstanding the fact that the nationalists were aware of progressive and regressive taxation, as they had raised a voice against regressive taxation much before 1938–9,[139] it is surprising that they resorted to

---

[137] *Khidmat*, 26 September 1952.

[138] Mullik, *My Years with Nehru-Kashmir*, 25.

[139] Presidential address delivered by Sheikh Mohammad Abdullah to the Sixth Annual Session of the Muslim Conference on 27 March 1938, cited in *Kashmiri Mussalmanun ki Siyasi Jadujahad*, 461–2.

such oppressive methods to tide over the financial crisis. Such tyranny did not obtain even in the rule against which they fought and which they defeated with mass support following their promise to build a 'New Kashmir'.

The National Conference government nationalized transport and trade, apparently to eliminate exploitation by middlemen, traders, and monopolists in the interests of the common people, artisans, and producers. But the practice proved quite contrary. For example, the purpose of the government in nationalizing transport on the Srinagar–Pathankot road and many other routes[140] was not to give relief to the people. On the contrary, private transport companies were forced to stop their business so that the Transport Department of the government could maximize the charges. The maximization of freight charges sent prices skyrocketing. 'If today only petrol is derationed', declared Aodhya Nath, secretary, Chamber of Commerce Jammu, on 5 May 1952,

> and the private bus owners get petrol in sufficient quantity as does the State Transport Department then the same goods for which the government has fixed Rs 5 per maund from Srinagar to Jammu would be carried at Rs 3 per maund. Recently when the Government carried the goods at Rs. 6 per maund some private bus owners offered to carry the same at Rs 3 and did carry at this rate.[141]

Similarly, the KCC submitted a memorandum in early 1953 to the government in which it asserted that the

> absence of free competition in transport has enabled the government to charge arbitrarily, thereby increasing the price level by about 50 per cent more so far as freight is concerned and throwing scores of people depending on private transport to penury starvation. The president of the Chamber described the miserable plight of the people of Kashmir as a result of the dreadful controls and state trading run by certain departments of the government.[142]

[140] Korbel, *Danger in Kashmir*, 216.

[141] Bazaz, *History of Struggle for Freedom in Kashmir*, 423.

[142] Korbel, *Danger in Kashmir*, 216–17.

To be sure, the basic commodities provided through service cooperatives were not comparatively cheaper than the market rates, as the institution of cooperatives in a socialist system may lead one to believe. Instead, the commodities sold through these cooperatives were much higher in cost than in a free-market economy. 'In only one seer of sugar the government is appropriating a profit of five annas,' protested *Martand* in November 1952.[143] The practice of profiteering by the government through controlled trade is also substantiated by official documents. For example, the total payments made during the year 1949–50 for the purchase of salt and other incidental charges including freight amounted to Rs 2,026,427, and the sale proceeds realized on this account and credited into the treasury amounted to Rs 4,544,789[144]—more than 100 per cent profit. This is the main reason that the people pleaded, though vainly, to be given freedom from the cooperatives.[145] The bitter memories of state monopoly of trade, especially of essential commodities, during the period under review have survived in a saying born out of the experiences of the time, signifying the positive relationship between monopoly of trade and price rise: *nun til gachhi ni akey wana gachhun* (the salt and oil should not be sold by one shop only), or *yeth chhu gumut nun til kune wana* (it is like salt and oil being sold by only one shop), meaning exorbitant prices due to monopoly.

After independence, internal customs were abolished throughout India. It was abolished even in PAK. But the Kashmir government refused to wind up the Customs Department in the state. Worse, the customs duties were exorbitantly high. As a result, goods imported from outside were cost-prohibitive. While there is no evidence of a demand made by stakeholders from the valley to

[143] *Martand*, 18 November 1952.

[144] Government of J&K, *Administrative Report of the Jammu and Kashmir* (13 April 1949–12 April 1950) (Srinagar/Jammu: Ranbir Government Press, 1952), 27.

[145] Government of J&K, *Report of the Inquiry Committee Appointed to Examine the Working of Land Reforms, Price Control, etc.* (Srinagar, 1953) (henceforth *Wazir Committee Report*), 33–5.

abolish customs duties, Jammu raised a strong voice. 'The time has come when the Kashmir Government should abolish the customs cordon,' demanded the president of the Chamber of Commerce, Jammu. He told a press conference on 3 April 1952 that 'continuance of the practice of charging internal customs duties on the goods entering from India has pushed up the cost of living. Within the state customs duties in some cases are as high as 200 per cent of the price of the article.'[146]

By prohibiting private trade in handicrafts, the government assumed the role of a monopolist, purchasing goods at cheap rates and selling them at exorbitant prices through emporiums that were established in different parts of India. The money thus earned did not go to the state treasury, nor did the artisan benefit. Instead, the money was spent on the salaries of employees who, as was true of this government, were either relatives or supporters and friends of National Conference leaders. They were paid high salaries even though they were not qualified for their posts. Most of the jobs were sinecures. Since the high overhead charges had made the prices of commodities sold at the emporiums cost-prohibitive, the government surprisingly 'lowered the wages and privileges of artisans',[147] instead of reducing the overhead charges. State monopoly ultimately proved devastating for the handicraft industry. The trade through emporiums survived during the period only because of the support of the Indian government and rich and politically conscious Indians. But such support cannot survive for long. And this is what happened ultimately.[148] The government of the day claimed that it sold goods worth lakhs of rupees annually through emporiums, but it remained silent about the comparative volume and value of exports of these commodities prior to 1947, when the trade in these goods was free. The silence is intriguing because the exports were much larger during the free trade days,[149] even though after 1947 the emporium system was patronized by the Indian state and by the politically conscious class in India for political reasons.

[146] Bazaz, *History of Struggle for Freedom in Kashmir*, 424.

[147] Bazaz, *History of Struggle for Freedom in Kashmir*, 419–20.

[148] Bazaz, *History of Struggle for Freedom in Kashmir*, 419–20.

[149] Bazaz, *History of Struggle for Freedom in Kashmir*, 419–20.

The nationalist government also levied new taxes or enhanced the old ones. Up to 1947, education had been free in government institutions. However, the 'popular government' levied a fee on education for the first time, generating student protests which were curbed with a heavy hand.[150]

Following the devastating famine of 1950, the government instituted a Flood Sufferers Relief Fund. Apart from collecting money from employees and increasing the prices of controlled items, National Conference workers also extorted money from the people, and a large number of well-off persons were beaten up on refusing to pay the money demanded from them. The collections continued for more than 18 months. The people became so exasperated by the atrocious methods by which money was extorted from them that even *Khidmat*, the official organ of the National Conference, had to admit editorially that such things could not happen even under the previous autocratic regimes of the Dogras.[151]

Forced Procurement of Food and Inadequate Rationing

It is a commonplace fact that rice has been the staple crop as well as staple diet in the Kashmir Valley. However, because of the geographical conditions of the valley as well as the dismal per unit productivity prior to the Green Revolution, the mass of the peasantry was not in a position to fulfil its needs even for a few months. In these circumstances, it was extremely oppressive to force the peasant to pay a portion of the land revenue in kind, to say little of the state's insistence on taking a portion of his produce to feed the non-agricultural population. The National Conference government no doubt took some substantial steps to increase paddy production, but the valley was yet far from becoming self-sufficient in rice production. Without considering this ground reality, the government pursued the policy of fulfilling its food needs internally by robbing the deficient peasantry of their produce. It also

[150] Department of Education, File 1651 Ed-726-US/1948, Archives Reference Library, Srinagar.

[151] Cited in Bazaz, *History of Struggle for Freedom in Kashmir*, 425.

banned private trade in foodgrains and itself became a grain trader. This policy ultimately proved disastrous for the people and for the government of the day.

During the Dogra period and even prior to that, the state used to realize a part of the land revenue in kind, known as *mujwaza*.[152] The rates fixed by the state used to be invariably lower than the market rates, so as to sell food at cheaper rates to the vocal city and urban population for manufacturing their consent. By supplying cheap food to the urban population and paying low prices to farmers by levying heavy subsidies on food, the state intervention, in the words of Lipton, 'urbanized the benefits and ruralized the costs'.[153] Although the National Conference had given its word to the peasants that this system would at least be rationalized, and had even established food committees during the Quit Kashmir movement under the leadership of Bagum Abdullah demanding the abolition of mujwaza,[154] it however dared not touch the system fearing unpopularity of the measure among the urban population.

What was an even more oppressive policy innovated by the National Conference government was its arbitrary assessment of the peasant surplus that the peasantry had necessarily to sell to the government on its unilaterally fixed low rates. This exaction was ironically called *khush kharidi*, meaning purchase by the willing consent of the seller. The question of determining the surplus was left to the petty revenue and police officials who made free use of their discretion to engage in high-handedness.[155] The 'surplus' so determined was forcibly taken away from the peasants by paying them officially fixed low rates. The cooperative and private agencies employed for making collections made the producer purchase paddy on cash payment from the black market at the rate of Rs 25 or Rs 30 per *khirwar* (83 seers), to be handed over to the government at the rate of Rs 9. Even this low procurement price was paid in easy instalments over a period of a year, forcing the peasant to

---

[152] Lawrence, *Valley of Kashmir*, 407.

[153] M. Lipton, 'Agriculture, Rural People, the State and Supplies in Asian Countries', in *Rural Transformations in Asia*, eds J. Breman and S. Mundle (Oxford: Oxford University Press, 1991), 112.

[154] Ishaq, *Nida-i-Haq*, 215–16.

[155] Qasim, *My Life and Times*, 46; Korbel, *Danger in Kashmir*, 213–14.

mortgage his lands and property to raise money for the purchase of paddy from the black market to be handed over to government as well as for his own requirement.[156] It may be mentioned that the number of rich peasants who really possessed a surplus was minuscule; the largest segment of the peasants was facing deficiency of food.[157] Thus, it has rightly been said, what was given to the peasant with one hand was snatched away from him by another, through this iniquitous and harsh system of food procurement.[158]

The grains so procured from the peasants were then sold to the urban population. However, the monthly ration distributed among the urban population used to be invariably insufficient to meet the needs of a family. Since the government did not favour the import of foodgrains, the leadership persuaded people to make up for the deficiency of the staple food—rice—by relying on substitute foods, rather than aspiring for sufficient cheap rice at subsidized rates. According to Sheikh Abdullah, he himself set the example by taking *watt* (a maize meal) once a day, and he also advised people not to hesitate to consume potatoes as a substitute meal in the interest of their economic prosperity and political autonomy.[159] To be fair to Sheikh Abdullah, this policy of urging citizens to give up rice and wheat, whose imports were draining out the wealth needed for industrial development, and to adopt 'substitute' foods was part of the post-colonial project of the Indian political leadership to realize self-reliance.[160]

---

[156] Government of J&K, *Jammu and Kashmir (August 53–August 54): A Review of the Achievements of Bakhshi Government* (Directorate of Information and Broadcasting, Jammu and Kashmir, 1954), Archives Reference Library, Accession no. 506, 8.

[157] One of the significant passages from the Wazir Committee report reads, 'Persons who could not produce from their lands enough to eat even for three months of a year were subject to the levy of *mujawaza*. No consideration was given to leaving enough food for families according to their strength or the size of the holdings.' *Wazir Committee Report*, 111.

[158] *Jammu and Kashmir (August 53–August 54): A Review of the Achievements of Bakhshi*, (1954), Archive Reference Library, p. 8.

[159] Abdullah, *Ātash-i-Chinar*, 497, 500–1.

[160] Benjamin Siegel, 'Self Help Which Ennobles a Nation: Development, Citizenship, and the Obligations of Eating in India's Austerity Years', *Modern Asian Studies* 50, no. 3 (2016): 975–1018.

Clearly, while on the one hand, the government policy subjected the peasants to oppression, on the other it also left the urban population a disaffected lot.[161] The shortage of rice was felt everywhere. In July 1951, a hungry mob led by Khawaja Abdul Khalique, a member of the legislative assembly (MLA), looted the government food depot in Sopore.[162] A month later, another food riot took place in Kupwara where military camps were burnt down. A strategic bridge was also blown up. Scores of suspects were sentenced to imprisonment.[163] In 1952, inferior quality rice was sold at Rs 34 a maund, and paddy at Rs 24 a maund. These were record prices for the two commodities since the great famine of 1870.[164] The scale of rationing in the towns was 6 seers of paddy per month per head, as against the fourfold demand, that is, demand of 24 seers of paddy per head. The situation was so grim that even well-to-do families had to be content with a one-time rice meal. The general public was living on maize, while in the villages the number of those living on fruit was by no means small.[165]

Commenting on the food policy of his own government, Syed Mir Qasim writes:

I visited my home town Doru. What I saw there was revolting: a 70-year-old man was made to crouch in a hospital compound with a heavy boulder on his fragile back. I was told he was being punished by the local revenue officer for not delivering the forced levy on paddy. I at once freed the old man and sympathetically asked him why he could not give the levy. He told me he did not have enough food even for his family.[166]

The same heart-wrenching accounts have been given by Korbel, Munshi Ishaq, and Bazaz.[167] It was through these 'plundering

---

[161] Ishaq, *Nida-i-Haq*.

[162] Saraf, *Kashmiris Fight for Freedom*, vol. II, 1228.

[163] Saraf, *Kashmiris Fight for Freedom*, vol. II, 1228.

[164] Saraf, *Kashmiris Fight for Freedom*, vol. II, 1228.

[165] Saraf, *Kashmiris Fight for Freedom*, vol. II, 1228.

[166] Qasim, *My Life and Times*, 45–6.

[167] Ishaq, *Nida-i-Haq*, 215–16; Bazaz, *History of Struggle for Freedom in Kashmir*, 30; Korbel, *Danger in Kashmir*, 213–14.

raids' and extortions that the government reduced the budget deficit from Rs 37 million in 1948 to Rs 4.711 million in 1952, providing an opportunity to Bakhshi Ghulam Mohammad later on to exploit this to his advantage.[168]

Corruption

In a special article published in *Khidmat* on 25 May 1952, Afzal Beg, the revenue minister, stated, 'There is hue and cry in our country against bribery. It is estimated that the corruption has assumed an epidemic form, which has taken the whole state in its grip. Despite the efforts of the government it has not been satisfactorily remedied.'[169] The *Hindustan Times* wrote in its issue of 8 June 1953, 'Graft of all kinds seem to flourish in all these altitudes.'[170] Josef Korbel quoted the government's own admission in the summer of 1953 that 'the cooperatives completely collapsed because of corruption and maladministration' of government officials.[171] The Wazir Committee wrote in its report: 'During our tours in the mofussils of Kashmir province, we were greatly distressed at the almost unanimous demand from the people at large, including unsophisticated peasants at almost every place in huge public gatherings, that all that they wanted was that they be relieved from the tyranny of cooperatives.'[172]

The prevalence of rampant corruption during the period is distressing especially given the context that the entire political movement in J&K had been built upon the promise of radical reforms, including eradication of misuse of power. The basic reason for this 'top to bottom' corruption was the political instability which made Sheikh Abdullah depend on some corrupt elements and the depraved party cadre for political survival. Far from taking action, he became complacent with regard to their malpractices, and even frowned upon those who pointed it out.[173] He, for example, knew

---

[168] Saraf, *Kashmiris Fight for Freedom*, vol. II, 1221–3.

[169] *Khidmat*, 25 May 1952.

[170] *Hindustan Times*, 8 June 1953.

[171] Korbel, *Danger in Kashmir*, 216.

[172] *Wazir Committee Report*, 49.

[173] Ishaq, *Nida-i-Haq*, 237.

very well about the corrupt practices of his deputy prime minister Ghulam Mohammad Bakhshi, but he deliberately remained
unmoved because Bakhshi helped him in maintaining 'order' with
his carrot-and-stick policy. Abdullah admits to this, though cloth-
ing it in a 'political' language:

> Why had I, of all my colleagues, chosen him [Bakhshi Ghulam
> Mohammad] for the position [of deputy prime minister]? The answer
> to the question lay in the conditions confronting us at that time.
> The entire state was in turmoil. ... The administration was falling
> apart. ... We needed men who could come to grips with this kind of
> situation—men imbued with self-confidence, courage and capacity.
> Such were the virtues possessed to a great degree by Bakhshi. ...
> There is no doubt that I made great use of his administrative abili-
> ties, but it is also a fact that his weaknesses equalled his virtues. ... I
> knew about his weaknesses. These would surface on occasion even
> while our movement was on. Even the Maharaja had an inkling of
> his weaknesses and had offered him material inducements in the
> shape of a car and other gifts which he had quietly accepted. When
> the cat was out of the bag my colleagues made accusations against
> him. However, I let the matter rest there.[174]

The havoc that National Conference workers-cum-officials
wrecked on the poor peasantry has attracted the attention of all
contemporary observers of Kashmiri society. Due to paucity of
space, it is sufficient here to quote one observer of the time, Josef
Korbel:

> Government officials are harassing cultivators and petty landhold-
> ers to procure *mujawaza* and surplus food grains. ... From the com-
> plaints which are being received from the *kisans* [peasants] of vari-
> ous areas, it is evident that in procurement of *mujawaza* ... and
> surplus paddy, the poor *kisans* are being subjected to inhuman treat-
> ment. Sometimes they are forced to sell all of their belongings in
> order to pay government taxes. A number of complaints depict very
> horrible conditions. It appears that human feelings and gentlemanli-
> ness are being sacrificed at the altar of barbarity as if law is helpless

---

[174] Abdullah, *The Blazing Chinar*, 323.

before these corrupt and barbaric officials. Money is being illegally extorted from poor people in every town and village. The local officers were resorting to all sorts of irregularities in order to deprive petty landowners of their produce. *Mujawaza* has been realized from *kisans* of Budgam district twice or thrice. The Revenue officials have made the life of the cultivators extremely miserable. ... The police have even surpassed the Revenue authorities in barbarities. In one case a *tehsildar* [district official] had issued warrants of arrest for a person and when on payment of *mujawaza* his warrants were cancelled, the police refused to let him go without receiving [a] bribe which the poor fellow could not pay and had to remain in prison. In Kulgan tehsil one old women had to sell her cow in order to pay *mujawaza*. ... The government is not paying any attention to redress the grievances of the people whose demands have failed to evoke any interest in government quarters.[175]

Intolerance of Opposition

Those who fought for twenty years against injustice, oppression and undemocratic style of administration imposed by Dogra Rule, the very people, after having gained power, did not hesitate in suppressing with heavy hand any voice that was raised against their misdeeds.

—Sanaullah Butt[176]

The administration seemed to rest on the props of tyrannical laws and their heartless implementation. A minor provocation was enough for petty officials to arrest and torture people like slaves in olden days. To top it all, the people were not allowed to express their views on the current state of affairs. Public places showed notices asking people not to talk politics.

—Mir Qasim[177]

The National Conference, which had been at the forefront of the struggle for freedom, claimed to have synthesized the communist

175 Korbel, *Danger in Kashmir*, 213–14.
176 Butt, *Kashmir in Flames*, 49.
177 Qasim, *My Life and Times*, 47.

ideology with democracy and liberal humanism.[178] However, the hopes of the people were belied when the new government became a dictatorship of the National Conference party, party and administration rolled into one, and the National Conference bruited about with pride its slogan—'One Party, One Programme, and One Rule'. People belonging to opposition parties were persecuted; all voices other than the voice of the government were muzzled; and a reign of terror and tyranny was let loose by National Conference workers who were also appointed as government officials.

Balraj Puri says:

> As a member of the NC, I had raised the issue of the separation of the party from the administration in the party forum. Abdullah, however, rejected my demand. ... he preferred the Soviet model in which the party controlled every branch of the administration. ... I warned Nehru that identification of the government with the National Conference would lead to the setting up of a totalitarian regime.[179]

However, the centre never showed any interest in curbing the authoritarian and dictatorial activities of those whom it considered useful instruments to serve its interests.

Those who know Sheikh Abdullah are of the opinion that by temperament he was authoritarian, unable to tolerate any opposition. The existence of multiple voices in a politically volatile environment further hardened his attitude towards enforcing one-party rule. All those leaders who advocated accession to Pakistan were put in jails or pushed into Pakistan, or they were externed from the state.[180] Alongside using the repressive machinery to

[178] Abdullah, *The Blazing Chinar*, 218.

[179] Balraj Puri, *Kashmir towards Insurgency* (New Delhi: Orient Longman, 1993), 43–4.

[180] Some of the leaders who were thrown out of the state or put behind the bar were Chawdhary Ghulam Abbas, Aga Saukat Ali, Maulana Mohammad Noorudi, Khawaja Ghulam Nabi Gilkar, Maulvi Mohammad Abdullah Shopyani, Maulvi Abdul Rahim, Khawaja Abdul Gani, Pandit Prem Nath Bazar, Khawja Abdul Salam Yatu, Mohammad Yusuf Saraf, Mohammed Akram Lone, Noor Mohammmad, Sham Lal Yechha, Allah

inflict heartbreaking physical and mental torture on their political opponents,[181] the National Conference party would unabash-
edly raise in every party gathering and publish in their official
organ the slogan: 'One Party, One Programme, and One Leader'.[182]
Listening to Radio Pakistan was prohibited; anyone suspected
of listening to it was arrested along with the radio set.[183] The
Statesman, a prominent Indian newspaper which always favoured
Sheikh Abdullah, observed on 1 March 1949, 'There are signs of
the establishment of a police state—futile notices in restaurants
forbidding political conversations when everybody talks politics;
more "Public Safety" prisoners than are necessary.'[184] Sir Owen
Dixon wrote in September 1950 that 'the state government was
exercising wide powers of arbitrary arrest.'[185] Subscribing to the
view of other contemporary sources, Korbel wrote, 'all observers
who have had the opportunity to visit Kashmir recently agree that
the rule of political persecution still continues and that the police
terror perpetuates itself, though fighting has been stopped for six
years.'[186]

All the 218 radio sets installed in different parts of Kashmir
Valley and 150 sets installed in Jammu under the Community
Broadcasting Centre were 'tuned to radio Kashmir, fixed and
sealed'.[187] All the dailies and weeklies published from the state
were either government controlled or under strict censorship.[188]
And to cap it all, as we shall see on the following pages, the
trend of conducting farce elections began with 'Awami Raj', as
the National Conference government was called by the National

---

Rakha Sagar, Mir Abdul Aziz, among others. Butt, *Kashmir in Flames*,
46–7; Bazaz, *History of Struggle for Freedom in Kashmir*, 408.

[181] Bazaz, *History of Struggle for Freedom in Kashmir*, 413–14; Qasim,
My Life and Times, 47; Butt, *Kashmir in Flames*, 46–7.

[182] Bazaz, *History of Struggle for Freedom in Kashmir*, 402.

[183] Butt, *Kashmir in Flames*, 49.

[184] *Statesman* (Calcutta), 1 March 1949.

[185] Cited in Korbel, *Danger in Kashmir*, 208.

[186] Cited in Korbel, *Danger in Kashmir*, 208.

[187] Korbel, *Danger in Kashmir*, 208–9

[188] Korbel, *Danger in Kashmir*, 209.

Conference itself.[189] Josef Korbel very aptly sums up the National Conference government under Sheikh Abdullah: 'The Kashmir people had every right to expect that their popular leader would live up to his promises. To them he was "Lion of Kashmir"; to detached observers, a democrat and socialist. However, as years went by, his policy cast increasing shadows over the enlightened scene of his theory.'[190]

Ignoring the Security Council Resolutions,[191] the Government of J&K decided in 1951 to hold elections for convoking the Constituent Assembly. People were to elect 75 members—45 to represent Kashmir and Ladakh, and 30 for Jammu. In Kashmir nobody could afford to incur the wrath of the National Conference by contesting against its candidates. Two candidates who had filed nomination papers without any serious thought withdrew under pressure.[192] Thus, there was no polling in Kashmir. All the 'forty-three candidates from the valley were elected unopposed one week before the election date'.[193] In Jammu, the nomination papers of 30

---

[189] For details, see Bazaz, *Democracy through Intimidation and Terror*; Bose, *Kashmir: Roots of Conflict, Paths to Peace*; Gauhar, *Elections in Jammu and Kashmir*.

[190] Korbel, *Danger in Kashmir*, 206.

[191] On being approached by Pakistan, the United Nations Security Council passed a resolution in late March 1951 'reminding the government and authorities concerned of the principles embodied in the Security Council resolutions of 21 April 1948, 3 June 1948 and 14 March 1950 and the United Nations Commission for India and Pakistan resolutions of 13 August 1948 and 5 January 1949, that the final disposition of the state of J&K will be made in accordance with the will of the people, expressed through the democratic method of a free and impartial plebiscite conducted under the auspices of the United Nations.' The resolution further warned that 'the convening of a Constituent Assembly as recommended by the General Council of the All Jammu and Kashmir National Conference and any action that Assembly might attempt to take to determine the future shape and affiliation of the entire state, or any part thereof, would not constitute a disposition of the state in accordance with the above principle.' United Nations Security Council Resolution, S/P. V. 539 of 30 March 1951.

[192] Korbel, *Danger in Kashmir*, 222.

[193] Korbel, *Danger in Kashmir*, 222.

opposition candidates were rejected on flimsy grounds. Genuinely agitated by this arbitrariness of the National Conference government, the Praja Parishad boycotted the elections. Thus the National Conference 'won' elections without any contest. 'No dictator could do better,' says Josef Korbel.[194] 'The manner in which this election was contested', says Sumantra Bose, 'made a mockery of any pretence of the democratic process and a grim precedent for future free and fair election in J&K.'[195] More shocking is the role played by Nehru in encouraging this farcical election. Interestingly, Nehru stated that he was 'sure that the way people have voted showed clearly that they were with the National Conference and with India'.[196] When Balraj Puri, during the preparations for these elections, organized a small show of anguish to agitate for fair play, Jawaharlal Nehru advised him to help Sheikh Abdullah as the 'luxury of democracy' could not be allowed in J&K. Nehru also told Balraj Puri, 'India's Kashmir policy revolves around Abdullah, and therefore nothing should be done to weaken him.'[197]

Ghulam Ahmad Mahjoor, the nationalist poet who had, during the freedom movement, portrayed a rosy picture of an anticipated independent Kashmir in his famous and commonly sung melodious poetry, which played an important role in the mass mobilization of Kashmir around National Conference discourse,[198] was so disillusioned with the authoritarian rule of the National Conference that he wrote the famous satirical poem, 'Azadi' (Freedom). It is worth quoting some verses of the poem:[199]

---

[194] Korbel, *Danger in Kashmir*, 222.

[195] Sumantra Bose, *The Challenge in Kashmir: Democracy, Self-Determination and Just Peace* (New Delhi: Sage, 1997), 31.

[196] *Hindu*, 1 November 1951.

[197] Puri, *Kashmir towards Insurgency*, 49.

[198] For details see Ghulam Ahmad Mahjoor, *Kalam-i-Majhoor* (Srinagar, n.d.), vol. X.

[199] Ghulam Ahmad Mahjoor, 'Azadi' (Freedom), in *An Anthropology of Modern Kashmiri Verse (1930–1960)*, trans. and ed. Trilokinath Raina (Pune: Sangam Press, 1972), 74–7.

Now raise a song of thanks and praise for freedom has entered our
   homes:
After ages indeed has freedom unveiled its face to us.
This freedom rains mercy on the soil of Western lands,
But to our land its gift is merely a dry thunder.
Poetry, destitution, helplessness, despair and tongues and lips sealed—
Freedom has descended on us casting these benign shadows!
This freedom is a hourie of Paradise and cannot roam from home to
   home:
Confined to a select few homes is the flutter of its charming dance.
This freedom proclaims the end of capitalism and exploitation.
Even as it drains its own people dry by garnering their wealth.
People moan and whine as rulers repose in contentment like
   bridegrooms,
Hugging the bride of freedom to their breasts in seclusion.
Nabir Shaikh alone, whose wife was whisked off, can divine the subtle
   meaning of freedom.
He went to file a complaint while his wife was delivered of freedom in
   somebody else's home.
Seven odd times did they rummage through her secret parts for a hand-
   ful of rice.
And then the poor vegetable vendor turned home, covering the basket
   with her head scarf.
All feel suffocated as agony rends their hearts:
'We would lay bare our woe', they say, 'but freedom will thrash and
   flog us.'[200]

Political Crisis

Lack of coherence between individual regions of the state now found
a serious outlet in Kashmir Politics.

—Josef Korbel[201]

Immediately after the Kashmir-based and Muslim-backed
National Conference assumed power, the contradictions and

---

[200] I am thankful to Professor G. R. Malik, former Professor of English,
University of Kashmir, for translating these verses for this work.

[201] Korbel, *Danger in Kashmir*, 226.

incongruities between the ethno-cultural identities of the artificially created state, and the mutually opposite ideas of Nehru and Sheikh Abdullah on the status of Kashmir, found an explosive outlet which threw Kashmir into a welter of chaos and culminated in the dismissal and imprisonment of the Sheikh, with enduring implications.

To be sure, Abdullah was independent-minded but, in the words of Korbel, 'his patriotism was too shallow to resist the temptations of power.'[202] Towards the twilight of his life, Sheikh Abdullah used to discuss 'freely and openly' the 'whole universe' with the governor, B. K. Nehru, at the Rajbhawan over 'endless cups of tea'. And what impression about Sheikh Abdullah Governor Nehru gathered from these discussions, he recorded in his autobiography, of which the following is noteworthy: 'He [Sheikh Abdullah] had chosen to opt for India rather than Pakistan at the time of partition because he knew well that he would have received short shrift from the *Quid-e-Azam*. ... He perhaps even hoped that he could over a period of time succeed in peacefully seceding from the Union altogether.'[203] This is undoubtedly true. Although Abdullah had already made up his mind to stay 'autonomous' within the Indian Union in lieu of which he supported Kashmir's accession to India, immediately after his release from jail on 29 September 1947, he convened a series of meetings in different parts of Kashmir wherein he demanded 'freedom first, accession after'.[204] In many public speeches and press conferences in October 1947, Abdullah said, 'Accession is of little importance. Freedom is more important. We do not want to join either Dominion as slaves.'[205] In the given fluid political atmosphere of Kashmir, it was difficult to guess which way Kashmir was heading through the haze of time. But the tribal raid hastened the march towards the goal

[202] Korbel, *Danger in Kashmir*, 207.

[203] Nehru, *Nice Guys Finish Second*, 594.

[204] *Khidmat*, 2 October 1947; Puri, *Jammu and Kashmir*, 56.

[205] *Hindu* (Madras), 16 June 1953; Korbel, *Danger in Kashmir*, 70–71; *Khidmat*, 9 October 1947; Statement released through the Associated Press of India on 10 and 21 October 1947, cited in Korbel, *Danger in Kashmir*, 72; Puri, *Jammu and Kashmir*, 57.

which Abdullah had set for himself, though in the ideal environment of 'freedom before accession' Sheikh Abdullah would have had better bargaining power, at least, to ensure a sustainable economic basis for the 'autonomous' state. The tribal raid culminated in Kashmir's accession to the Indian Union. The mutually agreed upon Instrument of Accession empowered Kashmir to legislate on all matters except defence, foreign affairs, and currency. However, the Indian government made it quite clear that the final solution of the Kashmir issue was subject to the will of the people, and the accession would be 'temporary' till then.[206]

Abdullah told Michael Davidson of the *Observer* that 'it would be improper to accede to either of the dominions. He liked to live in peace with both and for that a middle position with economic cooperation of both was adjudged the right alternative. However, the independent Kashmir's territorial integrity must be guaranteed not only by India and Pakistan but also by members of United Nations.'[207] Similar statements made by Abdullah were also published by the United Press of America.[208] Later, under pressure from Sardar Patel, Abdullah retracted these statements, saying he was merely 'thinking aloud'.[209] However, the independent-mindedness of the Sheikh reached such a critical mark following the intensification of the agitation by the Praja Parishad that he stubbornly refused to be persuaded by the conciliatory offers made by Nehru.[210]

Sheikh Abdullah had consistently to face the question: why did he favour Kashmir's accession to India rather than Pakistan? The most important arguments he advanced in defence of his decision were Indian: secularism and socialism.[211] Sadly, he was

---

[206] Text of Lord Mountbatten's reply dated 27 October 1947 to the Kashmir ruler signifying his acceptance of the Instrument of Accession, cited in Lakhanpal, *Essential Documents and Notes on Kashmir Dispute*, 57.

[207] Reproduced in *Montreal Daily Star*, 6 May 1949. Cited in Das, *Jammu and Kashmir*, 194.

[208] *Ranbir* (Jammu), 5 May 1949.

[209] Puri, *Jammu and Kashmir*, 93.

[210] See Chapter 1, this volume and following pages of this chapter.

[211] Government of J&K, Jammu and Kashmir Constituent Assembly, Opening Address by Sheikh Mohammad Abdullah the Hon'ble Prime Minister of Jammu and Kashmir, Srinagar, 5 November 1951.

disillusioned on both these premises. The Praja Parishad, a communal organization in Jammu, succeeded in mobilizing all the disgruntled elements who had been affected by the abolition of the princely order. The Praja Parishad also roped in the support of the Buddhist leader of Ladakh, Kushak Bakula, and the communal parties of India. The nerve centre of the Praja Parishad was eastern Jammu. The leading section of the Hindu majority in this part of Jammu was 'frankly communal'.[212] Economically, it represented that group of wealthy people who, as the maharaja's active supporters, constituted the privileged section of government officials, landlords, and businessmen. The land reforms carried out by the Sheikh government deprived them of jagirs, mukarares, and big land estates. Being mainly moneylenders, the cancellation of debt led to their incurring huge losses; the nationalization of transport and trade plus demobilization of the army particularly affected the rich section of Jammu Hindus, especially Dogra Hindus. Politically, however, the Praja Parishad found its 'principal support in the great majority of the non-Muslims'.[213] The propaganda alleging that their co-religionist benefactor had been replaced by a communal, rabidly anti-Hindu leader who was going to establish an independent country worked effectively as a mobilization tool. The appointment of Muslims to important positions, making Urdu the official state language, the bifurcation of Hindu-dominated Udhampur district, the refugee problem, and the closure of some Ayurvedic dispensaries also provided grist to the communal mill.[214]

Ever since the transfer of power from the maharajas to the National Conference led by Sheikh Abdullah, the restive Praja Parishad had led several demonstrations, met with police action and arrests, which kept adding fuel to the fire. The manipulated elections of 1951 made matters worse. In February 1952, the situation became extremely critical. The frustrated Sheikh made the famous public speech at Ranbirsinghpura in Jammu on 10 April

[212] Das, *Jammu and Kashmir*, 194.

[213] Korbel, *Danger in Kashmir*, 226.

[214] Behera, *State, Identity and Violence*, 109–10; Chowdhary, *Jammu and Kashmir*, 181.

1952, in which he criticized India for communalism and warned against any attempts towards the full application of the Indian constitution to the state. Describing these attempts as 'unrealistic, childish, and savouring of lunacy', he stated that

> no one can deny that communal spirit still exists in India. Many Kashmiris are apprehensive as to what will happen to them and their position if, for instance, something happens to Pandit Nehru. ... As realists, we Kashmiris have to provide for all eventualities ... if the special status for Kashmiri was not granted in the Indian Constitution, how can we convince the Muslims in Kashmir that India does not interfere in the internal affairs of Kashmir? ... we have acceded to India in regard to defence, foreign affairs, and communication in order to ensure a sort of internal autonomy. ... if our right to shape our own destiny is challenged and if there is resurgence of communalism in India, how are we going to convince the Muslims of Kashmir that India does not intend to swallow up Kashmir. ... such developments might lead to break in the accession of Kashmir to India.[215]

The pressures, both from the Praja Parishad as well as from Sheikh Abdullah, led to what is known as the Delhi Agreement of July 1952.[216] In Srinagar, the immediate reactions against the Delhi

[215] *Hindu* (Madras), 12 April 1952; *Khidmat*, Srinagar, 13 April 1952; Korbel, *Danger in Kashmir*, 223; Puri, *Jammu and Kashmir*, 99.

[216] By virtue of the Delhi Agreement, the residuary powers (powers other than those mentioned in the three lists) were given to the state; on the issue of citizenship it was agreed that in line with Article 5 of the Indian constitution, persons who have their domicile in the state of J&K would be treated as citizens of India. But the state legislature was empowered to confer special rights and privileges on the state's permanent residents. It provided that the state would have its own flag; however, it would not be a rival to the national flag. Hereditary rulership was abolished and it was to be replaced by the *sadr-i-riyasat* (governor, in other states) who shall be a state subject to be elected by the state legislature and will be recognized by the president of India. With regard to the provision of dealing with emergency, Article 352 dealing with internal emergency was to apply to the state in the case of external aggression only, and in cases of internal disturbances, the provision was to be applied only on the approval of the state government. Article 356 dealing with the imposition of president's

Agreement were so vehemently critical that a workers' meeting was 'hurriedly convened to favourably interpret its provisions'. Still, a wave of despair went round the valley.[217] But the most disappointed was the Praja Parishad, because the Delhi Agreement abolished monarchy and ratified the special position of the state. Their demand for extending the jurisdiction of the Supreme Court to Kashmir was not satisfied. The Praja Parishad demanded the full merger of the state with India or the creation of a separate state of Jammu as an integral part of India. The Parishad succeeded in winning the support of some members of the Indian Parliament, prompting the Sheikh to announce that the whole non-communal structure of the state was in danger. In December 1952, another wave of demonstrations precipitated the arrest of nearly 500 persons.[218]

Not only Jammu Hindus, but the Ladakhi Buddhists also clamoured for separation from the state. According to Josef Korbel, Sheikh Abdullah was trying to bring some measure of improvement to the 'almost hopeless condition of its people'.[219] A high school and a few elementary schools were opened at Leh, telephone communications were established, and forests were taken away from monasteries. Plans were prepared for the construction of roads and for irrigation, and basic medical services.[220] However, because of financial problems, these plans remained on paper. Prices of basic necessities like kerosene oil were very high. Although the Sheikh government did not introduce land reforms in Ladakh with the same zeal as in the Kashmir Valley and Jammu to placate the Buddhist clergy, still the Buddhist leadership joined hands with the Praja Parishad. Resenting being governed from

---

rule in the states and Article 360 dealing with financial emergency were not made applicable to the state. With regard to the jurisdiction of the Supreme Court, only the original jurisdiction under Article 131 dealing with disputes between the centre and the state or between one state and another or between a group of states and others was extended to the state.

[217] Saraf, *Kashmiris Fight for Freedom*, vol. II, 1204.

[218] Korbel, *Danger in Kashmir*, 228.

[219] Korbel, *Danger in Kashmir*, 229.

[220] Korbel, *Danger in Kashmir*, 229.

Srinagar, Kushak Bakula, the head lama of Ladakh and member of the Constituent Assembly, promoted the slogan, 'Ladakh for Ladakhis'. Basically afraid of land reforms, being himself a big landlord and a beneficiary of the land property of the monasteries, Kushak Bakula exerted pressure through coy hints such as 'across the border lies Tibet—perhaps here lies their destiny.'[221] However, when towards the end of 1949 the communist shadow slid over Tibet, Kushak Bakula 'began to realize his position between the "devil" of land reform and the "deep blue sea" of a communist rule in Tibet that might engulf him'. He began to place increased emphasis on the claim of autonomy within India. But should this fail, he warned in June 1952, Ladakh may seek political union with Tibet, 'as a last course left to us'.[222]

The year 1953 began with a renewed resolve by both Jammu and Ladakh to exasperate Abdullah completely and bring about his downfall. The Praja Parishad resorted to demonstrations, violence, sabotage, and disruption of work in various towns of Jammu by recruiting the support of Jan Sangh volunteers from Punjab.[223] It raised emotionally charged slogans rejecting the special position of Kashmir.[224] Nehru termed the movement 'most pernicious and malignant' in its 'narrow, bigoted, reactionary and revivalist approach'.[225] The *Times* (London) reported that the movement engineered and spearheaded by Praja Parishad 'has gone deep into the rural areas where the masses take part in Parishad processions and demonstrations and stubbornly endure police baton charges and tear gas attacks'.[226] 'Even in the seventh week of its agitation, planned protest meetings and processions, in defiance of the government ban, were a common sight.'[227]

---

[221] Korbel, *Danger in Kashmir*, 230.

[222] *Christian Science Monitor*, 29 June 1952.

[223] Mullik, *My Years with Nehru-Kashmir*, 34.

[224] The slogan *Ek Pradhan, Ek Nishan, Ek Vidhan* (One President, One Flag, One Constitution) became the battle cry of the Parishad, demanding abrogation of the special status of J&K.

[225] *Hindu*, 26 April 1953.

[226] *Times* (London), 24 January 1953.

[227] Saraf, *Kashmiris Fight for Freedom*, vol. II, 1207.

In this situation of total breakdown of law and order (in spite of police detachments having been called from Punjab to assist the local police), the head lama of Ladakh raised the ante and vociferously demanded separation from Kashmir. In fact, it was from December 1952 that we find him raising his pitch. He maintained that Ladakh had no bond with Kashmir. There was only one bond that linked Ladakh with Kashmir, and that bond was the maharaja. With the abolition of the hereditary monarchy, the bond was automatically broken.[228] He categorically mentioned that in no way—racially, linguistically, or culturally—did Ladakh have affinity with Kashmir. Interestingly, though, in respect of these cultural attributes, Ladakh is also equally distinct from Jammu, but the lama preferred to be a part of Jammu.[229]

In his usual way, Sheikh Abdullah blew both hot and cold. However, he was so pressurized by the whirlwind of unending agitations that he offered both Jammu and Ladakh autonomous positions within the proposed 'Autonomous Federated Units of the Republic of India'—the plan prepared as a solution to the centrifugal demand of the Praja Parishad and the head lama. But this reform too could not satisfy them, for by now they had succeeded in graduating their movement to a pan-Indian movement. The three communalist parties—Hindu Mahasabha, Jan Sangh, and Ram Rajya Parishad—threw their lot in with the Praja Parishad agitation and started a nationwide satyagraha for the complete integration of Kashmir with India. Significantly, according to Joseph Korbel, even Jawaharlal Nehru 'could not ignore their voice altogether, more so . . . the National Congress has not remained entirely deaf to their cries'.[230] The death of Shyama Prasad Mukherjee from a heart attack in custody in Srinagar added fuel to the fire. Nehru who had initially looked askance at the agitators now changed his tone and expressed his sympathies with them. Immediately after these expressions by Nehru, the agitation was called off on 7

[228] *Kashmir Affairs*, Rawalpindi, no. 49, 6 December 1952, quoted in Korbel, *Danger in Kashmir*, 233.

[229] *Times of India* (Bombay), 30 April 1953.

[230] Korbel, *Danger in Kashmir*, 234.

July 1953, but 'not before it had caused irreparable damage to the state's organic unity'.[231]

Regarding the factors that led to the calling off of the agitation by the Praja Parishad, versions vary. Mullik, while taking the credit for controlling the situation, says that he was sent by Prime Minister Nehru with a message to Praja Parishad leaders that in view of the external context of the Kashmir issue, a separate status for Kashmir had to be maintained for some more years. However, 'these special privileges would gradually disappear and Kashmir would be like any other state of India.'[232] Josef Korbel feels that 'something' behind the scenes was taking place which convinced the Praja Parishad to call off the agitation. This 'something' was perhaps the assurance given by Nehru through the director of the Intelligence Bureau that the Sheikh would be removed soon. This seems true, should we consider that by this time Bakhshi was in close contact with New Delhi, so much so that when the Sheikh declined to visit New Delhi in July 1953 to show his open resentment, Bakhshi went. So did the sadr-i-riyasat. The agitationists explained their reason for calling off the protest saying that the 'movement has achieved the purpose of impressing on the government and on the country the problem in an objective manner'.[233] 'There is no proof', says Korbel, 'of any connection between the behind the scene happenings and this sudden change in their policy, but the coincidence is conspicuous.'[234] Balraj Puri gives an entirely different explanation. According to him, it was in response to the press statement made by Prime Minister Nehru, and its endorsement by the state government, that the Government of J&K was considering giving autonomy to its regions, particularly Jammu, while framing the constitution of the state, that the Praja Parishad called off the agitation.[235]

The Praja Parishad's agitation, supported by its sister organizations in India, as well as its successful prodding of the Buddhist

---

[231] Das, *Jammu and Kashmir*, 203.

[232] Mullik, *My Years with Nehru-Kashmir*, 30.

[233] *Hindustan Times* (Delhi), 8 July 1953.

[234] Korbel, *Danger in Kashmir*, 237.

[235] Puri, *Jammu and Kashmir*, 116–17.

leaders of Ladakh to join hands with them, convinced Sheikh Abdullah that his notion that India was a secular country was nothing short of utopian. 'The communal happenings have shaken the foundations of this (Indo-Kashmir) relationship,' he said.[236] Mullik, who was unhappy with the Sheikh, and even poisoned Nehru by saying that the Sheikh 'is communal',[237] frankly admits that the agitations by communal forces changed Sheikh's views about India:

> The Praja Parishad agitation in Jammu and the Jana Sangh agitation at Delhi for complete integration of Jammu and Kashmir with India and the support which this agitation received from certain other quarters in India gave Sheikh Abdullah the much-needed excuse to openly challenge India's professed secularism and he used this as an argument to prove that the integration of Jammu and Kashmir, and particularly of the Kashmir valley and certain adjoining parts, which had Muslim majority, with Hindu India would not be in the interest of the Muslims in Kashmir; and from then onwards he started openly propagating for a special status for Kashmir.[238]

Except for the fact that the most powerful sections of Dogras were affected by the abolition of despotic, feudal, and sectarian rule, there was no justification for them to launch such a fierce struggle against Abdullah. Admittedly, there was a reason to get agitated over the manipulation of elections in 1951, but the agitation had begun much earlier. More importantly, the Praja Parishad even rejected the proposal of a federal polity offered by the Sheikh in April 1953. A document, *Is Abdullah Government Anti-Hindu?*, brought out by Hakim Parshuram Nagar, the National Conference district president, Jammu, provides details about community-wise representation in state services as well as the amount spent in Jammu on development works (see Table 2.3). According to this information, the numerical strength of non-Muslim officials was much larger than Muslims, though the latter formed the majority of the population. Also, in terms of development works

---

[236] *Hindustan Times*, 24 July 1953.

[237] For details see Mullik, *My Years with Nehru-Kashmir*, 28–34.

[238] Mullik, *My Years with Nehru-Kashmir*, 35.

Table 2.3 Creedwise Representation of Gazetted and Non-Gazetted State Government Employees in Various Departments

Category	Gazetted	Non-Gazetted	Menials
Muslims	163	4,943	2,090
Hindu	361	9,281	2,156
Sikhs	23	769	268
Others	11	436	155

undertaken in Jammu province, the statistical information does not show any discrimination towards the province.[239] As per the document, three out of the eight cabinet ministers were Hindus, as were two out of three deputy ministers. The strength of various communities in the services was as shown in Table 2.3.[240]

The document lists 22 key posts, including chief justice, principal secretary, chief engineer, home secretary, director industries, secretary external affairs, and so on, that were held by Hindus. Also, Jammu Hindus held 11 out of 18 inter-provincial posts.[241] The expenditure under different heads during 1948–49 stood as in Table 2.4.[242]

Convinced that Indian secularism was merely a façade behind which was hidden the real India about which he felt worried,[243]

[239] See Hakim Parshuram Nagar, *Is Abdullah Government Anti-Hindu?* (Jammu: National Publishing House), Archives Reference Library, Srinagar, Accession no. 533/GACC.

[240] Hakim Parshuram Nagar, *Is Abdullah Government Anti-Hindu?*, 2.

[241] Hakim Parshuram Nagar, *Is Abdullah Government Anti-Hindu?*, 2–3.

[242] Hakim Parshuram Nagar, *Is Abdullah Government Anti-Hindu?*, 5–6.

[243] 'Hereafter, all his reference to India were in hostile terms and contained strong criticism of any action of the Central or any State Government or any event anywhere in India which he could malign as communal in nature and he used all these instances to prove that India's profession of secularism was only on paper. He made several speeches with direct or indirect allegations on these lines and took advantage of the Friday gatherings at the Hazratbal Mosque to spread this poisonous propaganda.' Mullik, *My Years with Nehru-Kashmir*, 35.

Table 2.4 Budgetary Allocation to Jammu and Kashmir Provinces Under Different Heads

Head	Kashmir Province (Rs)	Jammu Province (Rs)
Irrigation	500,315	1,085,635
Roads	458,095	1,970,538
Bridges	2,974	14,731
Buildings	5,701	108,604
Water supply	50,441	201,101
Defence works	2,723,000	5,168,000
Transport works	80,000	176,000

the Sheikh decided not to look back, and instead to work for a permanent solution; but in any case no more to remain a part of India. On 18 May 1953, Abdullah called a meeting of the Working Committee of the National Conference to deliberate upon a permanent solution to the Kashmir problem. The committee appointed a sub-committee,[244] and on 9 June 1953 the sub-committee suggested four alternatives: (1) overall plebiscite for the state; (2) independence for the state; (3) independence for the whole state with joint India–Pakistan control on foreign affairs; (4) Dixon Plan with independence for the plebiscite area. Sheikh Abdullah intensified his campaign for obtaining the independence of Kashmir. In a speech in mid-June 1953, he reminded the public what he had pleaded for in 1947: 'freedom before accession'. As he had done before, he again insisted on the idea of an independent Kashmir.[245] In August 1953, Sheikh Abdullah threw a bombshell by saying that 'Kashmir's initial accession to India was forced on him because of India's refusal to give any help without the state's accession.'[246]

Indeed, Sheikh Abdullah was too deeply hurt to be won over by Nehru again. He did not find any scope for the survival of Kashmir's autonomous position, nor any future for Kashmiri

[244] The members of the sub-committee were: Sheikh Abdullah, Maulana Masoodi, Mirza Afzal Beg, Bakhshi Ghulam Mohammad, G. M. Sadiq, Sardar Girdharilal Dogra, and Pandit Shamlal Saraf.

[245] *Hindu* (Madras), 16 June 1953.

[246] *Hindu Weekly Review* (Madras), 10 August 1953.

Muslim interests within the Indian Union. He made these charges in a letter to Nehru on 4 July 1953 and to Maulana Azad on 16 July 1953.[247] Nehru's repeated invitations to the Sheikh for talks failed to move Abdullah. Nor could Maulana Azad, who was sent by Nehru to persuade Abdullah to modify his position. According to Mullik, Azad advised Nehru to 'dismiss Sheikh Abdullah before he committed any more mischief'.[248] And Rafi Ahmad Kidwai, who had also met with same rebuff from Sheikh, suggested the same.[249] Meanwhile, the Indian intelligence agencies had developed close contact with most of the cabinet colleagues of the Sheikh, who for ideological and personal reasons were not pulling on well with the Sheikh. The Intelligence Bureau had succeeded in creating a support structure within the cabinet that could help India in saving the situation. Mullik makes a veiled reference to this:

> In India many people had expressed doubt about Pandit Nehru's hold on the people of Kashmir, but this internal tussle, during which Sheikh was doing his best to undermine Pandit Nehru's influence, showed how strong the latter's hold was. Even in the Constituent Assembly, whose membership had been handpicked by the Sheikh, who had not allowed any opposition to grow, he could not carry the majority with him when he took an out-and-out anti India stand.[250]

From Mullik's *My Years with Nehru-Kashmir*, it becomes abundantly clear that D. P. Dhar, Bakhshi Ghulam Mohammad, Girdharilal Dogra, Sham Lal Saraf, and Karan Singh were already hand-in-glove with New Delhi. Sadiq and Mir Qasim also joined the bandwagon subsequently.[251] They were even more impatient to see the Sheikh dismissed from power than his opponents outside his government. Along with the director of the Intelligence Bureau, this group precipitated the deposition of Abdullah by

[247] For the draft of the letter to Maulana Azad, see G. Noorani, *Article 370*, 232–6.

[248] Mullik, *My Years with Nehru-Kashmir*, 37.

[249] Mullik, *My Years with Nehru-Kashmir*, 41.

[250] Mullik, *My Years with Nehru-Kashmir*, 40.

[251] For details see Mullik, *My Years with Nehru-Kashmir*, 24, 31, 33, 34, 36, 37.

spreading the rumour that he was planning to convene a session of Constituent Assembly which would declare the Valley of Kashmir independent.[252] But the most inflammatory rumour was that Sheikh Abdullah's quest for independence was backed by America. To substantiate this allegation, they quoted the meeting Abdullah had had with Adlai Stevenson, the Democratic leader of the United States, at Gulmarg during his visit to the subcontinent in 1953. Though the American ambassador to India, George V. Allen, and Adlai Stevenson himself had denied any interference in the Kashmir situation, Nehru declared later, 'it would not be correct to call it governmental interference, but individuals have not behaved properly, because again you must remember the basic fact that Kashmir is a highly strategic area.'[253] 'Events', says Korbel, 'moved swiftly towards a cloak-and-dagger denouement.' On 7 August 1953, three cabinet members led by Bakhshi submitted a memorandum to the sadr-i-riyasat accusing Abdullah of misgovernance and of creating conditions for rupture in the relations of the state with India and for the exploitation of the situation by foreign powers. The memorandum concluded by informing the Sheikh that the cabinet had lost the confidence of the people.[254] The memorandum was submitted to the sadr-i-riyasat (head of the state), who suggested an emergency meeting of the government. Abdullah refused the request and went to Gulmarg to spend the weekend there. The head of the state, avenging his father, dismissed Abdullah on 9 August 1953 and nominated Bakhshi as prime minister.[255] Nehru who, according to Mullik, had approved the plan,[256] played the diplomatic card, saying on 10 August 1953 in the Lok Sabha that the Government of India had been informed about the crisis in Srinagar 'but our advice was neither sought nor given'.[257]

[252] Korbel, *Danger in Kashmir*, 237.

[253] Jawaharlal Nehru's speech in the House of the People, 17 September 1953, quoted in Korbel, *Danger in Kashmir*, 238.

[254] *Times of India* (Bombay), 10 August 1953.

[255] Mullik, *My Years with Nehru-Kashmir*, 44.

[256] Mullik, *My Years with Nehru-Kashmir*, 43–4.

[257] Korbel, *Danger in Kashmir*, 242.

The news of the deposition and detention of Abdullah led to mass protests for about a month. According to the National Conference's version of the events, 1,500 persons were killed in police firing.[258] The official reports, however, put the number at 60.[259] About 10,000 people were arrested.[260] A complete 18-day strike was observed by traders, students, and transporters. Women's participation in the street protests was conspicuous. The demonstrators were demanding withdrawal of Indian troops and the holding of a plebiscite, besides raising slogans in favour of Sheikh Abdullah. According to Mullik, even when the violent disturbances subsided, the agitation for the Sheikh's release continued in the valley in one form or the other for a long time.[261] The contemporary writer-activist Balraj Puri says, 'By his confrontation with Delhi, he [Sheikh Abdullah] undoubtedly recovered whatever personal support he had lost during his years in power.'[262] This is also substantiated by the *New York Times*.[263]

* * *

The dawn of New Kashmir, which the people had long longed for and which the political leadership was zealous enough to usher in, was a huge disappointment in one respect or another to all sections of the society, including the rulers. The condition of the peasants for whom the New Kashmir Programme was basically formulated worsened further on account of the 'plunder' of their produce, exorbitant prices of essential commodities, corruption, and official tyranny. The urban craftsman also suffered equally because of state monopoly over trade in crafts, price rise, inadequate rationing,

---

[258] Saraf, *Kashmiris Fight for Freedom*, vol. II, 1214.

[259] Mullik, *My Years with Nehru-Kashmir*, 46.

[260] Saraf, *Kashmiris Fight for Freedom*, vol. II, 1214.

[261] Mullik, *My Years with Nehru-Kashmir*, 46.

[262] Puri, *Jammu and Kashmir*, 121.

[263] The *New York Times* correspondent reported from Delhi, 'Sheikh Abdullah is said to lean towards independent status and thereby strengthening his personal support which now appears to be failing.' Quoted by Satish Vashisht, *Sheikh Abdullah: Then and Now* (Delhi: Maulit Sahitya Prakashan, 1968), 96.

enhancement of existing taxes, and imposition of new taxes. The policy of nationalization and the closure of traditional routes hit the merchants, traders, and transporters hard. The dictatorship of the National Conference and the unruly behaviour of its cadres alienated the whole population. The educated middle class could neither give vent to their grievances, nor had any prospects in the environment of acute financial crisis which gripped Kashmir during this period. The political elite belonging to groups other than the National Conference was repressed into silence. Journalists could not breathe in the suffocating atmosphere of a media gag. While Sheikh Abdullah wanted freedom for himself, he denied the same to others. The Muslims were, according to Korbel, looking for 'unification with their Muslim brothers' and 'the Jammu Hindus and the Ladakhi Buddhists, clamouring for separation from Kashmir and for integration with India'.[264] The Sheikh, who was the real author of Naya Kashmir, was completely disappointed by the role of Hindu nationalists which militated with his idea of India—the idea which had formed the bedrock of Kashmir's accession to India. Indeed, only the National Conference workers were a happy lot, making hay in the sunshine of the 'quami haqumat' (nationalist government) as the National Conference rule was called by its leadership. This is what the disillusioned nationalist poet Mahjoor lamented, saying:

> Ye azadi che sorgich huur yi pheiri khana pata khanai
> Fakat kentsan garan andar chi maaraan graayi azadi
> (This freedom is a hourie of Paradise and cannot roam from home to home;
> Confined to a select few homes is the flutter of its charming dance.)
>
> —Mahjoor (1972), pp. 74–7.

Giving land to the tiller was no doubt a revolutionary reform. It had long-term substantial consequences in ameliorating the condition of subalterns. But in the immediate situation, it did not make any difference, not only because of the rampant corruption in the implementation of reforms that hit the common peasantry

---

[264] Korbel, *Danger in Kashmir*, 232.

hard, but more so because of the forced procurement of grains which devastated the peasantry, impoverishing them as they had to purchase grains from the black market and pay the same in the form of the most oppressive *mujwaza* and the so-called *khush kharidi*. The inhuman treatment meted out to the peasantry by the National Conference workers-cum-officials is simply heartbreaking. The Wazir Committee noticed with distress the wailings of the peasantry which, though deficient in foodgrains, was forcibly robbed of its scanty produce. Indeed, when surviving elders who were contemporaneous with the times under discussion hear the mention of 'mujwaza' and 'khush kharidi', they shudder with fear.

All in all, therefore, both friends and foes of Sheikh Abdullah are unanimous that the peasants were not happy over the land they got in ownership, because the land did not give them anything except misery. What was given with one hand, was taken away by the other. Daniel Thorner wrote on the basis of his personal observations, 'in practice any beneficial results [from land reforms] have been more than balanced by the government's food procurement policy, the cooperatives' reign of extortion, the implementation and redistribution through old bureaucracy and the sole political party's setting itself up as yet another privileged rural hierarchy.'[265] The significance of ownership of land lies in the ownership of its produce. The peasants of the Sheikh's regime became only the owners of land, but not of the produce. The state's claim over the ownership of the peasant produce was the oppressive innovation of the 'awami hukumat', as the National Conference government was boastfully called by its leadership. Indeed, it was during the period of Bakhshi, when the peasants were not subjected to the extortion of the forced procurement of foodgrains, that they became the proprietors of land as well as its produce. And it was then that the import of the policy of land to the tiller and the importance of its author (Sheikh Abdullah) was realized. It also needs to be mentioned that all the peasants of Kashmir were not landless peasants. Only that class of peasants was landless who tilled the land of landlords. The majority of

---

[265] Daniel Thorner, *The Shaping of Modern India* (New Delhi: Allied, 1980), 158.

the peasants were *milkiyati* (peasant owners). They had ownership rights over their land and were commonly called zamindars. Even the position of these peasants was worsened due to the procurement policy of the state.

A very significant measure was the scaling down of debts; but this reform too was a halfway house because, after the enactment of the law, the traditional moneylenders stopped giving credit, which proved more harmful to the peasantry in the absence of alternative credit facilities. In those times, money currency was scarce, and for payment of part of the land revenue in cash and to fulfil multifarious critical needs, the people generally turned to *sahukars* and *weddars*[266] for lending money in lieu of food or cash crop pledged to be paid at the time of harvest. Given this fact, the complaints published in the print media that the abolition/reduction of indebtedness in the absence of alternative rural credit caused greater problems to the people in general and the peasantry in particular, rather than solving them, are not flippant.

In terms of democracy, corruption, and tyranny, the post-colonial period was more disappointing than what obtained before October 1947. Although the leadership claimed that in the formation of the New Kashmir Manifesto they had 'combined the communist ideology with democracy and liberal humanism',[267] the reality was different during the regime. The National Conference rule was in essence the 'dictatorship of the National Conference'. All the contemporary observers knit their eyebrows over the regime's intolerance of opposition, not even tolerating healthy criticism: 'Public places showed notices asking people not to talk politics,' says Mir Qasim.[268] The National Conference leaders boastfully called their regime 'one party, one rule'. Having monopolized all means of information, education, cultural enlightenment, and

---

[266] *Weddar* was in common local parlance used for money lenders, who lent money in lieu of kind to be paid by the borrower at the time of harvest on the rates arbitrarily fixed by the money lender in advance at the time of lending money. Hence the rates used to be invariably much lower than the market rates.

[267] Abdullah, *The Blazing Chinar*, 218.

[268] Qasim, *My Life and Times*, 47.

entertainment, the government became a 'totalitarian dictator-ship',[269] to quote Korbel. As a matter of fact, while the Sheikh expected all—right from Security Council through New Delhi to Kashmir—to respect his wavering positions, he was an inveterate enemy of all those who held a contrary opinion. He seemed to love freedom for himself while hating it for others.

The prevalence of rampant corruption from top to bottom was admitted by the leadership too. The complaints against mistreatment by officials attracted the attention of almost all contemporary sources. Mir Qasim, who was associated with the government, and remained well disposed towards the Sheikh till his death, says, 'The administration seemed to rest on the props of tyrannical laws and their heartless implementation. A minor provocation was enough for petty officials to arrest and torture people like slaves in olden days.'[270] *Khidmat*, the official organ of the National Conference, wrote on 30 July 1952, 'It is a general complaint that some big state officers treat their subordinates and also the public in a way which is unbecoming. They resort to beatings and use vituperative language.'[271] 'The instances of administrative arbitrariness and excesses', writes Puri, 'had a similar reaction in all the regions of the state.'[272]

Robust state intervention was required to uplift Kashmir and the Kashmiris. Even a staunch enemy of communism like Josef Korbel, a contemporary observer of Kashmir, accepts that in the Kashmir-like conditions, there was no other alternative but state intervention in a big way to bring about worthwhile change in the otherwise wretched conditions of the people. Lauding the changes, especially in agriculture, as 'remarkable', Korbel justifies the idea that Kashmir needed powerful state intervention:

> the changes, especially in agriculture, have been remarkable. It would be wrong to judge them according to Western standards and principles of economic liberalism. It may well be that the task of

---

[269] Korbel, *Danger in Kashmir*, 218.
[270] Qasim, *My Life and Times*, 47.
[271] *Khidmat*, 30 July 1952.
[272] Puri, *Jammu and Kashmir*, 96.

raising the standard of living of a primitive society must be based upon active intervention by the state in economic and social affairs and the assumption of responsibility for planning and implementing production. Such a society often has no spontaneous incentive to improve living conditions.[273]

However, the role of the state as trader, and that too through corrupt National Conference workers, wreaked havoc among the poor masses. The state charged more than 100 per cent profit on some essential items in the name of 'fair price' cooperatives. Besides being a profiteering practice, the cooperatives became a den of corruption as the National Conference workers who ran them resorted to black marketing and nepotism. Theoretically, the cooperatives were supposed to sell goods at cheaper rates and serve mainly the poor common masses. In practice they did the opposite. The goods were sold at exorbitant rates, and those goods which were in great demand and of better quality were given to party people's kith and kin, and to those with political and money power. It is because of these two main reasons that even the government admitted that the cooperatives completely collapsed because of the 'corruption and maladministration of government officials',[274] and the people requested the Wazir Committee with folded hands to free them from the 'tyranny of the cooperatives'.[275]

The government followed the policy of 'self-reliance' for preserving political autonomy. But by pursuing 'self-reliance' without having sufficient internal resources, or the power to raise international finances and restore the geographical centrality of Kashmir so as to participate in global trade and tourism, the government proved to be nothing short of an incubus sucking the lifeblood of the people. By charging exorbitant rates for goods and services, and engaging in the forced procurement of food along with inadequate rationing, the government acted as a usurer and a plunderer of people's scarce resources.[276] To enlarge its scarce resources, it

[273] Korbel, *Danger in Kashmir*, 218.

[274] Korbel, *Danger in Kashmir*, 216.

[275] *Wazir Committee Report*, 49.

[276] *Amrita Bazar Patrika*, 'Sheikh Abdullah's Downfall', 17 March 1955, Archives Reference Library, Srinagar, Accession no. 185.

also imposed a tax on education, which had been free under the princely order.

The New Kashmir Programme also failed to unite people on basic and shared issues among all regions and communities. Although the progressive ideology espoused arraying people along the lines of exploiter versus exploited, it did not work in practice. The state became sharply divided on the lines of religion. True, the National Conference had rejected the two-nation theory and opted for Kashmir to accede to India, but the Jammu Rajput Dogras in particular and Jammu Hindus in general, including the Ladakhi Buddhists, rallied around religion and united to fight to finish the 'Kashmiri Muslim domination' led by Sheikh Abdullah, who had fought against their benefactor.

As a matter of fact, the period was riven by internal conflicts born out of ideologies rooted in region and religion. Sheikh Abdullah, the leader of the National Conference government, represented the interests of the marginalized sections of the society who were mainly Muslims, Scheduled Caste Hindus, and people living in remote areas. Also, he was independent-minded. The Dogra Rajputs whose material and emotional interests were not served by this ideology rose in revolt in association with Ladakhi Buddhist leaders and the Hindu nationalist forces of India, demanding abolition of Kashmir's special position. Kashmir became a battleground which, besides appropriating the precious time of the government, fuelled the financial crisis, as all the routes connecting Kashmir with India through Punjab were frequently sealed. The Sheikh put up a brave resistance against the demand of the Praja Parishad and insisted on not less than an independent position for Kashmir.

To be sure, the oppressive and dictatorial rule of the National Conference would have taken the edge off of Abdullah's mass appeal among the Muslims too, had the memories of his sterling role as leader of the freedom movement not been fresh in the minds of the people, and the wounds of the dark side of his governance not been blamed on some of his progressive steps. The whirlwind protests against the Sheikh in Jammu, led by the supporters of the erstwhile princely order, with communal overtones, provoked equal and opposite reactions in Kashmir, driving the discontent against the unpopular measures of the Sheikh government underground. Last but not least, Sheikh Abdullah's pro-autonomy utterances at the beginning and pro-independence outbursts subsequently, which were publicized with exaggeration by the well-knit party machine, 'recovered', in the words of Balraj Puri, 'whatever personal support he had lost during his years in power'.[277] Clearly, the Sheikh was a good fighter but a failed administrator, which is abundantly reflected in the governance of the period.

[277] Puri, *Jammu and Kashmir*, 121.

3

Patronage Government (1953–63)

Around midnight of 8–9 August 1953, Sheikh Abdullah was dismissed and arrested in a boardroom coup engineered by the central government[1] to eliminate the hurdle in its hidden agenda of fully integrating Kashmir with India. This came as a very rude shock to Abdullah,[2] who had proved a sincere friend to Pandit Nehru; so sincere that he ignored the basic structures—geography, history, culture, and economy—of Kashmir, and in preference to Pakistan had opted for Kashmir's accession to India. The shell-shocked Sheikh, who had not even in his dreams thought of any betrayal of the Accession Agreement, made no bones about the 'conditional accession' of Kashmir having cost him state power. Abdullah's uncere-monial exit was followed by massive demonstrations and protests across different parts of the valley. All contemporary observers—both friends and foes of Sheikh Abdullah—unanimously speak of the unprecedented outrage caused by the subversion of democracy

---

[1] See Mullik, *My Years with Nehru-Kashmir*, Chapter 3 'Fall from the Pedestal', 25–47. For the argument that the replacement of Sheikh Abdullah's government by a pro-India faction was part of a covert intelligence action, see also Gundevia, 'On Sheikh Abdullah', 107–12; Guha, *India after Gandhi*, 255.

[2] Abdullah, *The Testament of Sheikh Abdullah*, 40. For the reaction of Sheikh Abdullah to the integrationist moves by the Indian government, see Sheikh Abdullah's letter to Maulana Azad on 16 July 1953, Noorani, *Article 370*, 232–6.

in the state. The mass protests continued for many weeks, involving all sections of society across the valley.[3] Indeed, it was a very difficult situation, especially when the reaction of the Kashmiris found an echo in the hearts of many national and international audiences,[4] and that too at a time when Kashmir was a burning question being debated in international fora. Thus, the throne which the Indian government bestowed upon Bakhshi Ghulam Mohammad was not a bed of roses. His own legitimacy too was hanging by the eyelids. It needed astute statesmanship both at the centre and in the state to reverse the trend. And Bakhshi, with the full backing of the central government, proved equal to the task. Within two to three months, the tide had changed to the extent that Bakhshi was emboldened to say with pride that he commanded as much support among the people as Sheikh Abdullah had.[5] It is, therefore, instructive to look at the steps taken by Bakhshi to restore peace in trouble-torn Kashmir.

Materialistic Approach

It is really astonishing how Bakhshi Ghulam Mohammad and his colleagues in the ministry have, by their policy and hard work, changed

---

[3] Mullik, *My Years with Nehru-Kashmir*, 46; Bazaz, *History of Struggle for Freedom in Kashmir*, 575–6; Qasim, *My Life and Times*, 68–70; Saraf, *Kashmiris Fight for Freedom*, vol. II, 1212–13.

[4] Hafsa Kanjwal, 'Building a New Kashmir: Bakshi Ghulam Mohammad and the Politics of State-Formation in a Disputed Territory', PhD thesis, University of Michigan, 2017, https://deepblue.lib.umich.edu/bitstream/handle/2027.42/138699/hafsak_1.pdf?sequence=1&isAllowed=y (accessed 27 October 2017), 47.

[5] It is commonplace to hear the anecdote in Kashmir that when some journalist asked Bakshi Ghulam Mohammad about the position of his public support, he replied that four million people were with him. The journalist further asked how many people were with Sheikh Abdullah; Bakshi replied, four million. This was a strange reply, because the total population of Kashmir was only four million then. This is also quoted by Bose, *The Challenge in Kashmir*, 150.

the entire picture and outlook in the state within two months. They
have done so chiefly because of their economic approach.

—Jawaharlal Nehru[6]

Broadly speaking, Bakhshi used two opposite instruments to restore
'order' in Kashmir. These were: manufacturing consent, and the
use of the coercive apparatuses of the state. While the coercive
measures are dealt with in the next section, we will focus here on
those measures that Bakhshi adopted to gain public acceptance, if
not legitimacy. To put it briefly, the tide was reversed by the policy
of munificence, concessions, and development. Clearly, Bakhshi
and his mentors were ideologically convinced about the efficacy of
materialistic determinism. That is why they gave first preference
to addressing the economic aspirations of the people, and that too
with large-hearted generosity and the capacity to view things from a
total perspective so as to woo all sections of the society. They were
also conscious of the fact that mass support for them depended on
how 'expeditiously and effectively relief was provided for the people', to quote the *Amrita Bazar Patrika*.[7] It is, therefore, no wonder that immediately after taking over as prime minister, Bakhshi
promptly addressed the people on 9 August 1953 through radio and
emphatically reminded them of the economic crisis they had been
subjected to by the policies of his predecessor. He also emphasized
the benefits the people would get from the 'progressive economic
policies' of the Indian state. Bakhshi not only resorted to rhetoric,
but did something far-reachingly tangible.[8] In the same broadcast,
he announced a range of economic concessions which included:[9]

1. Abolition of the system of forced extraction of foodgrains from
    the peasantry.

---

[6] Jawaharlal Nehru, *Letters to Chief Ministers 1947–64*, vol. 3: *1952–
54*, gen. ed. G. Parthasarthi, Teen Murti, New Delhi, 1987.

[7] 'New Chapter Opened', *Amrita Bazar Patrika* (Calcutta), 19 March
1955, quoted in *Kashmir after August 9*, 17. (All articles from *Amrita
Bazar Patrika* are sourced from *Kashmir after August 9*).

[8] See Korbel, *Danger in Kashmir*, 244.

[9] Saraf, *Kashmiris Fight for Freedom*, vol. II, 1221.

2. Subsidized permanent ration system to the extent of 75 per cent of its cost for urban areas.
3. Grant of subsidies on the sale price of salt.
4. Free education from primary to university level.
5. Abolition of the monopoly of cooperatives.
6. Remission of cooperative debts.
7. Substantial increase in the salaries of all government servants and enhancement of wages of labourers.

In the preceding chapter, we have seen that the compulsory procurement of grain, through the two iniquitous institutionalized systems of mujwaza and khush kharidi, was a great source of tyranny for the peasantry in general and the poor peasantry in particular, not simply because they had to part with a substantial share of their produce at cheap rates to cater to the urban people, but more so because in view of the monocrop economy of Kashmir, small holdings, and extremely low per unit productivity, surplus production was unknown among the common peasantry. Their produce fulfilled their own food needs just for few months, and for the remaining months they had to subsist on subsidiary foods. The Wazir Committee appointed by the government in 1953 took serious note of this and reported, 'persons who could not produce from their lands enough to eat even for three months of a year were subjected to the levy of *mujawaza*. No consideration was given to leaving enough food for families according to their strength or size of their holding.'[10] This is the reason that the state had to use force for procuring grains from the wailing peasantry, giving rise to corruption,[11] and rendering the Sheikh's agrarian reforms practically meaningless. Rather, land became a liability, a source of misery. In fact, Walter Lawrence had in the late nineteenth century protested against the state policy of procuring cheap grains from the peasantry through coercive means for feeding idle urbanites at low rates. However, he could not succeed in pushing through his recommendation of abolishing the system,

[10] *Wazir Committee Report*, 111.

[11] Korbel, *Danger in Kashmir*, 213–14; Government of J&K, *Review of the Achievements of Bakhshi Government*, 8.

as the maharaja did not want to antagonize the vocal urban population.[12] The Sheikh, of course, did not want to oppress the peasantry, but his apprehensions of mortgaging 'freedom for a loaf of bread'[13] led him to ignore the fact that J&K did not grow enough food to feed the state's increasing population. The result of this misplaced policy of self-reliance created famine conditions everywhere. The rural population was robbed of the little stock they had just for a few months; and the urban population cried that the food rations allowed to them were acutely inadequate.[14] The situation not only led to antagonism between the people and the state, it also vitiated relations between the rural folk and the townsmen, as the former realized that the latter were fed at their cost.[15] Clearly, the procurement policy, which caused widespread resentment, was 'a source of profit to officials, of great loss to the state, and of misery and demoralization to the people'.[16]

To ride out the storm, the vexatious system of mujwaza was done away with to the great relief of the peasantry, more than the abolition of feudalism. The so-called 'khush kharidi' of the previous regime was really made into a 'sale by choice', as the rich peasantry, which possessed surplus stock, were now free to sell it to any agency. In case the peasant wanted to sell paddy to the government, its price was increased from Rs 9 to Rs 10 per *kharwar*.[17] Another rupee was paid to the peasant to meet the charge of transportation. All restrictions on the purchase and sale of foodgrains and their movement from one place to another within the valley were removed, and the ration scale in the city of Srinagar was raised to five *traks*.[18] About 150,000 maunds of rice, paddy, and wheat

---

[12] Lawrence, *Valley of Kashmir*, 438–40.

[13] 'New Chapter Opened', *Amrita Bazar Patrika*, 18.

[14] See Chapter 2 of this book.

[15] 'New Chapter Opened', *Amrita Bazar Patrika*, 18.

[16] 'Sheikh Abdullah's Downfall', *Amrita Bazar Patrika*, 14.

[17] Government of J&K, *Review of the Achievements of Bakhshi Government*, 9. Kharwar = 117 lbs approximately.

[18] Government of J&K, *Review of the Achievements of Bakhshi Government*, 9; *Kashmir Today* series II, 'A Reply to Critics', text of the speech by Bakshi Ghulam Mohammad in the State Legislative Assembly, 5 March 1955. Trak= 1. 32 lbs approximately.

were imported from India and supplied at reduced rates.[19] About 900 vehicles were utilized for bringing rice from Pathankot to Srinagar.[20] The sale rate of Jammu rice issued by the Food Control Department in Srinagar was reduced from Rs 25 to Rs 8, and that of wheat flour from Rs 25/10 to Rs 20 per maund.[21] The sale price of wheat *atta* (flour) and rice issued from government depots in Jammu city was reduced from about Rs 18 to Rs 12/8.[22] The rate of Kashmiri rice was also reduced from Rs 10/8 to Rs 8/8 per kharwar. Food rationing was also introduced in the towns of Baramulla, Sopore, Bandipora, and Bijbehara. Besides, the number of ration holders in Srinagar city rose by about 25,000 during the year 1954.[23]

To provide relief to the villagers in deficit areas, 70 relief centres were opened in Jammu and Kashmir (including Ladakh) where rice, wheat, flour, and maize were distributed among the people at the reduced rates.[24] Under the Bakhshi government, the cost of rice subsidies shot up to Rs 15 million per year as compared to Rs 1.9 million under Sheikh Abdullah.[25] Therefore, while the Sheikh came to be contemptuously called *ālubab* (father of potatoes), Bakhshi was elevated to the position of *bhatabab* (father of rice, the staple food of Kashmiris).[26]

[19] Government of J&K, *Review of the Achievements of Bakhshi Government*, 10.

[20] Government of J&K, *Review of the Achievements of Bakhshi Government*, 10.

[21] Government of J&K, *Review of the Achievements of Bakhshi Government*, 9.

[22] Government of J&K, *Review of the Achievements of Bakhshi Government*, 10.

[23] Government of J&K, *Review of the Achievements of Bakhshi Government*, 11.

[24] 'New Chapter Opened', *Amrita Bazar Patrika*, 17.

[25] G. N. S. Raghavan, 'Kashmir on the March', *Indian Express*, October 1956.

[26] Given the position of food production as well as food stocks in the state, it was no less than a miracle to be able to provide food to all, and that too at cheap rates. 'With the granaries of Food Control Department empty and the crop damaged by floods, the hoarders and black-marketers believed that the reform could never be successful.' However, with

The political importance of making rice—the staple food of Kashmir—available to Kashmiris in abundance, and that too at cheap rates, can be inferred from the contrast we find in the attitude of the Indian state urging its citizens elsewhere to consume substitute foods in place of rice and wheat to save foreign currency from being used for importing foodgrains.[27] Clearly, the policy of austerity, retrenchment, and switching over to substitute foods to save money for industrial development was incompatible with the political conditions of Kashmir. To win the loyalty of the people in the face of the disputed status of Kashmir and the lack of an emotional bond with India among its Muslim majority, there was unanimity among the Indian political leadership that compromises with national economic policies vis-à-vis Kashmir were politically correct if they served the larger political goals.

In those days when the roads remained blocked for six months, Kashmiris usually faced scarcity of salt, which was always a cause of concern for the state and the people, as the price of salt would soar to the extent of creating 'salt famine' (nun dreag).[28] Considering the general demand for salt and the low purchasing power of the masses, Bakhshi understood the political significance of reducing the cost of the commodity, especially when the Sheikh government had traded on the helplessness of the people by charging 100 per cent profit on its sale (for which Abdullah had been dubbed 'shylock'[29]); which is why the price of salt figured among the set of concessions granted by the state.

The remission of cooperative debts scaled down rural indebtedness from Rs 20 million to Rs 8.6 million. Bakhshi also redressed

---

generous aid from India and significant structural reforms within state structures, Bakshi made it possible. Government of J&K, *Review of the Achievements of Bakhshi Government*, 10.

[27] See Siegel, 'Self Help Which Ennobles a Nation'.

[28] Given the fact that during pre-modern times it was unimaginable to have stocks of salt, it is no wonder that the phrase 'horse-loads of salt' became idiomatic, denoting something which is promised to someone to entice them by deception. Thus, *nune gur te tile talav* (salt-loaded horses and ponds of oil).

[29] 'Sheikh Abdullah's Downfall', *Amrita Bazar Patrika*, 14.

a common grievance when he broke the monopoly of the cooperatives, which had become a 'symbol of tyranny'[30] for two reasons:
First, this institution had been monopolized by corrupt National
Conference workers, for whom they had become a source a of
loot,[31] and an instrument by which to impose themselves upon
the people and to harass those who were not ready to conform.[32]
Second, the monopoly of trade in essentials had caused their prices
to shoot up[33] so high that it gave birth to a saying in Kashmiri,
yeth chu gumut nun til kunise wanis, literary meaning that here
the trade in salt and oil is monopolized by one trader; technically
conveying that price rise was the natural consequence of monopoly of trade.

The political impact of the substantial enhancement of the
salaries of government employees and the wages of daily wagers
is easy to guess. The same is true of making education free up
to the university level, if we consider that the abject poverty of
the masses was a serious impediment in the way of popularizing
education.

To get an idea of how Bakhshi changed the course of events
in Kashmir, it is illustrative to quote the following excerpt from
the text of the speech that the prime minister gave in the State
Assembly on 17 March 1955:

> The House is aware that rice sold at Rs. 60/ to Rs. 70/ per *khirwar*
> (about two maunds) during the last six years and that the pay of a
> peon during this period was Rs. 12-1/3-15. Our only fault is that
> soon after our assumption of office, we raised a peon's pay scale to
> Rs. 25-1-30, plus Rs. 12/- as ration allowance and started issuing rice
> at Rs. 8/- per khirwar. We do not want to have the credit (for these
> reforms). Not only is rice selling at cheaper rates but other conces-
> sions also have been granted to them. For instance, education has
> been made free, customs tariff has been abolished and commodities
> are selling at cheaper prices. The peon could hardly purchase rice

---

[30] See Bazaz, *History of Struggle for Freedom in Kashmir*, 430–32; Puri,
Jammu and Kashmir, 129.

[31] Korbel, *Danger in Kashmir*, 216.

[32] See Chapter 2, note 102.

[33] Korbel, *Danger in Kashmir*, 216.

with his pay for only fifteen days of the month. But now he can purchase with his pay five maunds of rice. He understands very well the reality about their (opposition parties') slogans. He cannot be led astray by catchy slogans. He knows what could not be done for him during the period between 1948 and 1953 has been done during 1954.[34]

Indeed, Bakhshi Ghulam Mohammad and his advisors were convinced that the Kashmir crisis would be resolved by addressing the economic problems of the people. In his policy statement issued on 5 October 1953, Bakhshi reiterated his government's understanding of the Kashmir problem:

> By far the largest problem that confronts us, and to which I attach the greatest importance, is, as I have already stated, not of any political or military nature, but of eliminating the economic distress which has been on the increase since 1947 and is today at its worst. This crisis cannot be overcome by the termination of the State's association with India or by a merger of the State with Pakistan. Nor can it be avoided by an alignment with a foreign power. In the history of people, there are no short-cuts to success. The first and the foremost step to be taken by us will be to find out measures for arresting the economic degeneration. We are determined to go ahead with this plan. By the grace of God, our firm will and strong determination is sure to crown us with success.[35]

The policy of munificence, concessions, and development, however, needed oodles of money; and the Indian government, for which Kashmir 'strengthens the secular character of Indian polity' and thus helps in building a 'strong stable democracy' besides serving its 'defence interests',[36] was more than ready to meet all the

[34] 'Aid from India', speech by Bakshi Ghulam Mohammad in the State Assembly, 17 March 1955, 5.

[35] Government of J&K, *Unanimous Vote of Confidence in Bakshi Ghulam Mohammad: An Account of the Proceedings of the State Legislature on the Motion of Confidence in Bakshi Government (5 October 1953)* (Jammu: Naya Kashmir Orient Press, 1953), 30.

[36] 'Facts of Geography', *Amrita Bazar Patrika*, 8.

expenses on this count, besides providing the requisite technical support. Accordingly, the Indian Planning Commission advanced a loan of $14.9 million to the state government in December 1953.[37] The First Five-Year Plan was revised considerably after Bakhshi assumed office. Consequent upon this, the plan provisions went up from Rs 70 million to Rs 97.32 million in the central sector, and from Rs 30 million to 30.09 million in the state sector.[38] For the subsequent two plan periods also, the state received generous grants. Balraj Puri has given detailed statistics about the per capita grant-in-aid, special grants in the form of subsidies, and the additional central assistance for the development of border areas, state police, and border checkposts, which were much greater than for other states.[39] According to Puri, 'during the five year plan period from 1957–8 to 1961–2 Kashmir received the highest per capita grant-in-aid—Rs 41.7 or almost seven times the average for all the states which was Rs 6.'[40] The total transfer of funds to the state during the financial year 1959–60 was about Rs 1,211 lakhs.[41] It may be noted that at a time when all the states of India had deficit budgets, J&K had a surplus of Rs 38.016 million for 1959–60.[42] However, this was not due to the internal resource mobilization of the state, but because of generous funding from the centre. Indeed, between 1950 and 1970, nearly 90 per cent of the state's five-year plans were funded by the centre.[43] This amount excluded the funding that was spent on various centrally sponsored and operated schemes/programmes like the national highway, telegraphs and telephones, broadcasting, regional engineering and medical colleges, tunnels, regional research laboratories, and so on. A section

---

[37] Korbel, *Danger in Kashmir*, 245.

[38] Gupta, *Kashmir: A Study*, 397.

[39] See Balraj Puri, 'The Budget of Kashmir: What the Centre Means to the State', *Economic Weekly* (18 April 1959), 549–50; also see Balraj Puri, 'Central Aid to Kashmir: Effects of Finance Commission's Recommendations', *Economic Weekly* (May 1962), 811–12.

[40] Puri, 'Central Aid to Kashmir', 811.

[41] Puri, 'The Budget of Kashmir', 549.

[42] Puri, 'The Budget of Kashmir', 549.

[43] Prakash, 'The Political Economy of Kashmir', 2053.

of the political elite was sensitive about the implications of the departure from the policy of self-reliance to the policy of overdependence, and voiced its concern against this deviation.[44]

Considering the aforementioned huge financial assistance from the central government, it is no wonder that the Bakhshi government ushered in an era of all-round development. While due to the paucity of space, it is difficult to capture fully the multifarious developments that took place during the period of the Bakhshi government, the following pages delineate some of the important developmental schemes that ultimately turned the tide and restored 'cold peace'[45] in the state.

Agricultural Development

Bakhshi Ghulam Mohammad did not simply bank on importing large quantities of foodgrains and providing food at subsidized rates with the assistance of the central government. Doubtless, it was a temporary measure to respond to a crisis. He realized that the state needed to meet most of its internal demand locally by increasing agricultural productivity. Therefore, he not only maintained but rejuvenated the agricultural policy of his processor. Notwithstanding the sweeping agrarian reforms of the previous government which were carried forward by Bakhshi and successive governments, it was realized by all governments that overcoming the stark food deficiency of the state and improving the abject condition of the largest segment of the population presupposed increasing the per unit productivity substantially by introducing technological changes to supplement the institutional changes. The insistence on this aspect was reflected in the First Five Year Plan. Accordingly, the Bakhshi government executed many agricultural research schemes which were designed to:[46]

---

[44] See 'Aid from India', speech by Bakshi Ghulam Mohammad in the State Assembly, 17 March 1955.

[45] For the concept of 'cold peace', see International Crisis Group, 'India, Pakistan, and Kashmir: Stabilizing a Cold Peace', Asia Briefing, no. 51 (June 2006).

[46] Government of J&K, *Jammu and Kashmir: An Overview of Progress* (Srinagar: Department of Information), 1961, 63.

1. Introduce disease-resistant high-yielding crops.
2. Apply chemical fertilizers to ensure large production.
3. Establish demonstration farms.
4. Construct a soil research laboratory and a modern glass-house for research.
5. Establish private farms in each tehsil for the production of superior varieties of paddy and wheat seeds.
6. Provide additional machinery, material, and equipment for the control of diseases of crop and fruit trees.

As a follow-up measure, the Paddy Improvement Scheme in Kashmir was revised in 1954.[47] To determine the cause, control, and remedial measures of rai disease and to increase paddy production.[48] Ten seed plant nurseries were established during 1954 and 1955 to produce a 'double hybrid maize seed' and improved varieties of paddy; the number of seed multiplication farms increased to 15 during 1956–9; 100,000 maunds of pure seeds were distributed on a premium basis to the farmers; and large quantities of sulphate ammonia were given to the peasants on subsidized deferred payment, and even as free gifts.[49] Also, the Rice Research Scheme Programme was revised, with the objective of evaluation of varieties with high yield, response to heavy fertilization, easy threshing character, resistance to diseases, and so on. Besides, under the Coordinate Maize Breeding Scheme, Srinagar introduced sweet maize which fetched higher prices, and popcorn, suitable for popping, was introduced in 1958.[50]

To give a boost to agriculture, two important production programmes were started during the period of Bakshi—the Community Development Programme and the Intensive Agriculture Development Programme (IADP). With the aim of raising agricultural production by the extension of irrigation facilities, bringing as much land as possible under the plough and laying stress on improved agricultural practices, the Community

[47] H. S. Mann, 'Agriculture: An Analysis', in Saraf, *Jammu & Kashmir Trade Guide*, 150.

[48] Mann, 'Agriculture: An Analysis', 150.

[49] Aziz, 'Economic History of Modern Kashmir', 97–8.

[50] Mann, 'Agriculture: An Analysis', 150.

Development Programme was started in the state during the Second Five-Year Plan.[51] Similarly, the IADP, as proposed in the Third Five-Year Plan, was started during the year 1961–2 to step up agricultural production by intensive use of all the important farm inputs. Two districts, Anantnag and Jammu, were chosen for technological transformation under full state patronage through block development agencies.[52] Due to these programmes which ensured an adequate supply of improved seeds, fertilizers, pesticides, and credit, the two districts showed remarkable improvement from 1961–2 to 1965–6.[53]

However, the most remarkable development in agriculture was the ambitious project embarked upon by the government to bring dry land under paddy cultivation by excavating a network of canals and installing pump stations. Needless to say, half of Kashmir is under tablelands called *wuders* in Kashmiri and *karewa*s in geological literature. These wuders were either uncultivated or under rainfed crops with extremely low per unit productivity, and that too subject to timely rainfall. Even large low-lying areas were left either uncultivated or under rainfed crops for want of channelization of water resources. Since agriculture was the mainstay of the economy, the import of foodgrains was sapping the state's resources and the resources of the people. Given the fact that there were potentialities of agricultural growth through the extension of irrigation facilities, it is understandable that in the First Five-Year Plan, the highest priority was given to irrigation, for which the plan earmarked 16.5 per cent of the total plan allocation.[54] Indeed, irrigation had also been the priority of Sheikh Abdullah's government, but Bakhshi carried it forward vigorously with huge central assistance. This is evident from the impressive statistics available and reproduced in Table 3.1 on the number of canals excavated, lift irrigation schemes executed, and the wells dug up during the period, and the resultant huge area of land brought under

[51] Mann, 'Agriculture: An Analysis', 148.

[52] Mann, 'Agriculture: An Analysis', 148; Aziz, 'Economic History of Modern Kashmir', 99–100.

[53] Mann, 'Agriculture: An Analysis', 148.

[54] Aziz, 'Economic History of Modern Kashmir', 133.

Table 3.1 Canals Constructed in Kashmir Valley during the Regime of Bakhshi Ghulam Mohammad

S. No.	Name and Type of Irrigation Work	Year of Construction/ Improvement/ Remodelling	Land Irrigated in Hectares for Kharif Crop	Land Irrigated in Hectares for Rabi Crop	Location: Administrative Division
1	Bewerah Canal	1952–4	783	341	Islamabad
2	Shalatang Canal	1954	1,827	—	Sonawari
3	Dalna Lift	1954	230	2	Baramullah
4	Ladoora Lift	1954	484	—	-do-
5	Padgampora (Lift)	1954	1,493	156	Pulwama
6	Dagripora Lift	1954	870	191	-do
7	Barasoo Lift	1954	155	85	-do
8	Gorkha Canal	1955	613	68	Srinagar
9	Babal Canal	1955	3,361	—	Sonawari
10	Awneera Canal	1955	833	271	Shopian
11	Sindh Power Canal	1955	820	273	Srinagar
12	Melhora Canal	1955–6	752	49	-do-
13	Reshipora Canal	1955–6	1,333	321	-do-
14	Mehand Canal	1955–6	1,316	736	Islamabad
15	Rakh Litter Canal	1955–6	1,761	259	Pulwama
16	Kandi Tank	1956	155	—	Sopore
17	Hajan Markandal Lift	1956	2,422	—	Sonawari
18	Bonikhan Lift Station	1956	359	—	-do-

(Cont'd)

Table 3.1 (Cont'd)

S. No.	Name and Type of Irrigation Work	Year of Construction/ Improvement/ Remodelling	Land Irrigated in Hectares for Kharif Crop	Land Irrigated in Hectares for Rabi Crop	Location: Administrative Division
19	Hakbara Pump Station	1956	645	—	-do-
20	Galibal Pump Station	1956	576	—	-do-
21	Aegi Canal	1956	15,410	876	Srinagar
22	Khanchi Canal	1956	1,767	372	-do-
23	Lolab Storage Tank	1956	1,670	—	Sopore
24	Rajveer Tank	1956	586	—	-do-
25	Nathnisa	1956	124	—	-do-
26	Ompora Canal	1957	1,122	47	Srinagar
27	Heti Canal	1958	2,147	—	Sonawari
28	Paribal Lift	1959–60	603	—	-do-
29	Chandoosa Canal	1959–60	515	—	Baramulla
30	Kaniaz Lone Kuhl	1960	21	—	Sonawari
31	Krishi Nallahs	1960	776	—	Sopore
32	Tarzoo Veer	1960	1,530	—	-do-
33	Rakh Shalwat Pump Station	1960–61	1,146	—	Sonawari
34	Daslipora Pump Station	1960–61	733	—	-do-
35	Dehgam Kuhl [Lolab]	1961–2	2,579	—	Sopore
36	Padshah Kuhl	1962	4,535	1,490	Srinagar
37	Sindh Extension Canal	1962	1,382	195	-do-

38	Arhi Canal	1962	1,264	430	-do-
39	Haider Canal	1962	1,120	11	-do-
40	Baba Canal	1962	1,743	212	-do-
41	Lar Canal	1962	3,972	1,157	-do-
42	Dab Kuhl	1963	3,937	996	-do-

Sources: Abdul Rehman Mir, *Kashmir Mein Abpashi* (Srinagar: Shaheen, 1981).

Based on the information collected during filed study for the present work; Administrative Report, Government of Jammu and Kashmir, 1953-54, p. 45; Administrative Report, Government of Jammu and Kashmir, 1955-56, p. 199; John, Masarat, 'Jammu and Kashmir Under Bakhshi Ghulam' Mohammad, M.Phil Dissertation (unpublished), Department of History, University of Kashmir, 2012, pp. 86–8.

cultivation. Kashmir, indeed, underwent almost a revolution in terms of rice production. Its full economic and political import is realized when one undertakes a tour of the areas brought under cultivation by these canals and other means of irrigation, and talks with the people who were benefited by these canals. Clearly, the history of the progress of these villages dates from Bakhshi period, and it is no wonder that he continues to be remembered fondly as their saviour.[55] It is revealing to hear from economic historians that food production in the state increased from 8.256 million maunds in 1951–2 to 16.61 million maunds in 1964–5. According to them, this increase in productivity was mainly because of extensive irrigation facilities provided during the period.[56]

In the Kandi area of Jammu province, tubewells for irrigation and water supply purposes were sunk at Chheni Hamet, Samba, Bagocha Chak, Jetwell, Inlia, Gurah Salathian, Udhampur, and Sangway during the First Five-Year Plan at the estimated cost of Rs 4.41 million.[57]

The government also gave some economic relief to peasants and brought about a fundamental change in the status of tenants. Some concessions were granted regarding the levy of *abiana* (water tax) on different types of lands.[58] The canals built before 1948–9 did not entail the imposition of any tax on the lands irrigated by them; and those lands which were irrigated by canals built from the year

---

[55] In this regard, it is rewarding to talk to the peasants of Phak Pargana (the area near Hazratbal Srinagar). The pargana was for the first time irrigated during the period of Bakshi, who constructed what is popularly called the Bakshi Canal, or Bijli Kul. According to the peasants, they were living off rainfed crops prior to the Bakshi regime, and their condition was extremely pitiable. The same story is told by other peasants whose lands were for the first time irrigated during the Bakshi period.

[56] Aziz, 'Economic History of Modern Kashmir', 139–40.

[57] Government of J&K, *Review of the Achievements of Bakhshi Government*, 24.

[58] Government of J&K, *Five Months* (Srinagar: Bureau of Information, 1953); Ghulam Mohammad Khawaja, *Sheikh Abdullah ki Wazarat kay Zawal kay Asbab* (The Causes of the Downfall of Sheikh Abdullah Government) (Delhi: Darya Ganj, 1954), 76.

1948–9 were also assessed on *khushki* rates instead of *abi* rates.[59] Concessions of 25 per cent and 12 per cent, respectively, in the *abiana* rates were given to old irrigators whose lands were classed as *abi-awal* and *abi-daum*.[60] Besides, the State Tenancy Act was amended to place restrictions on the eviction of tenants from the land. According to this amendment, the tenant had to pay only a fourth of the produce to his landlord in the case of wetlands and a third in the case of dry land in respect of tenancy holdings exceeding 12.5 acres.[61]

In order to give a boost to animal husbandry, along with opening an institute for the manufacture of biological products to protect livestock and poultry against major diseases at Srinagar, intensive cattle development projects were launched both in Jammu as well as in Kashmir.[62] Two cattle breeding farms were set up, one each at Cheshmashahi, Srinagar, and at Belicharana, Jammu. Similarly, tehsil veterinary units, one mobile veterinary laboratory each at Jammu and Srinagar, and two artificial insemination centres at Srinagar were established. The cattle breeding process through artificial insemination in the state was started in 1954.[63] Sheep breeding also received great fillip during the period. By March 1956, large-scale sheep breeding had been undertaken in the state. Sixteen sheep and wool extension centres were organized. The most important development was the establishment of a full-fledged Directorate for Development and Research of Sheep Breeding during the Third Five-Year Plan.[64] Three sheep-breeding farms, one each at Anderwan (Sindh valley), Billawar (Jammu),

[59] Government of J&K, *Five Months*, 5.

[60] Government of J&K, *Five Months*, 5.

[61] Government of J&K, *Republic Day Series (January 1956–8)* (Srinagar: Press Information Bureau), 39.

[62] Shirazi, 'Animal Husbandry', in Saraf, *Jammu & Kashmir Trade Guide*, 153–4.

[63] Government of J&K, *Fifty Years of Animal Husbandry in Kashmir, 1947–98* (Srinagar: Animal Husbandry Department, Kashmir Division, 1998), 7.

[64] G. A. Banday, 'Sheep Breeding', in Saraf, *Jammu & Kashmir Trade Guide*, 158–9.

and Dachigam (Kashmir), were set up. Besides, 64 more sheep and wool development units were established. Training of personnel also received great attention, with officers being sent to other parts of India as well as foreign countries to get training in sheep and wool work.[65]

Flood Control

The occurrence of floods causing huge losses of men and materials had always been (and continues to be) a problem in Kashmir, as recovery from the devastation caused by floods took years. During the period of Bakhshi Ghulam Mohammad, Kashmir was subjected to two devastating floods in 1957 and 1959, which took a heavy toll of life besides damaging crops over about 200,000 acres of land and destroying thousands of houses covering 1,287 villages.[66] Bakhshi Ghulam Mohammad took steps to tackle the flood problem in an effective manner. He established a high-power State Flood Control Board in which engineers from the Central Water and Power Commission and representatives of central ministers functioned. The board was responsible for the overall planning of flood control proposals for the valley.[67] The government undertook an ambitious project of flood control and drainage at the estimated cost of Rs 25 million with the aim of not only protecting the city of Srinagar against flood but also saving approximately 45,000 to 50,000 acres of cultivated land from floods, and reclaiming about 25,000 acres of marshy and waterlogged land land by silting, over a period of 10 years.[68] Bunds on the river Jhelum from *Sangam* to Srinagar were strengthened to save the land from ordinary floods. The flood spill channel from Padshahi Bagh to Wallur was deepened and extended, and marginal bunds

---

[65] Banday, 'Sheep Breeding', 159.

[66] Saraf, *Kashmiris Fight for Freedom*, vol. II, 1229.

[67] *Kashmir: An Estimate of Progress* (Srinagar: Lalla Rookh Publications, n.d.), 44.

[68] Government of J&K, *Review of the Achievements of Bakhshi Government*, 25.

were constructed around marshy lands. For speedy drainage of floodwater in the upper part of the valley, floodgates, sluices, and drainage channels were constructed. Importantly, the prime minister himself was the chairman of the board which was set up to ensure that the work was properly planned and executed within the time limit.[69] The flood control measures taken by Bakhshi are a source of wonder to modern engineers of Kashmir, said an engineer friend to me.

Cooperatives

Bakhshi and his cabinet colleagues (the majority of whom were communists or Congress socialists) were staunch supporters of the cooperative movement. It is, therefore, not surprising that they put the cooperatives 'on a new footing to serve as a vital link with the people'. The important step in this direction was that the government broke the monopoly of the cooperatives, which had turned them into a 'symbol of tyranny'.[70] Now, unlike during the Abdullah government, the various facilities provided through cooperatives were not selectively available only to those who wore a badge of allegiance to the ruling party. Nor were commodities sold by cooperatives alone. The breaking of the monopoly brought prices down, to the satisfaction of the common people. A committee was appointed to examine the working of the cooperative movement and to suggest measures for revitalizing it on sound lines. The priority of the Bakhshi government was to strengthen service cooperatives, so that the benefits reached the people as quickly as possible. An Apex Bank was also established to make the cheap credit facilities of the Reserve Bank of India available to primary societies and ordinary cultivators. The working of industrial cooperatives was reoriented to ensure that craftsmen got their due share of the profits and were not exploited. In this regard, a land mortgage bank was established, which addressed a long-felt

---

[69] Government of J&K, *Review of the Achievements of Bakhshi Government*, 25.

[70] *Wazir Committee Report*, 49.

need in the state. This bank offered long-term loans to landholders on the mortgage of their land.[71]

The cooperative marketing societies offered marketing facilities to agriculturists. This system considerably reduced the dependence of farmers on middlemen and traders. Cooperative marketing societies marketed such commodities as paddy, wheat, maize, fruits, pulses, wool, ghee, and so on. The progress achieved in this direction can be gauged by the fact that the value of produce marketed by these societies increased from Rs 290,000 in 1956–7 to Rs 6,224,000 in 1959–60.[72]

Education

Though Sheikh Abdullah had considered land to the tiller and education as the two basic needs for a nation to attain prosperity,[73] he was constrained by scarcity of resources in fully implementing the education-related ideas envisaged in the Naya Kashmir Programme. With the full financial backing of the central government, Bakhshi completed the task initiated by his predecessor.

In his policy speech broadcast on Radio Kashmir immediately after he took over as prime minister, Bakhshi *inter alia* included measures for educational development in the list of his priority areas, and spelt out the steps he had decided to take in this direction. The steps included enhancement of salaries of teachers, abolition of tuition fee from primary to university level, providing textbooks to poor students, developing national languages (Kashmiri, Dogri, and Ladakhi), and promotion of education among the backward classes by providing scholarships.[74]

---

[71] Government of J&K, *Address to the Joint Session of Jammu and Kashmir Legislature* by Sadr-e-Riyasat Karan Singh, 9 February 1963 (Srinagar: Legislative Assembly Secretariat, 1963), 5.

[72] *Hamare Pyare Khalid-e-Kashmir* (Srinagar: Lalla Rookh Publications, n.d.), 37.

[73] Government of J&K, *Educational Reorganizational Committee Report* (Srinagar, December 1950), i.

[74] *Crisis in Kashmir Explained* (Srinagar: Lalla Rookh Publications, 1953), 10–11; Government of J&K, *Review of the Achievements of Bakhshi Government*, 11–16; 'New Chapter Opened', *Amrita Bazar Patrika*, 19–20.

This was an epoch-making development. A stumbling block in the way of mass education was removed. With very few exceptions, the poor masses were still illiterate; and the main reason was their inability to pay fees. What is more, the Abdullah government had increased the fee, provoking protests.[75] Considering that, unlike in the past, the Muslims were now showing a keen desire to receive education,[76] but their quest was frustrated by economic poverty, the political import of the inclusion of education-related concessions in the olive branch held out by Bakhshi to the restive Kashmiris is not difficult to appreciate. The policy of free education not only encouraged those 125,000 students[77] who were already in schools and colleges to continue their education; more importantly, it encouraged new aspiring parents to enrol their wards without any diffidence on account of economic poverty. It is, therefore, not surprising to see a spectacular growth in enrolment by the end of the Bakhshi regime. Thus, while in 1950, the total enrolment of students was 107,233, it had risen to 276,351 by 1960.[78]

As a matter of fact, right from the very inception of their entry into the politics of freedom, the National Conference leadership had realized the significance of education as an effective instrument of poverty alleviation. Further, passionate as they were to construct a Kashmiri personality after their own ideology, they could not find any better way than by shaping the educational policy with clear goals and guiding principles. For realizing the first objective, education figured among the priority areas in the agenda

[75] Government of J&K, Department of Education, File 1651 Ed-726-US/1948, Archives Reference Library, Srinagar.

[76] See the Foreword by G. M. Sadiq, Education Minister, J&K State, to Government of J&K, *Our Educational Policy: A Draft Statement* (1955), Archives Reference Library, Accession no. 657/GACC. Also see Government of J&K, File 1414,16/Schools/1949, and File No. 1410-nil/1955, Department of Education, Archives Reference Library, Srinagar, for the representations made by people of different areas for the opening of schools and colleges in their respective areas.

[77] Government of J&K, *Review of the Achievements of Bakhshi Government,* 11.

[78] S. L. Seru, *History and Growth of Education in Jammu and Kashmir 1872–1973* (Srinagar: Ali Mohammad and Sons, 1977), 150.

of the Naya Kashmir Programme and, as we shall see, Bakhshi took concrete steps to make education easily accessible to the poor. For the fulfilment of the second objective, the job of preparing the policy was assigned to progressive intellectuals headed by the committed communist education minister, G. M. Sadiq. It may be mentioned that as the Naya Kashmir Programme had informed the education policy of the Abdullah government too, there was, therefore, no difference in the substance of the aims and objectives of the educational policies of the two governments. The fundamental objectives of the education policy were: to provide the easiest possible access to education; to produce a 'productive citizen who is educated as well as able to earn his living'; to connect education with the realities of daily life and people's immediate surroundings; to create a scientific temper to help build a progressive and prosperous Kashmir; to develop a sense of belonging and patriotism for the motherland (Kashmir) as well as for India; to develop an appreciation for cultural heritage; to construct a personality in tune with New Kashmir; to cultivate a secular and liberal personality; and to promote extracurricular activities to 'productively' channelize the energy of the youth in the politically unstable conditions of Kashmir.[79]

For translating these objectives into reality, a large number of educational institutions were established, and emphasis was laid on technical and vocational education, the establishment of 'activity schools' at the primary level where children would learn by doing things,[80] teaching in the mother tongue and in the context of the immediate surroundings, establishment of 'multipurpose' schools after primary schooling to learn trades,[81] encouragement to students to opt for a career in trade and industry after matriculation,[82] patronage to the creamy layer of the talent pool

---

[79] For details, see G. M. Sadiq, *Our Educational Policy* (Srinagar: Ranbir Press, 1955). Also, see Sadiq, Foreword, in Government of J&K, *Our Educational Policy*.

[80] For details about the nature and character of activity schools, see G. M. Sadiq, *Our Educational Policy*.

[81] G. M. Sadiq, *Our Educational Policy*.

[82] G. M. Sadiq, *Our Educational Policy*.

for pursuing higher studies within and outside the state,[83] and promotion of extracurricular activities (art, music, drama, theatre, sports, and social service) under the centrally sponsored 'Youth Welfare' scheme to depoliticize the youth.[84] Texts were rewritten to appropriate the social sciences for popularizing the ideas of tolerance, amity, fellow feeling, human values, critical thinking, and the history of intimate relations between India and Kashmir.[85]

The financial integration with the Indian government helped Bakhshi to complete the process started by his predecessor. In 1950, the budget for education was 6 per cent of the state's total revenue; it had shot up to 12 per cent by 1956.[86] To realize the first ideal of New Kashmir's educational programme, the government focused on universal primary education, followed by the establishment of secondary and higher secondary schools and colleges.

During the first year of the Bakhshi government, the state witnessed an increase of over 26 per cent in the number of educational institutions,[87] and of about 30 per cent in the budget allocation for education.[88] The number of primary, high, and higher secondary schools rose from 1,115 in 1950–51 to 2,859 in 1960–61 and

---

[83] Government of J&K, *Review of the Achievements of Bakhshi Government*, 13.

[84] Sadiq, *Our Educational Policy*. Also see Sadiq, Foreword, in Government of J&K, *Our Educational Policy*.

[85] For details, see Kanjwal, 'Building a New Kashmir', 123–37.

[86] Seru, *History and Growth of Education*, 176.

[87] During the first year of the Bakshi government, the number of educational institutions rose from 1,245 to 1,820 and the budget from 5.40 million to 7.03 million. Further, one college, 30 high schools, 35 middle schools, 60 central schools, 331 primary schools, and 243 *maktab*s and *pathshala*s were opened. Government of J&K, *Review of the Achievements of Bakhshi Government*, 12. *Maktab* is an Arabic word meaning elementary school. These were primarily used for teaching of reading, writing, grammer and Islamis studies to children. *Pathshala* means school. Path means lesson/chapter/recitation and shala means hall/ house/school. Traditionally it was the name given to Hindu school were children were taught in Sanskrit.

[88] Government of J&K, *Review of the Achievements of Bakhshi Government*, 12.

from 55 in 1950–51 to 250 in 1960–61, respectively.[89] One education officer was appointed in each tehsil for effective inspection
of the schools.[90] Equal attention was paid to higher education.
An arts college was established at Jammu, the Islamia College
of Science and Commerce in Srinagar, and the Gandhi Memorial
College of Srinagar was converted into a science college.[91] The
Government College for Women, Nawakad, which had been
ordered to be upgraded from a high school to college by Sheikh
Abdullah in 1952, ultimately started functioning as a college in
1961. Similarly, the Government Agricultural College at Wadura,
Sopore, was established in 1960 to give a boost to the agricultural
sciences in the state.

As already discussed, Sheikh Abdullah had stopped or reduced
the grant-in-aid to many private institutions, resulting in their
extinction. The Bakhshi government restored all such cuts from
13 April 1953 at the annual cost of over Rs 200,000, benefiting 6
high schools, 29 middle and primary schools, 1 girls' college and
8 girls' schools.[92] Grant-in-aid was also sanctioned to many private institutions, notable among them Gandhi Memorial College
Srinagar.[93] The Jammu and Kashmir University, which was by then
only an examining institution, was converted into a teaching university. Eminent professors were brought from India to run some
of the teaching departments of Jammu and Kashmir University;
besides, some of the professors were deputed for higher studies in
India and abroad.[94] Women's education received special attention.
Separate primary, secondary, and high schools were established
for girls. While in 1952–3 the total number of girl students in all

[89] Government of J&K, *Jammu and Kashmir: 50 Years* (Srinagar:
Department of Information, 1998), 347.

[90] Government of J&K, *Review of the Achievements of Bakhshi
Government*, 12.

[91] *Daily Khidmat*, 7 November 1961.

[92] Government of J&K, *Review of the Achievements of Bakhshi
Government*, 12.

[93] Government of J&K, *Review of the Achievements of Bakhshi
Government*, 12.

[94] 'New Chapter Opened', *Amrita Bazar Patrika*, 20.

categories was 15,753, it rose to 19,350 by 1954–5.[95] The number of primary, middle, and high schools for women rose from 175 to 542, 15 to 69, and 9 to 38, respectively. Scholarships were provided to students belonging to the backward classes;[96] mobile schools were started for the benefit of nomadic people, namely, Gujjars and Bakarwals.[97] Technical education received a great impetus; the state got its first medical college, a regional engineering college, two agricultural colleges, two polytechnic colleges, and one Unani and Ayurvedic college.[98] This is besides a number of institutions established to train ancillary personnel for hospitals and dispensaries. Moreover, arrangements were made for reservation of seats in different technical and professional institutions in India for students of the state to receive education in medical, engineering, and other technical and professional fields.[99] While making nominations to these seats, special attention was paid to backward classes and communities, provincial representation, and representation of women candidates.[100] A scheme of granting interest-free education loans was introduced,[101] as a result of which a crop of technical personnel became available with adequate know-how in medicine and surgery, agriculture and animal husbandry, electrical, mechanical, and civil engineering, and for postgraduate studies in various disciplines.

[95] 'New Chapter Opened', *Amrita Bazar Patrika*,19.

[96] Government of J&K, *Review of the Achievements of Bakhshi Government*, 14; Suresh K. Sharma and S. R. Bakshi, *Encyclopedia of Kashmir*, series 4 (New Delhi: Anmol, 1995), 98.

[97] Government of J&K, *Review of the Achievements of Bakhshi Government*, 12–13; Government of J&K, *Republic Day Series*, 29.

[98] Government of J&K, *Address to the Joint Session* by Karan Singh, 8 February 1960, 6.

[99] Government of J&K, *Review of the Achievements of Bakhshi Government*, 14; *Daily Khidmat*, 25 July 1952.

[100] Government of J&K, *Review of the Achievements of Bakhshi Government*.

[101] Government of J&K, *Report of the Jammu & Kashmir Commission of Inquiry* (Gajendragadkar Commission Report), Srinagar, December 1968, 66.

It is also important to mention that in the girls' schools, the students were also imparted training in different vocations, namely, tailoring, needlework, laundry, cooking, painting, health and hygiene.[102] The government also laid much emphasis on extra-curricular activities. Students were encouraged to participate in sports, youth camps, and the Junior Red Cross Movement to shape their mindsets and promote among them the sense of social service. The construction of an impressive stadium called the Bakhshi Stadium[103] stands as a monumental testimony to Bakhshi's zeal to popularize the culture of sports in Kashmir.

Health

Health services were extremely poor at the time Bakhshi Ghulam Mohammad assumed power. There were only 2 hospitals and 87 dispensaries in the state. In order to meet the growing demand for effective health services and to bring within their orbit the far-flung areas of the state, an elaborate health programme was chalked out with the help of outside experts. To translate this programme into practice, Bakhshi increased the budget for health services by more than 50 per cent.[104] As a result, we find that the number of hospitals increased from 2 to 19 and there was also a spectacular increase in the number of dispensaries, which rose to 401 by 1960–61.[105] Medical schemes under community development projects were sanctioned for Ladakh, Budgam (Kashmir), and Mansar (Jammu).[106] Moreover, a vaccination campaign was launched in both provinces with the help of UNICEF to eradicate various deadly diseases.[107]

---

[102] Government of J&K, *Administrative Report*, 1959–60 (Srinagar/Jammu: General Administration Department, 1960), 17.

[103] The Bakhshi Stadium is one of the few stadiums in the entire state.

[104] *Daily Khidmat*, 25 December 1955; 'New Chapter Opened', *Amrita Bazar Patrika*, 20.

[105] Government of J&K, *Jammu and Kashmir: 50 Years*, 346.

[106] Government of J&K, *Review of the Achievements of Bakhshi Government*, 17.

[107] Government of J&K, *Review of the Achievements of Bakhshi Government*, 16.

Along with this, operation and X-ray apparatuses were considerably expanded and a deep X-ray was established in SMHS Hospital, relieving people of the trouble of going outside the state for this purpose.[108] One standard laboratory and a blood bank were established both in Jammu and Srinagar hospitals.[109] A mobile medical unit was also established to provide medical aid and advice to villagers at their doorsteps. To improve the standard and efficiency of the staff, a number of doctors were deputed for various training programmes both in the country and abroad.[110]

An organization of 6 technicians and 12 disinfectors under the control of a trained epidemiologist was set up to eradicate typhus from the state.[111] An anti-malaria scheme involving an expenditure of Rs 98,000 was launched in Jammu. A malaria unit was established in Jammu by the Government of India at an estimated cost of Rs 783,000.[112] Further, for the eradication of venereal diseases in Jammu, a scheme was executed at an estimated cost of Rs 94,000. This scheme was subsidized by the World health Organization, which supplied PAS (para-aminosalicylic acid) and penicillin free of cost for women and children.[113]

Both curative and preventive measures were taken to fight tuberculosis. A mass BCG (Bacillus Calmette–Guérin) campaign was launched in November 1954 all over the state. On the curative side, a sum of Rs 230,775 and Rs 244,000 was sanctioned for the expansion of the tuberculosis departments in Jammu and Kashmir, respectively. The bed strength of Jammu hospital was raised from 17 to 50, and that of Kashmir from 80 to 150.[114] The sanatorium at Batote was reopened for patients belonging to Jammu province,

---

[108] Government of J&K, *Review of the Achievements of Bakhshi Government*, 16.

[109] Government of J&K, *Review of the Achievements of Bakhshi Government*, 16.; *Daily Khidmat*, 25 December 1955.

[110] Government of J&K, *Review of the Achievements of Bakhshi Government*, 17.

[111] 'New Chapter Opened', *Amrita Bazar Patrika*, 20.

[112] 'New Chapter Opened', *Amrita Bazar Patrika*, 20.

[113] 'New Chapter Opened', *Amrita Bazar Patrika*, 20.

[114] 'New Chapter Opened', *Amrita Bazar Patrika*, 20–21.

and its bed strength was raised from 18 to 30. Likewise, the sanatorium at Tangmarg in Kashmir was improved to accommodate 85 beds.[115]

One of the most important steps taken by the government was to abolish the operation fee levied in the state hospitals.[116] An appropriate grant was sanctioned to state hospitals for the provision of patent medicines and antibiotics to indigent patients free of cost. Besides, free breakfast was provided to indoor patients at the two main hospitals at Jammu and Srinagar, and the diet scale of chest disease hospitals was increased. What is more, the government established state-owned medical shops to provide people standard medicines at reasonable rates.[117]

Industrial Sector

Developing the industrial sector partly by fostering and protecting the public sector was envisaged in the Naya Kashmir Programme and incorporated in the Jammu and Kashmir constitution under Section 14. Following this policy, Bakhshi established three industrial estates at Gandhinagar (Jammu), Barzulla (Srinagar), and Anantnag at a total capital outlay of Rs 5.26 million.[118] Besides this, a good number of industries were established. Of these, mention may be made of the drug manufacturing factory at Srinagar, surgical instrument factory (Srinagar), cement factory at Wyun (Kashmir), ceramic factory (Gandhinagar, Jammu), mechanical plate for making bricks and roofing tiles, the Government Paints and Mineral Factory, Pampore Plywood Industry, and a match box factory. Also, drug farms extending over 2,500 acres of land were established at Chakrohi, Manasbal, Bonora, Zainpura, Yarikah, and Katra.[119] For industrial development as well as for electrification,

---

[115] 'New Chapter Opened', *Amrita Bazar Patrika*, 21.

[116] Government of J&K, *Review of the Achievements of Bakhshi Government*, 16.

[117] Government of J&K, *Administrative Report*, 1959–60, 16.

[118] *Hamare Pyare Khalid-e-Kashmir*, 41.

[119] Government of J&K, *Administrative Report*, 1959–60, 10.

power generation was increased from 4,000 KWs to 31,000 KWs during the period of Bakhshi.[120] No less attention was paid to the development of small-scale industries. This can be gauged from the fact that in the Second Five-Year Plan, Rs 34.12 million was earmarked for village small-scale industries and mining. By 1955, a sum of Rs 1,500,000 was provided to craftsmen, artisans, and petty traders in cities and towns. A sum of Rs 120,000 was distributed by way of loans among the boatmen of Srinagar.[121] The State Industries Department was reorganized to enable it to start various small-scale local industries, and to provide employment and guidance to talented young men.[122] A carpet industry and a *namda* industry were established in Srinagar,[123] three tweed-producing centres at Sopore, Pampore, and Srinagar, a woollen demonstration-cum-production centre at Kishtwar, a centre for woodwork in Srinagar, a handmade paper and cardboard industry at Miran Sahib, and also a bamboo furniture manufacturing centre at Basholi. The pashmina industry was rejuvenated by establishing a Pashmina Syndicate at Ladakh and a Pashmina Centre at Basholi.[124] The cotton and woollen industry also received patronage by the state. To give a boost to the khadi industry, the development programme of the Khadi and Village Industries Commission was started in the state in 1961.[125] A khadi centre was set up at Hiranagar, a demonstration-cum-production centre at Samba, and a network of woollens centres throughout the state.[126] Apart from manufacturing quality goods in embroideries, shawls, carpets, and pashmina, the production centres established

[120] Kanjwal, 'Building a New Kashmir', 71.

[121] 'New Chapter Opened', *Amrita Bazar Patrika*, 21.

[122] Government of J&K, *Review of the Achievements of Bakhshi Government*, 18.

[123] *Namda* is a traditional Kashmiri carpet made by felting that orginated in eleventh century.

[124] *Kashmir: An Estimate of Progress*, 25.

[125] G. M. Zargar, 'Khadi and Village Industries', in Saraf, *Jammu & Kashmir Trade Guide*, 139.

[126] *Jammu and Kashmir: A Review of Progress*, Department of Information, Governemnt of Jammu and Kashmir, 1956, p. 8.

under state patronage also produced crafts such as woodcarving and papier mâché.[127] To respond to new tastes and demands, a research centre was established in 1957 to evolve new designs.[128] To give a boost to the small-scale industrial sector, a chain of 20 industrial estates with 275 industrial sheds were constructed at a cost of Rs 10 million during the Second and Third Five-Year Plan periods.[129]

Besides maintaining the existing policy of opening exhibition and sale centres in different cities of the country under the auspices of the government emporiums, Bakhshi succeeded in persuading the Central Cottage Industries Emporium, New Delhi, and the Small Scale and Cottage Industries Emporium, Bombay, to purchase Kashmiri handicrafts. To give a boost to the industrial sector, the State Financial Corporation Act, 1951 (LXIII of 1951), was extended to J&K in November 1959, and under it a State Financial Corporation was established with the objective of developing industry in the private sector.[130]

The government constituted a Labour Committee to recommend measures for improving the condition of labour and fostering a spirit of cooperation between employer and employee. Accordingly, the Industrial Disputes Act and the Trade Union Act were amended to address the felt inadequacies. Besides other things, 15 days' leave with pay was granted in a year to every labourer.[131] Work was also provided to the section of labourers who were out of employment before 9 August 1953.[132] The system

---

[127] Government of J&K, *Review of the Achievements of Bakhshi Government*, 19–20; Jammu and Kashmir, *A Review of Progress*, January 1956, 8.

[128] Government of J&K, *Administrative Report, 1958–59*, 183.

[129] For details see O. P. Modi, 'An Entrepreneur's View', in Saraf, *Jammu & Kashmir Trade Guide*, 135.

[130] Government of J&K, *Budget Speech of Bakhshi Ghulam Mohammad*, J&K Legislative Assembly Secretariat, Srinagar/Jammu, 1960–61, 6.

[131] Government of J&K, *Review of the Achievements of Bakhshi Government*, 27.

[132] Government of J&K, *Review of the Achievements of Bakhshi Government*, 27.

of contributory provident fund for labour was also introduced in all government industrial concerns.[133]

Roads and Buildings

Bakhshi Ghulam Mohammad is rightly called 'Bakhshi the Builder'. His decade was the decade of considerable infrastructural development. Apart from constructing educational institutions, hospitals, hotels, rest houses, dak bungalows, buildings for government departments including the two new secretariat buildings—one for Jammu and one for Srinagar—hydroelectric projects, housing colonies (Nursing Garh, Bal Garden, Jawahir Nagar, Gandhi Nagar), and stadiums.[134] Bakhshi Ghulam Mohammad showed special interest in establishing a network of roads within the state, as well as improving the connectivity of Kashmir with the rest of the country. A whopping Rs 24.54 million was spent on the project during the First Five-Year Plan,[135] and Rs 66.5 million during the Second Five-Year Plan. In fact, the road networking of Kashmir was for the first time zealously pursued during this period. Of a large number of new roads constructed during the period connecting villages with major urban centres, some are mentioned in Table 3.2. Bakhshi pursued this programme so zealously that ambitious projects that had been unimaginable until then were conceived and accomplished. Among them were the construction of the Banihal tunnel, Doda–Thathri Kishtawar road, Srinagar–Leh road (commenced in 1962, completed up to Kargil in 1964),[136] Srinagar–Teetwal road, and Baramullah–Uri road.

[133] Government of J&K, *Review of the Achievements of Bakhshi Government*, 27.

[134] D. D. Thakur, *My Life and Years in Kashmir Politics* (Delhi: Konark, 2005), 148. For additional information see, John, Masarat, 'Jammu and Kashmir Under Bakhshi Ghulam Mohammad', 2012, Appendix 5.3, pp. 116-19.

[135] *Hamare Pyare Khalid-e-Kashmir*, 12.

[136] Anwar Karim, 'New Ladakh', in Saraf, *Jammu & Kashmir Trade Guide*, 216.

Table 3.2   Roads Constructed during the Bakhshi Regime

Names of Roads	Total Length
Srinagar Division	—
Arigam Arihal Charpathri Road	3 miles
Awantipora Tral Road	7 miles
Budgam Chadoora Road	8 miles
Bemina Budgam Road	5 miles
Charari-Sharief Pakharapora Remu and Kanidjan Yus Road	19 miles
Chadoora Budgam Road	4.5 miles
Chadoora Syrasyar Road	8.5 miles
Extension of boulevard along Dal Lake	75 chains of 54 ft width
Humhama Raithan Road	11.5 miles
Humhama Naithan Road	12.0 miles
Improvement of Beehama Tullamulla Road	6.2 miles
Kakpora Romu Pulwama Nou Road	14 miles
Kangan Wangath Road	7 miles
Khunmukh Khrew Wahab Khar Road	10 miles
Metalling Beehama Safapora Road	7 miles
Nagam Rambagh Charari-Sharief	18 miles
Narkara Budgam Road	1.5 miles
Natipora Neu Road	12 miles
Pampore Ladhu Road	5 miles
Pampore Awantipora Road	8 miles
Pulwama Rajpur Road	5 miles
Pulwama Romu Road	5 miles
Pulwama Tahab Road	3 miles
Rambagh Nagam Charari-Sharief Road	16 miles
Rambagh Natipora Neu Road	12 miles
Road from N.M. to Khanaquo Maulla at Pampore	25 chains
Saloora Qamar Sahab Road	1½
Anantnag Division	—
Achabal Kokernag Viloo Road	7.07 miles
Anantnag Dooru Verinag Road	5 miles
Anantnag Town Road	1.5 miles
Approach Road from Kulgam Bazar to Tehsil	22 chains
Bawan Achabal Road	4 miles
Construction of Shopian Town Road	21 chains

Damhal Hanjipora Aharbal Road	10 miles
Dooru Town Road	46 chains
Fair Weather Road from Dooru to Kokernag	5 miles
Improvement to the Bejbehera Langanbal Road	21 miles
Kachduru Kaprin Gopalpora Road	3.12 miles
Kaimoh Kardar Road	6 miles
Khanbal Anantnag Achbal Road	8 miles
Khanbal Pahalgam Road	27 miles
Khandroo Shanzas Road	2 miles
Pahalgam Chandanwari jeep track	8 miles
Pahalgam Circuit Road	4 miles
Qazigund Kulgam Road	10 miles and 4 chains
Remodelling Mattan Road	1.5 miles
Shopian Bijbehara Road	15.5 miles
Shopian Khazanbal Aharbal Road	8 miles
Shopian Kulgam Road	13 miles
Baramulla Division	—
Baramulla Baba Reshi Road	18.1 miles
Construction of Baramulla Khadinyar Road	2.5 miles
Construction of Sumbal Tarzoo Road	13 miles
Handwara Langat Road	18 miles
Handwara Zachaldara Nichhama Road	14 miles
Improvement of Hajin Markundal Road	5 miles
Improvement of Hjiwara Magma Road	5 miles
Jeep track from Deyar Bali to Panipilla	—
Kupwara Sogam Chandigam Road	13 miles
Sogam Hayatpora Dilbagh Dardpora Road	—
Tangmarg Baba Reshi Road	22.5 miles
Jammu Division	—
Bajpura Gurah Salathain Sector	3 miles
Bajpur Ramgarh Road	6 miles
Bari Bhahmana Gurah Salathain Road	6 miles
Bari Brahmana Purmandal Road (Kaluchak)	13 miles
Bari Brahmana to Bishna	2 miles
Black topping of R.S. Pura Road	43 miles
Chak Mohd Yar to Chak Roi Road	4 miles
Dablahar to Arina Sector	4 miles
Dablahar to Maralin Road Distributory	5 miles
Hiranagar Link Road	3 miles

(Cont'd)

Table 3.2 *(Cont'd)*

Names of Roads	Total Length
Jammu Gajansu Road	8 miles
Jammu Ghaumansan Road	6½ miles
Main Sahib Deoli Road	6 miles
R.S. Pura to Dablahar Road	3 miles
Residency Julaka Mohalla Road	1 mile
Sambha Mansar Road	13 miles
Ujh Khor Punno Road	8 miles
Yakh to Raya Sector	3 miles
Jammu National Highway Division	—
Ramban Damakund Salal Reasi (Sector Ramban to Damakund)	10 miles
Widening of miles 108,109, Jammu Srinagar Road near military checkpost	—
Special Sub-division Kathua	—
Dialachak Ding Amb Road	6.5 miles
Dialachak Hiranagar Link Road	3 miles
Link Road Ujh Khor Punno Road	9.5 miles
Doda Division	—
Doda Thathri Road	19 miles
Udhampur Division	—
Sukethar Kacha Pind Jindrah Road	15 miles
Special Sub-division Poonch	—
Poonch Mandi Road	7 miles
Rajouri Budal Arnas Guldam Kund Road	4 miles

Source: John, *Jammu and Kashmir under Bakhshi Ghulam Mohammad*, 2012, pp. 112–15; 120–1.

A scheme for widening the roads of Srinagar and the beautification of the city was also launched. It necessitated the demolition of a large number of private houses. No protracted acquisition proceedings were resorted to; instead, Bakhshi Ghulam Mohammad himself went from door to door, doing a little haggling with the owners, and ultimately paying on the spot, often more than the market value.[137]

[137] Saraf, *Kashmiris Fight for Freedom*, vol. II, 1222.

Table 3.3 Bridges Constructed during Bakhshi Regime

Srinagar Division	Baramulla Division
• Budshah Bridge	• Baramulla Bridge
• Cement Bridge	• Kontra Bridge
• Construction of 1–45 span strutted bridge on Dudganga Nallah at Chatabal	• Seloo Bridge
	• Shoolora Bridge
• Rajbagh Bridge	• Sisri Bridge
• Wangat Bridge	• Sopore Bridge
• Zero Bridge	

Anantnag Division	National Highway Division
• Arwani Bridge	• Construction of Bailey Suspension Bridge over River Chenab near Ramban
• Badrani Bridge	
• Chandanwari Bridge	
• Ganeshpoora Bridge	• Julla Bridge at Thathri
• Kanganbal Bridge	• Mohalla Light Suspension Bridge
• Khanbal Bridge	
• Khudwani Bridge	**Jammu Division**
• Mau Bridge	• Bahlol Bridge
• Naninar Bridge	• Pikhoo Bridge
• Nohama Bridge	

• Padgampore Bridge	**Udhampur Division**
• Remodelling of Akhad Bridge No. 1	• Arch Bridge
	• Cheneni Bridge
• Remodelling of Akhad Bridge No. 2	• Saloh Nallah Bridge
• Remodelling of Seligam Bridge	
• Seer Bridge	
• Somgloo Bridge	
• Sonawam Bridge	
• Tangri Nallah Bridge	

Source: John, *Jammu and Kashmir under Bakhshi Ghulam Mohammad*, 2012, pp. 112–115; 120–1.

The construction of a network of roads produced a series of consequences with economic, cultural, and political import. I have outlined the commercial importance of the road networks in the following pages. Suffice it to say that besides giving a fillip to trade and commerce, the enormous expansion of the work done by the

Department of Roads and Buildings generated employment opportunities for engineers, ancillary staff, coolies, drivers, conductors, contractors, and a host of dealers connected with the building industry. The improvement in road connectivity had cultural and political implications too. Ideas and ways of life move along roads and through different communication systems, leading to two opposite consequences: syncretic cultures and revivalism. And we find both making rapid strides side by side. The same is true of competing political discourses. Also, the Indian government has always been seized of the political import of physical connectivity between Kashmir and the rest of India. The construction of the Banihal tunnel was the first major effort in this direction.

Trade and Commerce

We have seen that in accordance with the Naya Kashmir Programme, the policy of the Sheikh Abdullah government towards trade and commerce was one of state monopoly and control. This policy was obviously quite detrimental to the interests of the business community, which is why the Kashmir Chamber of Commerce had been pressing hard for the abolition of state trading and rigid control.[138] We have also seen that because of exorbitant rates charged by the service cooperatives, the system had become oppressive and, therefore, widely abhorred. Considering that the abolition of state trading and control would meet the demands of traders, transporters, and consumers, introduction of a market economy was among the very first steps taken by the Bakhshi government to manufacture public consent. Bakhshi also removed the customs barrier between Kashmir and the rest of India. The customs barrier symbolized, more than anything else, Kashmir's separate political identity, almost bordering on its semi-independent position. By removing this barrier, Bakhshi also wanted to increase the stakes of traders within India, as increased business interests within India would mean expanding the pro-India constituency in Kashmir. Bakhshi also reduced the toll tax by 50 per cent to

---

[138] Korbel, *Danger in Kashmir*, 216.

bring down the prices of goods, which, besides giving relief to the people, increased their demand.

The network of roads and the Banihal tunnel constructed during the period constitute a milestone in the growth of Kashmir's internal and external trade. While the network of roads within the three regions increased the movement of people and goods with enormous economic, social, cultural, and political implications, the construction of the Banihal tunnel is the second major milestone in the history of surface transport in Kashmir, after the Baramulla cart road constructed in the late 1880s. After the closure of the Baramulla route in 1947, Kashmir remained cut off from the external world for about six months a year, besides it being difficult to traverse through the 9,000 foot high Banihal pass. The construction of the Banihal tunnel became an enabling factor for quick mobility of people and goods between Kashmir and the rest of India, promoting trade and tourism. The increase in exports from Rs 15 million to Rs 165 million testifies to the immediate results produced by the Banihal tunnel and the removal of the customs barriers. That during the Bakhshi regime, the strength of private sector transport rose from 1,872 vehicles in 1947 to 6,325 vehicles, corroborates the remarkable growth in the movement of goods and people during the period.[139]

Besides the promotion of trade with different parts of India, internal trade also received a great impetus following the development of rural transport and communication infrastructure. The vast network of roads and the vehicular traffic considerably enhanced the mobility of goods, and that too at reduced costs, encouraging increasing demand for those rural products that, for want of quick and cheap transport, were either traded in small quantities or could not be commercialized. True, the real history of the inroads made by the market into the self-sufficient village economy began a few years after Bakhshi assumed power. This was further helped by the comparatively better purchasing power of the village community following the liberal economic measures and massive investments by the Bakhshi government in

---

[139] Kanjwal, 'Building a New Kashmir', 70.

rural infrastructure, providing work to rural folk—men, women, youth.[140]

Tourism

Tourism was an important source of income for the state during the colonial era. According to official estimates, around 30,000 tourists used to visit Kashmir every year providing a source of income to 200,000 people.[141] Besides its economic value, which itself is an important means of peace building, the tourist flow has been consistently regarded as an effective political instrument, in that it projects normalcy in Kashmir to Indian and international constituencies,[142] and also supposedly brings Kashmir closer to India through people-to-people contact. It is therefore understandable to hear a sympathetic Indian journalist saying, 'every tourist who goes to Kashmir this summer (1949) will be rendering as vital a service to Kashmir—and to India—as a solider fighting at the front.'[143] This is the reason that, right from 1947, one of the most important priority areas of governments in Delhi as well as state governments has been to promote tourism in Kashmir (though New Delhi has been more interested in promoting domestic tourism in the Kashmir Valley). Accordingly, the Bakhshi government

[140] The correspondent of the *Amrita Bazar Patrika*, who visited Kashmir in 1955, wrote: 'in the course of my tours in Jammu and Kashmir, I found that work on digging new irrigational canals, remodeling "kuhls" and on the lift-irrigation projects was going apace and at present nearly eighty-thousand labourers were working on different projects.' *Amrita Bazar Patrika* (Calcutta), 17 March 1955.

[141] Korbel, *Danger in Kashmir*, 217.

[142] A correspondent for the *Times of India* explained the meaning of increasing tourism in Kashmir in these words: 'the steeply rising figures of visiting tourists are a reliable index of the sense of security and political stability now obtaining in the state, which is humming with constructive nation building activity, providing increasing employment to the people and building up better and higher standards of living for them.' Quoted in Kanjwal, 'Building a New Kashmir', 81.

[143] K. A. Abbas, 'The Enchanted Valley', *Swatantra*, 23 April 1949.

showed keen interest in restoring the past glory of Kashmir tourism, which was hit hard by the disturbances in the subcontinent in 1947.[144] Some of the important steps taken in this direction were: marketing the tourism industry of Kashmir by reorienting its publicity policy and setting up publicity wings all over the country;[145] abolition of the permit system and customs barrier between the state and the rest of the country;[146] improvements in air service, construction of the Banihal tunnel and a network of roads within the state; the tourist reception centre at Srinagar; and the construction of 14 dak bungalows, 24 rest houses, 13 dormitories, 111 huts, 1 youth hostel, 2 government hostels, and good hotels.[147] To accommodate the tourists, the Visitors Bureau had licensed about 300 houseboats and over 20 hotels in Srinagar by 1954, besides 10 huts in Pahalgam, 1 dak bungalow and 1 serai at Katra, 10 huts at Cheshmashahi, as well as dak bungalows at Poonch, Kud, and Doda.[148] Ration cards were issued to the tourists on their first visit to the Visitors Bureau, and all necessaries of life were made available to them in abundance at competitive rates. To encourage more tourists to visit Kashmir, the bus fare from Pathankot to Srinagar and back was reduced from Rs 40 to Rs 27.[149] To maintain the flow of passenger traffic and essential goods from and to the state, the government organized a Transport Department with a fleet of about 500 vehicles, including a series of deluxe buses.[150] Regular bus service was made operational in

---

[144] Korbel, *Danger in Kashmir*, 217.

[145] M. A. Pandit, 'Facilities for Tourists', in Saraf, *Jammu & Kashmir Trade Guide*, 201; *Hamare Pyare Khalid-e-Kashmir*, 46.

[146] *Kashmir: An Estimate of Progress*, 24.

[147] Government of J&K, *Review of the Achievements of Bakhshi Government*, 18; Pandit, 'Facilities for Tourists', 201; *Hamare Pyare Khalid-e-Kashmir*, 46.

[148] Government of J&K, *Review of the Achievements of Bakhshi Government*, 18; Pandit, 'Facilities for Tourists', 201; *Hamare Pyare Khalid-e-Kashmir*, 46.

[149] Government of J&K, *Review of the Achievements of Bakhshi Government*, 18.

[150] Government of J&K, *Review of the Achievements of Bakhshi Government*, 19.

the cities of Jammu and Srinagar.[151] As a result of these steps, the tourism industry received a great fillip. In July 1954, Indian official sources reported that up to that date, 21,000 tourists had already visited Kashmir.[152] By 1960 the number had risen to 74,560 (63,370 domestic tourists and 11,190 foreign tourists).[153]

Power

The power sector, though very critical for the development of the state, continues to remain the weakest component of the state economy. However, the Bakhshi government tried its best to develop this sector. This is evident from the plan outlay of Rs 32.92 million earmarked for power generation during the Second Five-Year Plan and the actual spending of Rs 37.1 million during the plan period. Of the important power projects established during the period were the Sind Hydroelectric Project, with a generating capacity of 6,000 KWs, the Hydroelectric Power House established at Udhampur with an installed capacity of 800 KW, Gandarbal Power House, Salal Project (45,000 KW), Chenani Project (14,000 KW), Nichihama Project (10,000 KW), Kalakot Thermal Power House (22,500 KW), and Jogindranagar Power Project.[154]

Cultural Project

For manufacturing popular consent, it was no doubt considered important to give first priority to the fulfilment of the material needs of the people; but policy planners were aware that the material project, though a prerequisite, was not sufficient per se to overcome the political crisis. It was considered necessary to supplement it by a cultural project to convert the collective mentality

[151] Government of J&K, *Review of the Achievements of Bakhshi Government*, 19.

[152] Korbel, *Danger in Kashmir*, 217.

[153] Mirza Nazir Ahmad, *Management of Tourism in Kashmir* (New Delhi: Dilpreet Publishing House, 2010), 45.

[154] 'New Chapter Opened', *Amrita Bazar Patrika*, 22.

from rebellion to rejoicing,[155] to promote a liberal, tolerant, and modern outlook, to render the environment inhospitable for 'purist' ideologies, to create a bond between India and Kashmir, and to project normalcy in the state. This is what we know from the official records and state-sponsored organs,[156] though in his policy statement of 1956 Bakhshi deliberately omitted mentioning the politics driving his cultural project:

> Now that we have earnestly launched schemes for the economic and social regeneration of the people of our State, it is necessary that adequate attention should be paid to our cultural heritage so that these traditions are nourished and carried forward. We feel that our progress would be incomplete, if side by side with these material things, we ignore our cultural needs.[157]

It is, therefore, not difficult to understand that alongside the saga of material development, the Bakhshi regime is also distinguished for promoting the entertainment industry by cultivating traditional means of entertainment as well as inventing new ones under full state patronage; monopolizing the cultural production by co-opting the poets, litterateurs, and artists into his fold; and by institutionalizing the state's cultural project through the establishment of the Jammu and Kashmir Cultural Academy, a government publishing house called Lalla Rookh Publications, and the publication of Kashmiri, Hindi, Dogri, Urdu, and English journals.[158] Indeed, the establishment of cinemas, state-sponsored festivals, orchestras, operas, dancing, and *mushaira*s (poetic symposia) formed the salient features of the Bakhshi government's cultural governance. The most popular, however, was the state-instituted Jashn-i-Kashmir (Festival of Kashmir), organized annually at different towns in Kashmir. This festival was an innovation of Bakhshi, and

---

[155] See 'Srinagar Dairy', *Kashmir Today*, 1, no. 1 (Srinagar: Lalla Rookh Publications, September 1956).

[156] 'Srinagar Dairy', *Kashmir Today*; Government of J&K, *Message by G. M. Bakshi, 8 October 1956*, Archives Reference Library, Accession no. 105/GACC.

[157] *Daur-i-Jadid* (weekly), Kashmir, October–November 1956, 4.

[158] For details see Kanjwal, 'Building a New Kashmir', 173ff.

it was held for the first time in 1956—around three years after he assumed office. It was marked by a variety of cultural activities—theatre, music, poetry, dance, sports, and so on—from Kashmir as well as from various Indian states. It drew large crowds for entertainment, and also attracted tourists and senior leaders from India, including Prime Minister Jawaharlal Nehru. 'I am glad to say', said Bakhshi 'that these celebrations have evoked widespread interest outside the state and mass enthusiasm in Kashmir. ... its success has surpassed our expectations.'[159] Apart from hunting for talent from within the state, musicians and dancers were also invited from outside. While the professional qawwals and Kathak dancers from India enthralled the Srinagar audience at the Bakhshi Stadium and Tagore Hall, *band patheer* (folk theatre of Kashmir), *chakri* and *rouf* (folk singing and dance), mushaira, and drama formed the general features of the Jashn-i-Kashmir and other festivals. In addition to Jashn-i-Kashmir, many other state-patronized festivals were innovated and organized. These were Jashn-i-Bahar (Spring Festival) and Shab-i-Shalimar and Shab-i-Nishat.[160]

For promoting cultural activities, Bakhshi Ghulam Mohammad constructed the Tagore Hall for producing and staging plays. And on 25 October 1958, the J&K Academy of Art, Culture, and Languages was established to promote the cultural heritage of the state. State patronage of cultural activities and the prime minister's personal presence in most of these functions[161] made entertainment a *reference culture*, as is evident from the popularity of organizing orchestras and staging open-air plays and folk dramas in every nook and cranny of Kashmir.[162]

---

[159] Kanjwal, 'Building a New Kashmir', 173ff .

[160] Saraf, *Kashmiris Fight for Freedom*, vol. II, 1223.

[161] In the words of J. N. Sadhu, special correspondent, *Indian Express*, 'He (Bakhshi) is fond of music and at times forgoes anything for Kashmiri *Kalaam*. ... In the words of a famous Indian poet, Bakhshi sahib has a taste of a Mughal king.' Cited in *Bakshi Ghulam Mohammad—A Study*. Directorate of Information and Broadcasting, Government of Jammu and Kashmir, September 1957. Archive Reference Library, Srinagar, Accession No. 538/GACC.

[162] Based on oral history.

Apart from patronizing artists and actors, Bakhshi also appropriated writers and poets. The effort to control culture by patronizing poets, writers, as well as the religious class was also made during the period of Sheikh Abdullah;[163] Bakhshi institutionalized and consolidated the process. Apart from establishing the Cultural Academy to nurture poets, litterateurs, and artists who were willing to conform to the ideology of the government, Bakhshi disbanded the Cultural Congress established by Sheikh Abdullah and invited its members to join the newly created organization, Koshur Markaz. He also offered them government jobs and funds for writing projects, besides the routine job of writing poems for state-sponsored festivals, programmes, and government publications.[164] Ultimately, the progressive writers who had become a force to reckon with, especially after 1947, and whose support was sought by all political leaders to strengthen their position, were co-opted by Bakhshi's largesse. When asked only to bend, many chose to crawl.[165] It was on account of state patronage to writers and poets that many organizations, namely, the Cultural Forum, Cultural and Research Society, Bazm-i-Urdu, and the Literary Society, came into existence[166]—part of the policy to generate a complex array of groups and institutions to create, enlarge, and strengthen the support structure. Bakhshi also patronized the popular version of Islam in Kashmir—'shrine and saint worship'—by ordering the repairing, facelift, or constructing of shrines and visiting them frequently, which endeared him to the shrine-and-saint-minded masses and earned for him the support of the influential custodians of shrines[167]—shrines which attracted large crowds

---

[163] Sheikh Abdullah had created the Cultural Front and the Cultural Congress. *Tableeg-ul-Islam*, a religious organization, was affiliated with the National Conference.

[164] 'Bakhshi Number', *Sheeraza*, vol. 35, J&K Academy for Arts, Culture and Language, Srinagar; Kanjwal, 'Building a New Kashmir', 173–88.

[165] For encomiums written by the poets praising Bakhshi, see *Tameer*. Under the title 'Bakhshi Number', Vol. 5, 25 July 1960, Department of Information, Srinagar. Also see Kanjwal (2017), pp. 173–9.

[166] *Hamare Pyare Khalid-e-Kashmir*, 20.

[167] Based on oral history.

who returned brainwashed by oral sermons by the state-promoted shrine babas and pirs.

Towards Building an 'Equitable' Society

It goes without saying that the National Conference leadership was ideologically pro-poor and, like true socialists, they believed that for bridging the gulf between the rich and the poor, it was a prerequisite to extend special treatment and privileges to the marginalized sections. In Section 23, the constitution of Jammu and Kashmir lays down a directive principle that the state shall guarantee to the socially and educationally backward sections of the people special care in the promotion of their educational, material, and cultural interests, and protection against social injustice. Since the Muslims as a community were on the margins, and their share in state services and education was minimal despite being in a majority, all the post-colonial governments in Kashmir up to G. M. Sadiq extended special concessions to the community. In 1954, the government issued an order authorizing a community-based ratio for admissions to colleges, which was justified on the ground that being educationally backward, the Muslims would not be able to compete with the educationally advanced non-Muslims.[168] As the order gave Muslims 70 per cent and non-Muslims 30 per cent seats in Kashmir, and non-Muslims 70 per cent and Muslims 30 per cent seats in Jammu,[169] it led to a sharp reaction in the minority press in Kashmir.[170] As a community, the Kashmiri Pandits turned against Bakhshi and complained to the central government.[171] This was an expected reaction, because the Kashmiri Pandits were traditionally

[168] Government of J&K, Department of Education, File No. 1006-Edu-249-D/54/1954, Archives Reference Library, Srinagar.

[169] Government of J&K, Department of Education, File No. 1006-Edu-249-D/54/1954, Archives Reference Library, Srinagar.

[170] Government of J&K, Department of Education, File No. 1006-Edu-249-D/54/1954, Archives Reference Library, Srinagar.

[171] See Butt, *Kashmir in Flames*, 77; Saraf, *Kashmiris Fight for Freedom*, vol. II, 1223; M. L. Koul, *Kashmir, Past and Present: Unraveling the Mystique*, http://www.koausa.org/pastpresent/chapter11.html (accessed 9 November 2017).

almost the only educated community of Kashmir, and had a dominant presence in the colleges despite being only 4 per cent of the population.[172] Again, on the plea that the Muslims of the state and non-Muslims of Jammu were backward and therefore they required a special push, the Bakhshi government followed the same reservation policy in appointments to the service sector too. Bakhshi Ghulam Mohammad explained the rationale behind this reservation policy to the Gajendragadkar Commission of Inquiry:

> In evidence before us Bakshi Ghulam Mohammad unequivocally stated that merit could not be the only criterion in these matters, for if merit were made the sole criterion there would be no place for Muslims, Jammu Dogras and other backward people in the state. He, therefore, introduced communal and regional representation for filling government posts and making admissions to educational institutions. When the Public Service Commission was established in 1957, it was also advised to keep these considerations in view and to suggest an equal number of candidates from the Jammu and Kashmir regions and also an equal number of candidates belonging to the Muslim and non-Muslim communities for filling vacancies in government service.[173]

The reversal of the appointment policy of the Dogra state,[174] the establishment of a network of educational institutions, and the abolition of tuition fees signalled a second revolution (after the abolition of landlordism) in the history of upward mobility among Kashmiri Muslims in particular and backward communities in general, generating a tension between those whose traditional calling was state service, and the new aspirants and their supporters.[175]

[172] Kanjwal, 'Building a New Kashmir', 154.

[173] Gajendragadkar Commission Report, 74.

[174] The appointment policy of Dogra rulers was biased in favour of their own community. For details see Rai, *Hindu Rulers Muslim Subjects*, 249–58. For community-wise representation in different government departments on the eve of 1947, see Dar, 'Inter-community Relations in Kashmir', 107–23.

[175] For the ill-will which the Pandit community bore against Bakhshi, see Butt, *Kashmir in Flames*, 77; Saraf, *Kashmiris Fight for Freedom*, vol. II, 1223; Koul, *Kashmir, Past and Present*.

Meeting Regional Aspirations

In the preceding chapters, we have discussed in detail the restive Ladakh and Jammu regions during the period of Sheikh Abdullah. In the interest of the integrity and solidarity of the state, as well as to serve the ends of legitimacy, there was urgent need to pacify the two regions by attending to their pressing issues. Bakhshi declared, 'the rights and privileges that we secure for the state as a whole have to be shared in equal measure by the people of its different parts.'[176] Accordingly, immediately after assuming power, Bakhshi invited Kushak Bakula, the head lama of Leh, for talks. During these talks, besides evolving plans for the development of Ladakh and expansion of education in the region, Bakhshi invited the head lama to join his cabinet as deputy minister for Ladakh affairs, which the latter accepted.[177] Bakula on 25 November 1953 stated, 'As far as Ladakh is concerned we are determined to give our full support and co-operation to the new Prime Minister.'[178]

Ladakh's immediate problems were food, fuel, communication, and modern education. The Bakhshi government took various steps to address these problems. The Ladakh Affairs Department was created and headed by Kushak Bakula, and looked after the developmental and other activities of the region. The government deputed an expert to Ladakh to make an assessment of the preliminary work in connection with community projects, which were subsequently sanctioned.[179] Similarly, an expert was deputed to survey the small-scale industrialization in the region, starting with the setting up of namda, handlooms, and weaving industries.[180] During the Second Five-Year Plan, 1956–61, Rs 8.66 million was invested for the development of the region. During the Third Plan period, 1961–6 (part of which covered the Bakhshi period), the plan outlay was enhanced to Rs 14.73 million. Under

[176] Balraj Puri, *Simmering Volcano: Study of Jammu's Relations* (New Delhi: Sterling, 1983), 40; Kaul and Kaul, *Ladakh through the Ages*, 206.

[177] 'Ladakh and Jammu Get More Attention', *Amrita Bazar Patrika* (Calcutta), 21 March 1955, 24–5.

[178] Bazaz, *History of Struggle for Freedom in Kashmir*, 581.

[179] Kaul and Kaul, *Ladakh through the Ages*, 207.

[180] Kaul and Kaul, *Ladakh through the Ages*, 207.

non-plan expenditure, Rs 93.44 million was spent on transport and communication and Rs 1.49 million on 'Ladakh development'.[181]

Regarding Jammu, the crucial demand was the resettlement of refugees. The Bakhshi government took keen interest in meeting this demand. As a result, 6,227 non-camp displaced persons and 1,478 camp refugees were given land, a process that started immediately after Bakhshi took over in August 1953. By May 1955, about 20,000 acres of land had been allotted to these refugees.[182] Further, non-camp refugees were allowed rehabilitation facilities other than the allotment of land, like loans for building hutments, purchase of bullocks, and agricultural implements. Substantial concessions were given to refugee widows, orphans, and invalids. A colony was raised on the outskirts of Jammu city to accommodate around 4,000 families. Similar colonies were set up at Udhampur, Sunderbani, Rajouri, and Nowshera.[183] With a view to providing better housing facilities to displaced persons from the Nagrota Relief Camp, an additional sum of Rs 100 per family was provided. Further, primary and middle schools and dispensaries were set up in the areas inhabited by refugees. And where qualification and merit were equal, displaced candidates were given preference over others in matters of recruitment to services; displaced persons were also sanctioned to have bank deposits and such other dues payable to them immediately by the Panchayats and Public Works Department; and necessary facilities and assistance were provided to refugee students to enable them to carry forward their studies.[184]

Crisis in Governance

Bakhshi's policy of establishing hegemony was not free of serious flaws. He knew more than anyone else that he was pitted against

---

[181] Kaul and Kaul, *Ladakh through the Ages*, 216.

[182] 'Ladakh and Jammu Get More Attention', *Amrita Bazar Patrika*, 27.

[183] Government of J&K, *Review of the Achievements of Bakhshi Government*, 28.

[184] Government of J&K, *Review of the Achievements of Bakhshi Government*, 28.

the deep-seated political sentiment of the Muslims of Kashmir, which was represented by the Sheikh. Therefore, without taking any chances, he, alongside leading a benevolent state, took steps to create calm through corruption, coercion, and intolerance of dissent.

Narrating his eyewitness account of the regime, Prem Nath Bazaz writes: 'His government is remembered largely for two salient attributes, both of which became his lasting legacy to Kashmir politics; one is the pervasive corruption, with ministers and bureaucrats looting the public exchequer with impunity; and second crude, mafia-style authoritarianism, with the slightest political dissent forcibly stifled by police and gangs of organized thugs.'[185]

In the preceding chapter, I referred to the corrupt practices of Bakhshi as deputy chief minister during the Sheikh government. Now when he had been made prime minister by Delhi, with its full backing to change the tide in Kashmir by hook or by crook, he had the licence to give full expression to his corruptibility, nepotism, and administrative fraud. Syed Mir Qasim, who was in the cabinet of Bakhshi, says: 'Corruption had become rampant under Bakhshi's rule, aggravating the common man's misery. There was an atmosphere of loot and Bakhshi Sahib defended the looters.'[186]

Bakhshi embarked on the project of corrupting both his supporters as well as opponents. 'By giving tempting bribes in the shape of huge sums, government jobs or P.W.D. and forest contracts', says Bazaz, 'he corrupted political workers, journalists, religious leaders, and others both inside as well as outside the state. His route permit system under which a permit holder could earn thousands while sitting at home gained notoriety.'[187] No branch of administration remained untouched by the practice of open corruption. 'To be corrupt carried no stigma, no discredit for an official or a public worker.'[188] 'The workers of his National Conference, from the secretary of the organization down to the ordinary worker in

---

[185] Bazaz, *History of Struggle for Freedom in Kashmir*, 585; Bazaz, *Kashmir in Crucible*, 69–70.

[186] Qasim, *My Life and Times*, 82.

[187] Bazaz, *Kashmir in Crucible*, 70.

[188] Bazaz, *Kashmir in Crucible*, 70.

a mohalla or village got a bad name for him. Most of his workers, in liaison with the bureaucracy misused their position for their interests.'[189]

The Justice Ayyangar Commission which was set up in January 1965 to enquire into the allegations of corruption against Bakhshi Ghulam Mohammad reported:

> By the abuse and exploitation of the official position of Bakshi Ghulam Mohammad, undue pecuniary advantage was obtained by his relatives.[190]

> Bakshi Ghulam Mohammad and his relatives had by 1963 acquired vast assets and pecuniary resources and this process of acquisition was facilitated by the abuse by Bakshi Ghulam Mohammad of his official position or by the exploitation by his family and other relatives, with his consent, knowledge or connivance.[191]

Out of 38 allegations against Bakhshi Ghulam Mohammad, the Ayyangar Commission found 15 charges true. The commission also found that Bakhshi had received Rs 5.4 million as bribes, out of which 3.3 million had been received by him personally.[192]

Indeed, during the period of Bakhshi, the state possessed the attributes of both a benevolent and a predatory state,[193] therefore

---

[189] Manzoor Fazili, *Kashmir Government and Politics* (Srinagar: Gulshan Books, 1982), 46.

[190] Government of J&K, *Report of the Commission of Inquiry against Bakshi Ghulam Mohamed* (Ayyangar Commission), 30 June 1967, 287.

[191] Government of J&K, *Report of the Commission of Inquiry* (Ayyangar Commission), 698.

[192] Saraf, *Kashmiris Fight for Freedom*, vol. II, 1253.

[193] A benevolent state is one considered to be acting solely in the societal interest and that is equipped with the needed information, knowledge, and policy instruments to intervene in an optimal way. Whereas a predatory state is one that is seen to be subjected to the pushes and pulls of interest groups, whose main interest is in redistribution, rather than growth and development. See T. N. Srinivasan, 'Neoclassical Political Economy, the State and Development', *Asian Development Review* 3, no. 2 (1985); D. Lal, *The Hindu Equilibrium* (Oxford: Clarendon Press, 1988).

diluting both formations. It was predatory in that the state comprised a group of self-seeking individuals interested in maximizing their own utility at the expense of the welfare of the society. By interacting strategically with private agents and the bureaucracy, government policies and the budget became the mechanisms of redistribution and dispensing of subsidies overtly and covertly among the 'coalition partners'. The Justice Ayyangar Commission made *inter alia* one adverse comment—regarding the willingness of officers to do patently improper things at the behest of the political head. To quote Justice Ayyangar:

> I thought it would be in public interest, if by going into these details I brought out the manner in which the improprieties were committed and the machinery employed to put through schemes for self-favoritism. Some of them are no doubt ingenious but several are really clumsy as they involve apparent manipulations and alterations which are clearly seen in the documents to achieve their purpose. In this connection my analysis of the evidence would serve to show how when abuse starts from the top, demoralization sets in the permanent services, and even officers who by virtue of their status and position could normally be expected to take an objective view of matters coming up before them, succumb to the temptation of becoming subservient and winning tools for those under whom they serve.[194]

Making full use of the police force to suppress his opponents, Bakhshi virtually turned the state into a police state. He set up a special wing of police known as Special Staff headed by Ghulam Qadir Ganderbali, nicknamed by Kashmiris 'Qadir Natta', with special powers. Ganderbali unleashed a reign of terror, especially against anti-India elements. In police custody, he used such brutal methods as were not permissible under any law.[195] No wonder then that Qadir Ganderbali became synonymous with *zulum* (oppression) in Kashmir.

[194] Government of J&K, *Report of the Commission of Inquiry* (Ayyangar Commission), 712.

[195] D. N. Dhar, *Dynamics of Political Change in Kashmir: From Ancient to Modern Times* (New Delhi: Kanishka, 2001), 179.

Besides the 'Special Staff' for coercing people by employing third-degree methods, ragtags were organized into what was euphemistically called the 'Peace Brigade'.[196] The Peace Brigade was set up during Sheikh Abdullah's reign, comprising National Conference workers and all types of anti-social elements. It was practically an organization of thugs, the only difference being that they had government permission to do as they liked.[197] But during Bakhshi's time, this group was reorganized and used to suppress the opponents of the regime. This is also substantiated by another contemporary, the cabinet colleague of Bakhshi, Mir Qasim:

> the common man, under Bakshi's tyrannical rule, was denied even basic civil liberties. Political dissent was sought to be crushed ruthlessly. The government agents forced hot potatoes into the mouths of the opponents, put heavy stones on their chest; and branded them with red-hot iron. The Peace Brigade, which Bakshi Saheb had set-up, initially consisted of political workers, but later bad characters infiltrated into it, especially in Srinagar. They were free to harass the people and humiliate the women folk.[198]

The main job of the Peace Brigade was to disrupt 'peace' by attacking and disturbing any public function of any opponent group or party, and then to restore the peace of a graveyard. This naturally gave rise to an organized 'goondaism' by the ruling party. The hired hoodlums, disparaged as *kuntra-pandah* (literary meaning 29 and 15) for their monthly pay packet of 29 rupees and 15 annas, imposed Bakhshi's rule in tandem with the zealous police officers.

'The network of espionages and the fear of being reported as an individual harbouring anti-government views Bakhshi made any Kashmiri to suspect everyone he met. It was a deplorable state of affairs surely in no way conducive to the reconstruction of the moral and cultural life of the people so essential for the establishment of a democratic society,'[199] says Bazaz. Writing about the

---

[196] M. S. Pampori, *Kashmir in Chains* (Srinagar: Pampori Publishing House, 1992), 386.

[197] Saraf, *Kashmiris Fight for Freedom*, vol. II, 1189.

[198] Qasim, *My Life and Times*, 82.

[199] Bazaz, *Kashmir in Crucible*, 70.

policy of coercion adopted together with the manufacturing of consent, a contemporary journalist says: 'The intelligence network was introduced in jails too to keep a watch on the activities of the political prisoners. These detectives were deliberately put inside the jail in the garb of Sheikh Mohammad Abdullah's supporters and this way they could be in the inner circles of the political prisoners and report back to the government about their day-to-day thinking.'[200]

Like in the Sheikh's case, India was not prepared to let any sort of opposition against Bakshi to grow as long as his services were required for 'national interests'. Balraj Puri recalls:

Nehru warned me against being too idealistic and asserted that national interest was more important than democracy. He conceded that Bakshi used unscrupulous methods, but argued that India's policy on Kashmir now revolved around him, and despite all its shortcomings the Bakshi government had to be strengthened. He added, 'We have gambled at the international stage, we are there at the point of the bayonet. Till things improve, democracy and morality can wait.'[201]

Bakshi's policy towards the press was the same as had obtained during the period of Sheikh. Independence of the press was inconceivable. The press had either to be a collaborator or cease to exist. Mir Qasim writes:

Bakshi Saheb's attitude towards the press, too, was dictatorial. He likened newspapermen to snakes who must either be crushed or given milk to keep them on your side; but in no case should a snake be left free because it would bite somebody. The press, under such a Prime Minister, suffered tremendous hardships and risks. Any mention of Sheikh Saheb was considered an anti-Bakshi campaign. His anti-press stance also affected newspapers outside Kashmir. Kashmir became a very sensitive matter to report.[202]

[200] Butt, Kashmir in Flames, 66–7.
[201] Puri, Kashmir towards Insurgency, 46–7.
[202] Qasim, My Life and Times, 83.

The extent of goonda raj during Bakhshi's regime can be inferred from the following horrendous experience of Stephen Harper of the *Daily Express* upon his arrival in Srinagar: 'I had scarcely arrived in Srinagar, the capital last week when a mob swarmed around my car. They shouted murder him, we do not want British reporter here. Car door and canopy were ripped off. Hands grabbed and tore at my clothes. Little baskets of charcoal carried around for heat were poured over me and burned my face.'[203]

The policy of conducting farce elections initiated by the Sheikh government and characterized both by the rejection of nomination papers on flimsy grounds and creating circumstances for 'winning elections unopposed', continued to be followed by Bakhshi too. In the 1957 elections, if they can be so called, the National Conference 'won' 69 out of 75 seats. Of the 43 valley seats, 35 were 'won' by official National Conference candidates without any contest. In the whole state, 30 National Conference candidates, including Bakhshi, were returned unopposed. Another 10 candidates of the National Conference got elected after the nomination papers of their opponents were declared invalid. A token contest was allowed on eight seats in the valley. Out of these eight, the National Conference 'won' seven. A past master in declaring the nominations of opposition candidates null and void was Abdul Khaliq Malik, popularly known as Khaliq DC, a Bakhshi henchman. To this day, Kashmiris recall hilarious stories related to Khaliq DC. Interestingly, the candidates whose nominations papers were approved by him (who belonged to the National Conference only) came to be popularly known as 'Khaliq-made MLAs'. In fact, such was the extent of authoritarianism and political discontent prevailing in the state that, in 1958, two of the Bakhshi's closest colleagues, Ghulam Mohammad Sadiq and Mir Qasim, quit the ruling party, though temporarily, and floated their own party, the Democratic National Conference.

In the 1962 elections to the State Assembly, the National Conference got 68 out of the 75 seats. Out of the 43 constituencies in the valley, 32 were decided in favour of the National Conference without any contest. Out of these 32 seats, on 20 seats no other

---

[203] *Daily Express* (London), 5 February 1957.

candidate was allowed to file nomination papers; the non-official candidates who had filed papers on 8 seats withdrew before the polling date out of fear, and in the case of the other 4, the papers of opposition candidates were declared invalid on frivolous grounds. In Jammu, the National Conference won 27 out of 30 seats through dubious tactics. In protest against the massive malpractices and farcical process, many political parties like the Praja Parishad, Praja Socialist Party, and Akali Dal jointly held a mass demonstration in Jammu city. However, the prime minister of Kashmir dismissed their complaints as frivolous.[204]

As Akbar notes, 'Riled by adverse reports in the international press, even Nehru wrote to Bakhshi after the 1962 polls advising him to lose "a few seats" in the future, so that the image of the world's largest democracy would not be unnecessarily tarnished.'[205] He further wrote, 'it would strengthen your position much more if you lost a few seats to bona fide opponents.' Nehru wanted to caution Bakhshi not to use representative institutions to trample over the rights of opponents. But being a hardcore believer in undemocratic principles, Bakhshi replied: 'If you stick to democratic principles, you will never be able to have peace in Kashmir.'[206]

It was in the din of populist measures and the distribution of largesse coupled with the declaration of a sort of 'emergency' and the creation of a 'culture of fear' that Bakhshi implemented the agenda of expediting the process of integration of Kashmir with the Indian Union, which Abdullah had resisted. On 6 February 1954, the 'purged' Constituent Assembly ratified the accession. On 13 April 1954, the customs barrier was removed, making Kashmir economically an integral part of India. On 14 May 1954, the president of India issued a constitutional order by which most of the provisions of the Indian constitution were applied to the state of Jammu and Kashmir. An exception was made only with regard to the position and functions of the sadr-i-riyasat and the rights to

---

[204] Das, *Jammu and Kashmir*, 269–70.

[205] M. J. Akbar, *India: The Siege Within* (New York: Viking, 1985), 258.

[206] K. R. Sunder Rajan, 'The Challenge in Kashmir', *Tribune*, 9 July 1984.

acquire immovable property in the state as well as employment and settlement therein. This order practically annulled the 1952 Delhi Agreement and created circumstances for bringing an end to Article 370. In terms of financial and fiscal matters, Kashmir's relation with New Delhi was brought on par with other undisputed units of India. Union departments, namely, customs, central excise, posts and telegraph, civil aviation, All India Radio, and so on, had their operations extended to the state, just as in other states. The Indian Supreme Court now had full jurisdiction in Kashmir. The fundamental rights of citizens guaranteed by India's constitution were extended to Kashmir, but in the interest of 'security', these civil liberties could be suspended at any time with no judicial review of such suspensions. K. G. Kannabiran, a well-known Indian civil liberties activist, while analysing the impact of this legislation observes: 'What we in India experienced for a brief period . . . during Indira Gandhi's emergency regime, Jammu and Kashmir has suffered for . . . years. . . . we cannot deny a people rights that flow out of citizenship and then expect their allegiance.'[207]

Many more integrative measures followed the constitutional order of 1954. The functions of the comptroller and auditor general of India were extended to the state in 1958. In 1958, a constitutional amendment was also effected, whereby J&K was brought under the purview of the Central Administrative Services, which resulted in the employment of non-residents of J&K in central administrative agencies, economic enterprises, and banks based in Kashmir. 'Thus', says Sumantra Bose, 'while Kashmir's political arena was monopolized by corrupt, despised puppets installed at Delhi's behest through the calculated destruction of representative democracy, its day-to-day administration too gradually came to be dominated by people with no roots among the population.'[208]

---

[207] K. G. Kannabiran, 'The Slow Burn', *Illustrated Weekly of India* (New Delhi), 1 July 1991.

[208] Bose, *Challenge in Kashmir*, 33–4. In 1989, while Muslims comprised 65 per cent of the population in Indian-administered J&K (and Hindus 32 per cent of the population), Hindus mostly from outside the state made up 84 per cent of the high-level officers, 79 per cent of the clerical employees,

Clearly, Bakhshi enjoyed the full support of Nehru; and the support extended by the prime minister was conditional. Bakhshi had to accomplish what had been resisted by Sheikh Abdullah: facilitate the integration of Kashmir with India. Sumantra Bose succinctly describes this contractual relationship:

> Bakshi Ghulam Mohammed's term in office lasted a full decade, until October 1963. The sequence of events during that decade strongly suggest a contractual relationship between Bakshi and the government of India, whereby he would be allowed to run an unrepresentative, unaccountable government in Srinagar in return for facilitating IJK's 'integration' with India on New Delhi's terms. The result was twofold: a crippling of rule of law and democratic institutions in IJK; and an erosion of IJK's autonomy, achieved (as required by Article 370) with the 'concurrence' of IJK's government—which consisted of a motley clique of New Delhi client politicians.[209]

Although Bakhshi took all conceivable measures to change the collective political sentiment of the Muslims, he could not succeed in accomplishing the project. To his great shock, when under compulsion from Nehru (who himself was under pressure from the international community), he released Abdullah from jail on 8 January 1958, the lion of Kashmir was 'received like a victorious Roman hero', to quote the correspondent of the *Blitz*, in the *Bombay Weekly*.[210] The Kashmiri writers of the time, even those

---

and 73 per cent of low-level employees in the centrally operated services. A minuscule 1.5 per cent of the high-ranking officers in centrally owned banks in the state were Kashmiri Muslims. Only 25 per cent of Indian Administrative Service officials posted in the state in 1989 were natives of the province, and of 22 secretaries, the highest rank, a mere five were Kashmiri Muslims. The problem of inequitable representation extended to provincial-level administration organs and economic enterprises. Here, Hindus comprised 51 per cent of senior administrators and 47 per cent of top officials in economic enterprises, the tiny Pandit minority being disproportionately overrepresented. Cited in Bose, *Challenge in Kashmir*, 52.

[209] Bose, *Kashmir: Roots of Conflict*, 68.

[210] Quoted by Abdullah, *The Blazing Chinar*, 453.

who were not well disposed towards the Sheikh, also recorded it as an exemplary reception.[211] The rousing reception accorded to Sheikh, in spite of the undeclared emergency enforced by Bakhshi, shocked the central leaders.[212] Much more, it shattered Bakhshi.[213]

The desperate Bakhshi saw no other way to escape from the public fury but to re-arrest Abdullah who had been out for not more than 100 days. The situation came to such a pass that the opportunist Bakhshi ultimately found it politically prudent to co-opt some leaders of the Plebiscite Front to keep the pot of turmoil boiling.[214] Thus, when by means both fair and foul he could not succeed in securing his monopoly over power, he used the conflict as an industry to perpetuate his hegemony.

* * *

The liberal financial support, political backing, and policy inputs from the central government, together with Bakhshi's own dynamism, mobilization capacity, administrative knack, informal and contingent dealings with the local society, zeal and zest for development, profound understanding of Kashmiri society, and prompt response to the aspirations of the different sections of the people, helped the otherwise imposter premier to turn the tide, and that too within a few months of taking over the government. Even Nehru was surprised to see so quick a return of normalcy in Kashmir.[215] It must have shocked Abdullah to learn while behind bars how,

---

[211] Ishaq, *Nida-i-Haq*, 284; Butt, *Kashmir in Flames*, 72.

[212] One of the contemporaries of the time, Pandit Kashap Bandu, told another leader, Munshi Ishaq, that 'on the day Sheikh Sahib reached Srinagar, he was in Delhi with the Home Minister, Pandit Govind Ballabh Pant talking about Kashmir problem. In the adjoining room telephone rang and when Pandit ji returned, he was depressed so much so that it seemed as if he had lost some dear one. The reality was that the Sadr-i-Riyasat of the State, Dr. Karan Singh had informed him about the rousing reception accorded by the people to the Sheikh.' Ishaq, *Nida-i-Haq*, 284.

[213] Butt, *Kashmir in Flames*, 72–3; Ishaq, *Nida-i-Haq*, 284.

[214] Butt, *Kashmir in Flames*, 69–70.

[215] Nehru, *Letters to Chief Ministers*.

within two months following his dismissal, most of those who had solemnly owed allegiance to him turned coat one by one,[216] and in just a little more than a month Bakhshi was able to hold conventions of the National Conference in different towns in the valley, beginning with Srinagar. Certainly, within three years he had acquired a complete grip over affairs, and would proudly say that he was no less than Sheikh Abdullah in commanding a public following. In this, he was not making a flippant remark. His funeral procession was the biggest in living memory, and, for the first time, hundreds of women broke with tradition in accompanying the body to the graveyard.[217]

Bakhshi had his finger on the pulse of the people. He had no doubts about the fact that sacrifice might be the wish of a few, but comfort was the priority of all. It is, therefore, understandable that he approached the whole problem from a materialistic perspective. He firmly believed that it was possible to evaporate discontent among the people and convert them from foes to friends by addressing their basic economic problems and creating a saga of all-round development, including offering patronage and largesse to the influential sections of society. Not surprisingly, therefore, he inaugurated his rule by announcing a series of populist measures, followed by unprecedented infrastructural development, for which the central government extended its fullest cooperation and offered the maximum financial support. It should be mentioned that during the initial months when tempers were running very high, Bakhshi did not hurt the political sensibilities of the people. Instead, like the present Kashmiri-based mainstream political leaders, he sought the support of the people on the promise of improving their economic lot. Indeed, for restoring peace and winning over the people, Bakhshi followed an indirect route: by meeting the most pressing basic needs of the people. To quote M. Y. Saraf:

> On his first visit to Baramulla as Premier he told a select meeting of Muslim elders that he needs their cooperation; that if and when plebiscite was held, he knew they would vote for Pakistan but what

---

[216] Bazaz, *History of Struggle for Freedom in Kashmir*, 581.
[217] Saraf, *Kashmiris Fight for Freedom*, vol. II, 1253–4.

was the use of non-cooperation till then? He told that he had services and scholarships to offer, enormous development funds for improvement of their economic wellbeing and in return, he needed their support. This argument he must have advanced at other places too; whether it was a strategy or a practical plea, it did succeed.[218]

The financial integration of Kashmir with the centre, which the Sheikh had resisted but which was gladly accepted by Bakhshi, and Bakhshi's political skill in exploiting the situation created by the dismissal and imprisonment of Abdullah, resulted in the flow of liberal financial aid from the centre, ushering in the new saga of a subsidized economy and infrastructural development. Indeed, the per capita Rs 41.7 statutory grant-in-aid to J&K for five years from 1957–8 to 1961–2 was almost seven times the all-India average.[219] Per capita income was raised from Rs 188.41 to Rs 236 over this period. In the first two five-year plans, the state created 33,569 jobs; and the revenue of the government rose from Rs 52.3 million in 1953–4 to Rs 245.35 million at the end of the Bakhshi period.[220] In the year 1959–60 alone, a total of Rs 121.1 million was transferred to the state from the centre.[221] The importance of the assistance and the great development work undertaken was admitted even by those who favoured Kashmir's accession to Pakistan.[222] Indeed, there was hardly any sector which did not see a revolutionary development during the decade-long rule of Bakhshi. The economic progress achieved by the state was reported by important newspapers of the world, with such varied political orientations as the *New York Times*, *Economist* (London), *News Chronicle* (London), *Pravda* (Moscow), *Al Joumhouria* (Cairo), and *Stockholms Tidningen*.[223]

[218] Saraf, *Kashmiris Fight for Freedom*, vol. II, 1222.

[219] Puri, *Jammu and Kashmir*, 129.

[220] Government of J&K, *Some Basic Statistics* (Jammu: Ranbir Government Press, 1964), cited in Kanjwal, 'Building a New Kashmir', 75.

[221] Puri, 'The Budget of Kashmir', 549.

[222] C. B. Birdwood, *Two Nations and Kashmir* (London: Robert Hale, 1956), 195.

[223] For more foreign comments, see *Kashmir: An Open Book* (Srinagar: Lalla Rookh Publications, 1958); *Kashmir through Many Eyes*, 1957. Lalla Rookh Publications, Srinagar, Department of Information

The hustle and bustle of multifarious developmental, economic, social, and cultural activities—an unprecedented scene
for the people—and the opportunities it threw up for everyone to
improve his/her position subsumed the mass dissent (though temporarily), besides creating a spectacle of performativity, productivity, and prosperity. The mind-boggling state-led developments and
the transformation of the life of the people gave legitimacy to what
was otherwise rule by a usurper. Some poorer sections who were
greatly benefited by Bakhshi fondly remember him even today.

Bakhshi did not leave any strata of the society with its material demands unmet, and he used all the available practical strategies to accomplish the project of expanding hegemony. The largest
segment of the population—the peasantry—was won over by a
series of measures, namely, abolition of mujwaza and the forcible
extraction of foodgrains, lifting of the ban on free movement of
foodgrains, construction of a network of roads, canals, and dams,
introduction of lift irrigation schemes, sinking of tubewells, reduction of the grazing tax and irrigation tax (abiana), scaling down of
the land ceiling, construction of embankments to save paddy land
and the habitations situated around the Jhelum and Wular from
devastating inundations, provision of subsidized goods through
cooperatives, breaking of the cooperatives' monopoly, commis-
sioning of different projects and providing work to about 80,000
wage labourers in one year,[224] and expanding education in rural
areas. For the urban people, Bakhshi took the revolutionary step
of subsidizing rations and making them available in plenty. The
unprecedented increase in employment generation, enhancement
of salaries of employees and casual labourers who primarily came
from urban areas, reduction in the prices of essential commodities,
the establishment of a series of institutions, the construction of
roads and buildings, abolition of state monopoly in trade and elimination of the customs barrier between Kashmir and India, reduction in the rates of freight and fare which immensely increased

---

and Broadcasting, Accession No: 56206, Nehru Memorial Museum and
Library.

[224] 'New Chapter Opened', *Amrita Bazar Patrika*, 19.

the use of transport by the people, de-rationing of petrol,[225] granting bonuses to workers,[226] scaling down of debts of cooperative departments, grant of loans to traders and artisans, a 50 per cent reduction in road toll and octroi duty,[227] state patronage to crafts, the unprecedented fillip to commerce and tourism, new housing colonies, and above all meeting the people every morning to redress their grievances[228] and extending generous help to widows, orphans, and the destitute[229]—all this changed the tide of sensitive Srinagar in his favour.

To alter the network of potential alliances, Bakhshi embarked upon the policy of openly bestowing state favours on leaders and opinion makers—politicians, political workers, businessmen, the educated elite, journalists, religious leaders, poets and litterateurs, and other local power holders. They were given tempting bribes in the shape of huge sums, government jobs, admission to technical institutions, Public Works Department and forest contracts, licences, route permits,[230] and so on. Indeed, Bakhshi is known for his magnanimous displays of generosity and largesse. It is popularly said in Kashmir that Bakhshi used to say, 'one who would not be able to improve his economic position during my period, he would not be able to achieve it ever.'

To further strengthen his policy of political alliances, he entered into matrimonial relations with some influential political families.[231] What is no less important in this regard is that Bakhshi

---

[225] Government of J&K, *Review of the Achievements of Bakhshi Government*, 26.

[226] Government of J&K, *Review of the Achievements of Bakhshi Government*, 26.

[227] Government of J&K, *Review of the Achievements of Bakhshi Government*, 26.

[228] Ishaq, *Nida-i-Haq*, 265.

[229] 'Bakshi Ghulam Mohammad had promised to meet the expenses of a widow's daughter's marriage himself. . . . about four years after his resignation, he borrowed rupees three thousand from an officer to fulfill his promise.' Saraf, *Kashmiris Fight for Freedom*, vol. II, 1254.

[230] Bazaz, *Kashmir in Crucible*, 70.

[231] Ishaq, *Nida-i-Haq*, 265.

successfully exploited the deeply embedded anti-Abdullah sentiments among the pro-Pakistan sections of Kashmiris among whom
the Shias constituted the majority. By employing diplomacy and
patronage, a large section of this population joined the ranks of
Bakhshi.[232]

By deserting Sheikh Abdullah, collaborating in his dismissal,
taking his seat and putting him behind the bars, launching a tirade
against him especially questioning his nationalist credentials,[233]
ratifying the state's accession with India, facilitating the centre
to expand its jurisdiction over the state, and giving representation
to all regions and states in the cabinet and the administration,
Bakhshi earned the loyalty of otherwise estranged identities—
the Kashmiri Pandits,[234] the Hindu Dogras, and the Ladakhi
Buddhists who had not yet reconciled to the demise of Dogra rule
and the changes that followed thereof. However, Bakhshi's policy
of recruiting Muslims in the service sector in proportion to their
population—a compulsion of governance and politics—was not
taken well by the Kashmiri Pandits.

Bakhshi and his mentors were intelligent enough to realize that
alongside providing basic facilities to the people, meeting their
economic needs, and creating the opportunities for upward mobility, it was equally important to take proactive steps to promote
a mentality with a desire for pleasure and comfort rather than
the will to sacrifice. Thus he embarked upon a cultural project to

---

[232] Ishaq, *Nida-i-Haq*, 265.

[233] See the propaganda literature produced during the period against
the nationalist credentials of Sheikh Mohammad Abdullah. Some of this
literature is preserved in the J&K Archives.

[234] Upon being accused by Bakhshi of non-cooperation in front of
Nehru, one of the Hindus present stood up and with folded hands said,
'Bakshi Maharaj! It is unfortunate that of all people, *you* should have
accused *us* of non-co-operation. Since the imprisonment of Sheikh
Abdullah, you have been saying your Eid prayers in Jammu and since the
number of Muslims attending the prayers is very small, we have been
sending Hindus to these congregations who joined you in the prayers to
build up their size so that *you* may not be embarrassed!' Saraf, *Kashmiris
Fight for Freedom*, vol. II, 1223.

promote entertainment, sports, pilgrimages, and festivities, both secular and religious. He even invented state-sponsored festivities, besides sponsoring shrine festivals and the establishment of song and drama clubs. To institutionalize the state patronage to culture, he founded the Cultural Front and the J&K Academy of Art, Culture, and Languages. And to control and shape the collective thinking of the people, Bakhshi assumed the presidentship of Awqaf-i-Islamia, the Muslim religious body that controls most of the shrines and mosques in Kashmir, having removed Abdullah from that position. For establishing hegemony, 'a cultural project, wrote Gramsci, could not be some avant-garde movement imposed upon people, instead it had to be rooted in the "humus of popular culture as it is, with its tastes and tendencies and with its moral and intellectual world, even if it is backward and conventional".'[235]

No less striking a feature of Bakhshi Ghulam Mohammad as an administrator was his unique knack of establishing direct rapport with the common people and his quick decision making. His ability to mix with the people and attend to their problems instantly won him tremendous popularity. According to a British correspondent, Bakhshi's technique was simple indeed; he did not lecture his audience, he got his audience to participate and to talk back to him.[236] Interactions with a wide range of people, especially the subalterns, through informal, routine, and contingent dealings gradually but surely evaporated the public rage and antagonism against Bakhshi. Reading this style of functioning together with the all-round development, it is no wonder that between an anti-Indian wave on the one hand and the sweeping constitutional developments that struck at the very roots of Kashmir's separate autonomous status on the other, Bakhshi was able to secure a niche in the hearts and minds of the people of Kashmir. It is little wonder that Bakhshi's methods were recommended for emulation

---

[235] Antonio Gramsci, *A Selection from the Cultural Writings* (London: Lawrence & Wishart, 1985), 102; Steve Jones, *Antonio Gramsci*, Routledge, London and New York, 2006, p. 37.

[236] Special Representative, 'Kashmir: An Open Book', *Sunday Times* (London), 26 July 1955.

to inexperienced and less gifted politicians holding office, and he was lionized whenever he travelled in India.[237]

Informed by a socialist ideology and under pressure of political expediency, Bakhshi did not only carry forward the policy of propaganda initiated by his predecessor, Sheikh Abdullah, but gave it a further boost to meet many ends simultaneously. He had to satisfy the international audience, given the hullabaloo over the Kashmir dispute at international fora and the critical reportage on Sheikh Abdullah's removal in international newspapers. He had also to counter the campaign launched far and wide by Pakistan against India in Kashmir; more importantly, Pakistan and Azad Kashmir Radios were fondly listened to by Muslims of Kashmir. Internally also the state was reeling under severe pressure from the Plebiscite Movement demanding a referendum and attacking the government. And no less important was the fact that Bakhshi's own legitimacy was in question. To wean people away from the piquant mobilization strategies employed by opponents both within and without, the 'national interests' and the interests of the ruling clique demanded changing and remaking the mindset of the people. This crucial job of the state was assigned to the Jammu and Kashmir Department of Information, which utilized Radio Kashmir, local, national, and international media, information centres established throughout the state, and government publications including tourist guides.[238] The focus of the propaganda was to bring into the limelight the achievements of the government, through selective reportage, selected sources, and selected places to represent the government. The job of the Department of Information was to stop 'adverse' reporting, show Pakistan and PAK as behind on the road to progress, undermine the importance of internal dissenting voices, and show the popularity of the government and the contentment of the populace engrossed in pursuing their respective mundane interests and enjoying life in the wake of the increasing tourist flow. And, therefore, that Kashmir was as normal as any other place, and was fulfilling the requisite parameters of normalcy and modernization.[239]

---

[237] Bazaz, *Kashmir in Crucible*, 71.

[238] For details, see Kanjwal, 'Building a New Kashmir', 75–8.

[239] Kanjwal, 'Building a New Kashmir', 75–8.

Bakhshi was a shrewd politician. To remain in power he was not coy about using any means, fair or foul. In this regard, it is rewarding to quote his contemporary, the editor of the daily *Aftab*:

Inside Kashmir Bakshi Ghulam Mohammad worked successfully on three fronts. His first front was in New Delhi where with money and gifts he managed to create a pro-Bakshi lobby comprising Congress leaders including influential members of Parliament. And the members of this front worked for the Bakshi in New Delhi. The second front on which he worked was to create a proper political image for himself in the state and within three years he had tightened his grip on the political and administrative machinery of the state. The third front in which also he succeeded was to keep alive anti-India campaign in the state. He knew if the elements, who were struggling against India, turned weak it would result in weakening his power. ... Conscious of it he kept on financing, though secretly, some people of the Plebiscite Front and the Political Conference. Some other pro-Pakistani elements too received liberal financial help from him. He told me many times 'existence of such elements was necessary otherwise New Delhi will do anything here it liked'.[240]

Clearly, all was not rosy during the time of Bakhshi. He cultivated corruption, personalized the administration, and patronized goondaism to win loyalties and silence his opponents, which shattered the moral fabric of the society. Bakhshi was Machiavelli's 'half-man and half-horse'. He combined in himself human means but he also knew how to imitate beasts. While he left no stone unturned in ruling by manufacturing consent, he resorted to rampant corruption and used force and denied democracy to those elements who refused to give their consent. Since the most powerful collective sentiment in Kashmiri society—the sentiment of rights—was represented by those elements against whom coercion was used, Bakhshi's material and cultural project could only succeed in temporarily controlling the sentiment without, however, extricating it, a fact which was fully known to the policy planners at Delhi, more so to Nehru. And this became abundantly clear by the exemplary,

[240] Butt, *Kashmir in Flames*, 69–70.

rousing receptions people accorded to Sheikh after his release from Kud jail on 8 January 1958. That is why when Bakhshi succeeded in securing calm, Nehru thought of removing him for being a hard-liner in giving Sheikh Abdullah a space in Kashmir politics. Yet, Bakhshi's success in creating 'cool peace' exerted pressure upon the Plebiscite leadership to give concessions to the circumstances and soften its stand, as is evident from the Front's willingness to participate in elections and the interest Sheikh showed in opening channels of communication with Nehru.[241]

---

[241] Qasim, *My Life and Times*, 79–80, 82–3.

4

A Difference in Degree (1964–75)

History repeated itself in 1963. Bakhshi was shown the door; not, however, through the type of coup d'état which threw Abdullah out of power in 1953, but through a gimmick, using the facade of the Kamraj Plan. The cause behind the removal of both these first two prime ministers was, nevertheless, the same: Nehru's hunger to swallow up Kashmir. Like Abdullah, Bakhshi resisted the insatiate hunger of Delhi to make Kashmir like any other state of India. Clearly, Bakhshi was installed as prime minister only when the Indian government satisfied itself that he would be forthcoming regarding its plan of gradual merger of the state with the Indian Union. And for the same purpose it stood behind him through thick and thin. He kept his word, and rendered the prey half-dead. Article 370 giving Kashmir a special position was considerably eroded, as we have seen in the preceding chapter. Nonetheless, Bakhshi refused to bleed the state to death. He resisted the moves to change the nomenclature of prime minister to chief minister and sadr-i-riyasat to governor. He also declined to merge the National Conference into the Indian National Congress.[1] This was not acceptable to Nehru, and he employed the method of embarrassing Bakhshi and encouraging the rival group in the National Conference, headed by G. M. Sadiq. Mir Qasim, who was one of the important members of the rival group, spoke the truth (though

[1] Puri, *Jammu and Kashmir*, 153.

in the late stages of his life) about Indian diplomacy towards Kashmir:

> whenever New Delhi feels a leader in Kashmir is getting too big for his shoes, it employs Machiavellian methods to cut him to size. This it does by projecting a lesser leader as an alternative with the help of a two-way overt or covert campaign—one, by convincing the targeted leader that his position was unassailable and the lesser leader was conspiring against him; and second, by telling the lesser one that he was more popular than the main leader who had out-lived his utility. Tale-tellers, of whose black art one heard in Mughal courts, play a major role in this kind of conspiracy.[2]

And, like Sheikh Abdullah, when Bakhshi gauged the gravity of the machinations of the central government, he burst out and made a series of 'anti-national' statements precipitating the crisis.

> He [Bakhshi] reiterated that Kashmir had acceded to India only to the extent of Defense, Foreign Affairs and Communications and warned the Democratic National Conference [headed by G. M. Sadiq] opposition in the State assembly—which favoured extension of certain provisions of the Indian Constitution to the State—that he would not allow it 'to sell Kashmir to India'. He also asserted that Article 370 would be abrogated on his dead body. In an interview with *Kashmir Post*, he said: 'We prefer to be poor but independent rather than rich and under other's control.' From the very beginning, he was opposed to interference by leaders of Indian political parties 'in the internal affairs of the Jammu and Kashmir State'. Bakhshi also resisted moves to merge his National Conference into the Indian National Congress. For 'after all, the National Conference has its own tradition ... and its following has certain sentiments which needed to be respected.'[3]

The integrationist forces at the centre, who were waiting in the wings to install a more favourable group already encouraged/created by them to complete the unfinished agenda, got an excuse to

---

[2] Qasim, *My Life and Times*, 119.
[3] Puri, *Jammu and Kashmir*, 153–4.

exert pressure on the already convinced central government of India to make an 'appropriate' change about installing the rival group in NC headed by G. M. Sadiq. On 27 November 1963, Prime Minister Nehru and his colleagues in the government assured Parliament that further steps for gradual erosion of Article 370 would be taken 'in the next month or two'.[4] And as expected, the agency which had already been made ready to accomplish this project was the rival group in the National Conference government headed by G. M. Sadiq—the group which had shown an abundant proactiveness towards completing the process of integrating Kashmir with the Indian Union.[5]

However, Bakhshi, who commanded majority support in the Legislative Assembly, resisted the move to make Sadiq prime minister, and instead proposed the name of another colleague, Shamas-ud-Din, obviously to rule in his name. Nehru had to accept this, though not without demurring. However, within two and a half months after Shamas-ud-Din assumed office on 12 October 1963, a mysterious but highly consequential development occurred on 27 December 1963. This was the 'theft' of the *moe-e-muqqadas* (the hair relic of Prophet Muhammad preserved in the shrine-cum-mosque of Hazratbal, Srinagar), which caused such a massive upsurge in Kashmir that, for the first time, the Indian leadership and the concerned Indian agencies felt Kashmir slipping out of their hands. All the religious and political parties united, forming what is known as a *majlis-i-amal* (action committee) and converting a 'non-political demand' into a 'massive political agitation'.[6] The extremely worried Nehru[7] virtually handed over the

---

[4] 'Speeches in the Lok Sabha', *Keesing's Contemporary Archives*, 1963–4, cited in Puri, *Jammu and Kashmir*, 153.

[5] Qasim, *My Life and Times*, 82.

[6] For details on the moe-e-muqqadas *tehreek* (movement), see: Qasim, *My Life and Times*, 94; Mullik, *My Years with Nehru-Kashmir*, 119–42; Saraf, *Kashmiris Fight for Freedom*, vol. II, 1235–40; Bazaz, *Kashmir in Crucible*, 74–5; G. N. Gauhar, *Hazratbal: The Centre Stage of Kashmir Politics* (Srinagar: Gulshan Books, 1998); Butt, *Kashmir in Flames*, 85–92; Ishaq, *Nida-i-Haq*, 303–8.

[7] Mullik, *My Years with Nehru-Kashmir*, 122–65.

administration of Kashmir to CBI director B. N. Mullik and Home Secretary Vishwanathan, and kept himself informed at regular intervals each day.[8] As the demand was the recovery of the relic and identification of the culprits, Nehru's insistence on the immediate recovery of the relic was quite understandable. It was due to Mullik's dexterity that the relic was recovered on 4 January 1964 through a miraculous process which, according to Mullik, being 'an intelligence operation, [was] never to be disclosed'.[9] However, after its recovery, the Action Committee doubted its authenticity; and it was again owing to the statesmanship of the head of the Intelligence Bureau of India that the authenticity of the relic was endorsed by a group of Muslim religious leaders. It may be mentioned that B. N. Mullik, Vishwanathan, and S. Benerji had insisted on avoiding the risk of getting the relic authenticated.[10]

Towards a New Kashmir Policy

Though the 'extraordinary' feats of intelligence agencies succeeded in controlling the public rage in Kashmir, it convinced Nehru that the Kashmir policy called for a rethinking. Indeed, if Bakhshi Ghulam Mohammad's stick and carrot policy had kept the lid on mass political discontent, it burst into flames engulfing every hamlet and mohalla of Kashmir during the moe-e-muqqadas agitation. It was not in 1953 but now that the Indian government found itself on shaky ground despite having spent huge sums of money on 'buying Kashmiris', to quote the words of Wajahat Habibullah.[11] The massive street protests persuaded Nehru that there was an urgent need to revisit the Kashmir policy. To quote Mullik:

[8] Mullik, *My Years with Nehru-Kashmir*, 122–65.

[9] Mullik, *My Years with Nehru-Kashmir*, 142.

[10] Qasim, *My Life and Times*, 97–8; Mullik, *My Years with Nehru-Kashmir*, 157.

[11] Wajahat Habibullah, *My Kashmir: Conflict and the Prospects of Enduring Peace* (Washington, D.C.: United States Institute of Peace, 2008), 55.

The Prime Minister started by saying that, even after fifteen years of association, if Kashmir still remained in such an unstable state that even on a simple issue like the Moe-e-Muqaddas, the people could be so provoked as to rise in defiance of the government, then in his opinion, a new approach had to be made and a radical change in our thinking about Kashmir was called for. He said that he felt disappointed that after all that had been done for the people of Kashmir, they were apparently still dissatisfied and though much of this dissatisfaction was due to a certain amount of mis-government, all of it could not be ascribed to these causes. He also felt that Sheikh Abdullah still had a strong hold on the people of Kashmir and in the changed circumstances, no political settlement in the valley could be thought of without bringing him in.[12]

'In India itself', says Sheikh Abdullah, 'perceptive statesmen like Rajagopalachari, Jayaprakash Narayan and some members of Parliament underlined the need for my release . . . to end the disturbance prevailing in Kashmir.'[13] Evidently, Nehru had decided to reverse the repressive policy of Bakhshi, give dissent a fair hearing, check the misuse of power, and, more importantly, to make concessions with regard to the demands of Sheikh Abdullah. The hardliners in his cabinet were shocked by the sea change in the attitude of the prime minister towards the Sheikh, but before his imposing personality no one could muster the courage to differ with him.[14] For implementing the new thinking, Nehru's first preference was Sadiq, who, besides being a clean man, was a supporter of the idea of giving Sheikh Abdullah a fair deal. Sadiq was thus installed on 28 February 1964, regardless of the fact that the majority of the legislators were supporters of Bakhshi. It is worth noting here that all prime ministers (Sadiq was also prime minister up to March 1965) of Kashmir including Sheikh Mohammad Abdullah were installed by the central government.

[12] Mullik, *My Years with Nehru-Kashmir*, 172.

[13] Abdullah, *The Blazing Chinar*, 486.

[14] Abdullah, *The Blazing Chinar*, 486–7; Mullik, *My Years with Nehru-Kashmir*, 172–3. Mir Qasim says that when he tried to explain the negative consequences of Sheikh Abdullah's release, the colour of Mr. Nehru's face changed: 'You also talk of fears' he roared', Qasim, *My Life and Times*, 100.

Milieu

The period of the Sadiq government (28 February 1964–12 December 1971) was one of great stresses and strains owing to both endogenous and exogenous challenges. When he took over, Kashmir had virtually turned into a police state where police brutalities and the implication of innocent people in false cases had become the order of the day. Dissenting voices were ruthlessly suppressed, and corruption had seeped into the collective mentality of the ruling class and officialdom. On the other side, the government faced mass opposition spearheaded by the Plebiscite Movement, which asserted itself aggressively especially after the 1965 war.[15] The Pandit community of Kashmir was also restive owing to the state's policy of giving special privileges to the backward sections of society, who in Kashmir happened to be Muslims mainly. There was such pent-up anger in the community that a small incident of a Pandit girl marrying a Muslim boy led to a widespread agitation by the Kashmiri Pandits, who were supported by the communal forces of Jammu and other parts of India.[16] Jammu and Ladakh were seething with discontentment over what they termed as the discriminatory policy against them.[17] The unemployed educated Muslim youth, whose number by the mid-1960s had risen to the thousands, were motivated to reinforce the ranks of the Plebiscite Movement and radicalize the politics,[18] and, further, in 1965, thousands of Pakistani soldiers in civilian garb infiltrated across the ceasefire line into Kashmir to carry out subversive activities.[19] A section of Kashmiri youth became radicalized and began carrying out underground activities to overthrow Indian rule in Kashmir.[20]

[15] Qasim, *My Life and Times*, 113; Ishaq, *Nida-i-Haq*, 319–34; Butt, *Kashmir in Flames*, 128–50.

[16] Qasim, *My Life and Times*, 117. For more sources, see fn 110.

[17] See Chapter 1 of this book.

[18] Saraf, *Kashmiris Fight for Freedom*, vol. II, 1263.

[19] Saraf, *Kashmiris Fight for Freedom*, vol. II, 1147–60; Butt, *Kashmir in Flames*, 107–11.

[20] Saraf, *Kashmiris Fight for Freedom*, vol. II, 1271–5; Swami, *India, Pakistan and the Secret Jihad*, 49–75; Watali, *Kashmir Intifada*, 311–23. Also see the sub-section 'Student Movement and Underground Activities' later in this chapter.

And the Hindu nationalists were continuously pressuring the state and central governments to abolish the special position of Kashmir and suppress the dissenting voices.[21]

It was in this difficult scenario that the leftist but 'liberal' and unflinching centrist, a man of integrity and courage of conviction, G. M. Sadiq, had to function. Following in the footsteps of his predecessors, Sadiq pursued, though with greater zeal, the socialist policies envisaged in the Naya Kashmir Programme. He was, however, more 'liberal' in tolerating dissent. In fact his is considered the first 'liberal government' since the National Conference came to power.[22]

Policy of 'Liberalization'

In his policy statement on 1 March 1964, Sadiq declared that his government 'would do its best to ensure that the fundamental rights guaranteed to the citizens under the constitution become a reality so that opportunities of greater and freer participation in the economic development and social progress of the state are open to them'.[23] He said that 'it shall be the earnest endeavour of my government to ensure the rule of law and respect for the rights and liberties of the people.'[24] In pursuance of these aims, the government took the following steps:

1. Curbs on freedom of speech and assembly were removed.
2. All detenues belonging to different political parties were released.
3. The special police organization (Peace Brigade) that had gained notoriety as an instrument of repression was disbanded.

[21] Bazaz, *Kashmir in Crucible*, 78–81; Puri, *Jammu and Kashmir*, 153–4.

[22] Saraf, *Kashmiris Fight for Freedom*, vol. II, 1257.

[23] Government of J&K, *Ghulam Mohammad Sadiq, Prime Minister of Jammu and Kashmir* (J&K Information Department), Archives Reference Library, Srinagar, Accession no. 456, 1.

[24] Government of J&K, *Ghulam Mohammad Sadiq, Prime Minister of Jammu and Kashmir* (J&K Information Department), Archives Reference Library, 1.

4. The Preventive Detention Act was made more liberal and brought on par with the law in force in the rest of the country.[25]

Following the issuance of a statement by Sadiq on 5 April 1964, the conspiracy case against Abdullah was withdrawn on 8 April 1964, and he was released unconditionally on the same day. This ushered in a new era in the political history of Kashmir. Viewing the development against the backdrop of Nehru's unflinching support to Sheikh Abdullah vis-à-vis his adversaries, and the publicity that there had been a change in Nehru's policy towards Kashmir, even his (Sheikh Abdullah's) 'opponents' according to Mullik 'fell head over heels in welcoming him back . . . even the Praja Parishad'.[26] That the Sheikh received, in the words of Mir Qasim, 'a hero's welcome'[27] from the people especially in Kashmir, or that, in the words of Prem Nath Bazaz, 'the whole country [Kashmir] seemed to be flocking to hear them [National Conference leaders],' is quite understandable.[28] After all, Sheikh Abdullah and his comrades represented the collective will of the people.

The removal of restrictions on civil liberties ushered in a new era of freedom, accommodation, and tolerance. Within five to six months, three dozen new journals and newspapers were started in the state, representing various points of view.[29] To quote the editor of *Aftab*: 'The announcement initiated a new and historic period for Kashmir. . . . the 17 year old mansions of curbs and restrictions on civil liberties were demolished. The people saw their demands fulfilled.'[30]

Pursuant to his new approach to the Kashmir problem, Prime Minister Nehru invited Sheikh for talks to New Delhi, and according to Sheikh Abdullah he saw Nehru deeply serious in solving

---

[25] Government of J&K, *Ghulam Mohammad Sadiq, Prime Minister of Jammu and Kashmir* (J&K Information Department), Archives Reference Library, 1–2.

[26] Mullik, *My Years with Nehru-Kashmir*, 175.

[27] Qasim, *My Life and Times*, 101.

[28] Bazaz, *Kashmir in Crucible*, 78. For further details see also Abdullah, *Ātash-i-Chinar*, 541–3; Butt, *Kashmir in Flames*, 102.

[29] Government of J&K, *Ghulam Mohammad Sadiq*, 2–3.

[30] Butt, *Kashmir in Flames*, 101.

the Kashmir issue, and that too during his own lifetime. Sheikh recorded what transpired between the two:

> The time has come when India, like a generous elder brother, should make a beginning towards solving the Kashmir problem because it is this problem which has impaired the relations between the two countries (India and Pakistan). Panditji responded saying, I fully agree with your sentiments; and I want to accomplish this work in the twilight of my life—the work which should have been done much earlier. Perhaps . . . you would act as a bridge in this direction.[31]

As President Ayub of Pakistan as well as Prime Minister Nehru wanted to have a negotiated settlement on the Kashmir issue, both desired Sheikh Abdullah to visit Pakistan.[32] On reaching Rawalpindi, the Sheikh received a rousing reception,[33] and President Ayub agreed to come to Delhi for talks as was desired by Nehru. But as ill-luck would have it, Nehru passed away on 27 May, just three days after Sheikh Abdullah's visit to Pakistan. And with this, a momentous development that would have radically changed the course of history with a positive bearing on the life and conditions of the people of the subcontinent, did not come to pass.

From Liberalization to Repression and Coercive Integration

The situation now came back to square one. The policy makers at Delhi comprised the same category of people who had treated Sheikh as the enemy and were determined to frame him in a 'cast-iron case'.[34] These hawks were headed by Home Minister Gulzari

[31] Abdullah, *Ātash-i-Chinar*, 557.

[32] Abdullah, *Ātash-i-Chinar*, 557; Butt, *Kashmir in Flames*, 102; Qasim, *My Life and Times*, 101.

[33] Abdullah, *The Blazing Chinar*, 503–5.

[34] Abdullah, *The Blazing Chinar*, 561; Mullik, *My Years with Nehru-Kashmir*, 173.

Lal Nanda, who had a strong hatred for Sheikh.[35] The dominant influence of hardliners in New Delhi led by the home minister turned the 'new thinking' upside down, and they hotly pursued the same policy which Nehru had followed, and for which he had remorse towards the end of his life. Sheikh Abdullah sums up his disappointment:

> Shastri [Lal Bahadur Shastri, the new prime minister] was a noble and moderate person. Immediately after becoming Prime Minister he, in his policy statement, declared that he would follow in the footsteps of Jawaharlal Nehru. When I met him a few days after he assumed the office, I requested him to maintain the continuity of [new] Indo-Pak policy [initiated recently] and to revive the talks at an appropriate time and take up the thread where it was left owing to the sudden death of Jawaharlal Nehru. Shastri promised that the same would be done; but Gulzari Lal Nanda was a powerful influence in the Cabinet. He did not let the status quo to be maintained; instead he insisted on quickening the process of integration of Kashmir with India to the extent that he got many laws passed which eroded the special position of Kashmir as guaranteed by the Article 370. He used his power and influence to close all those channels which were opened [during the last days of Nehru] for talks to improve the bilateral relations between India and Pakistan. In the state of Jammu and Kashmir too he also tried to revive the practices of Bakhshi regime; and in this way began the preparations for blowing out the lamp of freedom, which the people of Kashmir had lit by their blood during the *mou-e-muqqadas* movement.[36]

Indeed, the celebrations following the announcement of the liberalization policy by Sadiq were still being held by the social democrats in particular and the people of Kashmir in general, when a strong tirade against the Sadiq government was launched by the Hindu nationalists, including many Congress leaders who were unable to tolerate the people of Kashmir giving rousing receptions to the Plebiscite Front leadership and expressing their suppressed sentiments.[37]

[35] Qasim, *My Life and Times*, 103.

[36] Abdullah, *Ātash-i-Chinar*, 570.

[37] Bazaz, *Kashmir in Crucible*, 80.

The result was that the well-wishers of Sadiq, who had paid glowing tributes to him for his policy of liberalization, were extremely disappointed after just nine months when they saw him reverting to the policies of his predecessor. On 28 June 1965, P. N. Bazaz wrote a letter to Sadiq in which he complained:

> You say that 'there has been absolutely no change in your basic stand on democratization and other issues' and that 'you do not impose any restrictions on freedom of expression even in the face of worst provocation'. This categorical statement can hardly prove reassuring when I know that the assertion is not borne out by facts and recent developments in Kashmir. Imprisonment of hundreds of political workers without trial under the provisions of infamous laws mostly for holding views not in agreement with your own, arbitrary suspension of journals not supporting your policy and ban on holding of public meetings in the valley do not corroborate your statement. Added to this is the high-handedness of the police force, local and imported, and the picture becomes quite bleak. Excesses are excesses and do not gain sanctity because they are perpetrated under men claiming to be liberal.[38]

Not only this, but the Hindu members of the National Conference, especially the Hindu Dogra members, made a concerted demand for merging the National Conference into the Congress, calculated to annihilate the individuality of Kashmir. The campaign against liberalization was intensified and was carried out within the camp of the Congress outside the state. 'It has been the old practice and tradition of the Congress', says Bazaz, 'to halfheartedly oppose a Hindu communal demand in the beginning, but mostly concede it when the passions are raised.' The fresh attack of the 'reactionaries was so concerted and furious that government of India considered wisdom in appeasement'.[39] On 20 November 1964, the Lok Sabha discussed the bill proposed by a private member urging that Article 370 be done away with. 'The Congress members vied with the Jan Sanghis to prove that they

---

[38] Bazaz, *Kashmir in Crucible*, 232.
[39] Bazaz, *Kashmir in Crucible*, 82.

did not lag behind in destroying the individuality of Kashmiris,' says Bazaz.[40]

The willing Congress buckled under pressure; so did Sadiq. Thus, it was decided to extend in successive stages all those articles of the constitution which did not apply to the state of J&K until that point. As a first step in this direction, on 4 December 1964, the union home minister, G. L. Nanda, announced the government's decision to apply Articles 356 and 357 to Kashmir. The manner in which the central government surrendered before the Hindu nationalist forces, on the one hand, encouraged Hindu communalism, and on the other weakened the position of Sadiq in the eyes of the people. What is more, the home minister in Parliament and the education minister in Jammu assured their respective audiences that 'Kashmir will be fully integrated with India.' Encouraged by these statements, the 'Hindu reactionaries' refused to be appeased by these measures and insisted on the repeal of the entire article. The Jammu Jan Sangh declared that the 'Dogras will not rest till this aim is achieved.'[41]

It is amazing that neither the Government of India nor the state government took the reaction of the state's people into consideration. Their sole desire was to appease Indian public opinion. 'The interference of the communal forces in Kashmir and the support which they got from the central government scared away the Kashmiri Muslims especially, in light of the communal riots in many towns of India namely Jabalpur, Jamshedpur and Calcutta in which Muslims were killed, looted or made homeless.'[42]

The Sadiq government played with the collective sentiment of Kashmiris repeatedly to appease Hindu nationalists, who were also taking refuge under the umbrella of the Indian National Congress. Thus, in an attempt to efface the separate political identity of Kashmir, Sadiq along with his comrades fulfilled the wishes of his adversaries and masters beyond the Kashmir Valley by dissolving the National Conference and replacing it by the Pradesh Congress Committee on 26 January 1965.

[40] Bazaz, *Kashmir in Crucible*, 82.
[41] Bazaz, *Kashmir in Crucible*, 85.
[42] Bazaz, *Kashmir in Crucible*, 85.

Not realizing that the National Conference was a symbol of Kashmiri Muslims' political and cultural achievements and national existence, reared by the people with great sacrifices, with this move Sadiq and his colleagues further distanced Kashmir from Delhi instead of bringing Kashmir nearer to India. It was a move to erase the separate political identity of Kashmir. Bakhshi, who was entrusted with the job of integrating Kashmir with India, had appropriated this historic party, which had spearheaded the freedom struggle in Kashmir, and thus diluted its special image. Yet he resisted converting it into a branch of the Indian National Congress.[43] Sadiq and his group in the National Conference had been testing the waters for a long time with regard to their quest to become members of the Indian National Congress and establish a branch of the Congress in J&K.[44] The new forces in the central government encouraged them to give a practical shape to the idea. Visualizing its implications for the special political identity of Kashmir, Sheikh Abdullah announced a social boycott (*tark-i-mawalat*) against Sadiq and his Congress party. The clashes between the two parties engulfed the whole valley. The social boycott took such an ugly turn that if a Congressman died, people would not attend his funeral. Mir Qasim, the founder-president of the Congress party in Kashmir, gives a vivid picture of this boycott:

> He [Sheikh Abdullah] called for a boycott of the Congress and ostracisation of its leaders. This caused immense difficulties for the Congress workers. The worst form of this ostracisation was the Sheikh's refusal to let the Congress supporters bury their dead in Muslim graveyards. It was during this boycott campaign that my aunt, who brought me up after my mother's death in my infancy died. Among those who visited my house to condole the death was Mr. Beg's brother Mirza Ghulam Qadir Beg. The supporters of the boycott at once informed Sheikh Abdullah who demanded an explanation from Mirza Beg. The latter pleaded ignorance of the Sheikh's edict, 'otherwise I would not have allowed Mirza Ghulam Qadir Beg to visit Qasim's house'. If they could do this to me, one could understand the plight of the common Congress worker. In Nawakadal

---

[43] Puri, *Jammu and Kashmir*, 154.
[44] Qasim, *My Life and Times*, 106–7.

(Srinagar) a worker's aged mother died. His neighbours burnt up the
bier (taboot) on which her body had been taken to the graveyard.[45]

The central government took serious note of the social boycott movement and directed the state government to take action
against the Plebiscite Movement, with the result that around
2,000 people were detained under the Defence of India Rules.[46]
Despite the social boycott, the Congress gained strength by enroll-
ing all sorts of riff-raff to swell its ranks.[47] This is another tragedy
of Kashmir politics—a direct result of conflict, which discouraged
men of integrity from entering the arena of politics. Consequently,
politics became the domain of people whose only objective was
advancing their own material and political interests.

Unmindful of the reaction of the Muslims of Kashmir to his integrationist policies, Sadiq in March 1965 amended the state constitution and rechristened the sadr-i-riyasat as governor and the position
of the prime minister as chief minister, thus bringing Kashmir in
line with other states. Reminding Sadiq of the promise he had made
a year ago but failed to keep under the pressure of Hindu nationalists, without realizing its baneful implications for the agenda of
Kashmiri's emotional integration with India, Bazaz wrote

> Indeed, you assured me [a year ago] that you will not be a party to
> the further impairment of autonomy enjoyed by the state people
> under Indian constitution nor to the replacement of the National
> Conference by the National Congress—the two aims that had been

[45] Qasim, *My Life and Times*, 107.

[46] Butt, *Kashmir in Flames*, 105; Abdullah, *Ātash-i-Chinar*, 580.

[47] While this was done by all parties, the patronage politicians did it
mainly. For example, writing about the practice of enlisting the services of
goons to build a support structure, Ghulam Ahmad writes, 'Mohammad
Shafi Qureshi, a political adventurer, an opportunist and once President of
Muslim Students Federation established a branch of Congress in Srinagar.
... Qureshi enlisted the service of a ruffian known as Noora *Taburdar* (axe-
man or woodcutter) and his friends and comrades in hooliganism to entice
and enroll all sorts of riff raff in the Congress fold to swell its ranks with
the blessings of the central party.' Ghulam Ahmad, *My Years with Sheikh
Abdullah: Kashmir 1971–1987* (Srinagar: Gulshan Books, 2008), 39–40.

set by Indian reactionaries for achievement. . . . Your surrender before Indian reaction has not added to your prestige nor has it in any way fulfilled the objective of Kashmiri's emotional integration with India. It has only produced widespread disaffection against your government which has been easily exploited by the fanatics, demagogues and enemies of democracy.[48]

Disgusted with the hard-line policies of the central government, Sheikh Abdullah left for Hajj (holy pilgrimage) via England, Egypt, and Algeria. In Algeria he met Chinese prime minister Chou En-lai, who was there to attend an Afro-Asian conference. The news that the Chinese prime minister and Sheikh Abdullah had met at a dinner and exchanged views on Kashmir raised a storm in India, as China was its 'abominable enemy', especially after the Indo-Chinese war of 1962. To douse the anger, the Government of India asked Sheikh Abdullah to return immediately after performing Hajj, or else his passport would be impounded. Sheikh Abdullah complied with the directions and returned on 7 May 1965. He was arrested at the Palam Airport and sent to Ootacamund (Tamil Nadu) for detention along with Afzal Beg. Afzal Beg was the point man of Sheikh Abdullah who after the arrest of Sheikh in 1953 launched the Plebiscite Movement in Kashmir and was its President. In May 1965 when Sheikh Abdullah was again arrested after his return from Hajj, Afzal Beg was also arrested as the government feared that he may launch a renewed agitation against the arrest of Sheikh.

As usual, the arrest of Sheikh Abdullah led to a mass agitation, which was ruthlessly suppressed. A large number of people were arrested. What is more, between 7 March 1965 and 10 May 1965, quite against the policy of liberalization announced by Sadiq in his policy statement, restrictions were placed on 10 newspapers.[49] The street protests were followed by a civil disobedience movement launched jointly by the Plebiscite Front, the Political Conference, and the Awami Action Committee on 6 June 1965. A programme of courting arrest in groups each day was chalked out,

---

[48] Bazaz, *Kashmir in Crucible*, 234.
[49] Butt, *Kashmir in Flames*, 106–7.

and thousands of people came forward and registered themselves for courting arrest.[50]

Operation Gibraltar and the Infiltration of Armed Personnel

It was in this restive situation that, in August 1965, rumours went around that Pakistani armed personnel had entered the Kashmir Valley and Poonch district of Jammu province. They were called *mujahid* in the valley and *razakara* in Poonch and Rajouri.[51] Almost all the infiltrators were from Pakistan administered Kashmir (PAK). The plan was to begin guerrilla-fuelled protests on 9 August, the day when Sheikh Abdullah had been deposed and arrested in 1953. This day was observed every year with a massive public meeting at which 'anti-India speeches and slogans were raised and an atmosphere of insurrection was created'.[52] In a protest note to the Government of Pakistan on 10 August 1965, India alleged the entry of 1,200 guerrillas.[53] According to one report, 5,000 armed men had infiltrated into Poonch district[54] (present-day Poonch and Rajouri districts of Jammu province). The majority of the armed personnel were from PAK. They were equipped with automatic rifles, stun guns, small machine guns, and other modern types of firearm.[55] The Plebiscite Front, however, dissuaded people from getting involved in the guerrilla war; and instead of organizing processions on the day as per routine,[56] the Front observed *hartal* (strike) in the city and other towns on 9 August 1965. The venue of the public meeting was changed from Khanyar to Mujahid Manzil to avoid infiltration in the public meeting.[57]

---

[50] Butt, *Kashmir in Flames*, 106–7; Ishaq, *Nida-i-Haq*, 323.

[51] Choudhary, *Kashmir Conflict and Muslims of Jammu*, 156–8.

[52] Qasim, *My Life and Times*, 108; Ishaq, *Nida-i-Haq*, 329.

[53] Saraf, *Kashmiris Fight for Freedom*, vol. II, 1156.

[54] Choudhary, *Kashmir Conflict and Muslims of Jammu*, 156.

[55] Butt, *Kashmir in Flames*, 110.

[56] Ishaq, *Nida-i-Haq*, 330.

[57] Butt, *Kashmir in Flames*, 108.

However, the night of 10 August 1965 saw sporadic firing, and on 11 August there was exchange of fire between the Border Security Force and the infiltrators at Bemina in Srinagar, which created panic and terror among the people. Security was beefed up at sensitive places and installations. Simultaneously, mini-wars were going on between the Indian security forces and the guerrillas at other places in the valley, namely, Beerwah, Ganderbal, Gulmarg, Yusmarg,[58] and so on. A propaganda machinery in the shape of a secret radio station called Sada-i-Kashmir (Voice of Kashmir) was also pressed into service to mobilize the Kashmiri people and to boost the morale of the infiltrators.[59] On 14 August 1965, following a decision taken in a meeting of top army commanders and the chief minister, the entire area of Batamallo (a politically sensitive quarter of Srinagar) was set on fire to finish off the guerrillas.[60] The situation was so precarious that many central leaders including the union home minister, defence minister, and the prime minister of India came to Kashmir in succession.[61] Besides, during the entire period, the army chief remained in Kashmir before he suddenly left for New Delhi on 1 September 1965, signalling the war between India and Pakistan.[62]

The contemporary evidence reveals that Operation Gibraltar was worked out in consultation with some leaders of the Plebiscite Front, though they were in a minority. The majority was against armed rebellion, at least at this stage.[63] Indeed, Maulana Masoodi, the leader of the Plebiscite Front, who like others had come to know of the infiltration of armed personnel as well as their plan to raise an armed insurrection with the help of the local people, in his address to the public on 9 August 1965 dissuaded the people from committing any 'mistake'.[64]

[58] For details see Saraf, *Kashmiris Fight for Freedom*, vol. II, 1147–57.

[59] Saraf, *Kashmiris Fight for Freedom*, vol. II, 1156; Butt, *Kashmir in Flames*, 108; Ishaq, *Nida-i-Haq*, 331–2.

[60] Butt, *Kashmir in Flames*, 109; Ishaq, *Nida-i-Haq*, 330.

[61] Saraf, *Kashmiris Fight for Freedom*, vol. II, 1156; Butt, *Kashmir in Flames*, 110–11.

[62] Butt, *Kashmir in Flames*, 210–11.

[63] Butt, *Kashmir in Flames*, 109–10. Munshi Ishaq also substantiates this in his memoir *Nida-i-Haq*, 329–33.

[64] Butt, *Kashmir in Flames*, 108.

In Poonch and Rajouri, the situation was entirely different because of the bitter memories of the massacre of Muslims in Jammu province in 1947 and the cultural affinity between the people of the area and the infiltrators called 'razakars' (volunteers) by locals.[65] They spread all across the vast area which had seen, besides the massacres, large-scale migrations and divided families in the aftermath of Partition. Zafar Choudhary captures the situation in Poonch and Rajouri areas on the basis of oral history. The following excerpt illustrates the difference between Jammu province and the Kashmir Valley in their response to Operation Gibraltar:

> The Razakaars mixed up too well with the locals, established control and expanded their areas of influence deep into interiors, extended to Budhal in north and Kalakote in the east. Hundreds of government employees left their jobs and joined ranks with the Razakaars who had soon established their own 'local governments' in the villages. . . . Many respondents in Rajouri and Poonch recalled that for few months there existed no signs of state authority.[66]

The people of Poonch and Rajouri had to pay a heavy price for the support they extended to the razakars. Bitter memories of the heartbreaking violation of human rights by the Indian Army have come down to us through oral history, pointing to the fact that the events of 1965 have become an important component of the collective consciousness of the people of the area.[67]

The Student Movement and Underground Activities

The prolonged democratic struggle, which had apparently borne no fruit except bullets, beatings, interrogations, and incarcerations, exasperated sections of the educated youth who were no longer contented with cool democratic struggle. Though they did not

---

[65] Choudhary, *Kashmir Conflict and Muslims of Jammu*, 155.

[66] Choudhary, *Kashmir Conflict and Muslims of Jammu*, 155.

[67] Choudhary, *Kashmir Conflict and Muslims of Jammu*, 159–60.

pick a quarrel with their leaders, they were nevertheless restive with the latter's style of struggle.[68] The unrest among the youth intensified owing to political education by the Plebiscite Front, repressive measures by the government, organized anti-Muslim riots in India,[69] unemployment among the urban educated youth, non-recruitment of Muslims into centrally administered departments, inadequate representation of Muslims in state services and professional educational institutions,[70] the arrogance of Indian civil as well as military officers,[71] propaganda by the electronic media of Pakistan and PAK, Algeria's successful struggle for freedom, the Vietnamese war against the mighty USA,[72] Pakistan's continuous support and encouragement of indigenous resistance, and the infiltration of a large number of armed guerrillas from across the Ceasefire Line. Given the treatment meted out to Sheikh Abdullah, it is difficult to disagree with the argument that Abdullah and those close to him tacitly supported the radicalization of the resistance.[73] Thus in March 1964, according to Praveen Swami, 'radical elements at the edges of the National Conference [read Plebiscite Front] set up the Students and Youth League (SYL) to push the case for separation from India more aggressively than their parent organization.'[74] Branches of the league were set up in

---

[68] Ishaq, *Nida-i-Haq*, 331–3.

[69] Mohammad Yousseff Bhat, *Prison Dairy: Kashmir Untold Story 1965–68* (Srinagar, 2017), 172.

[70] Mohammad Yousseff Bhat, *Prison Dairy: Kashmir Untold Story 1965–68*, 146, 188; Saraf, *Kashmiris Fight for Freedom*, vol. II, 1262–3, 1268–9.

[71] Mohammad Yousseff Bhat, *Prison Dairy: Kashmir Untold Story 1965–68*, 146–7; Saraf, *Kashmiris Fight for Freedom*, vol. II, 162–3.

[72] Saraf, *Kashmiris Fight for Freedom*, vol. II, 162–3. The politically conscious youth used to keep themselves abreast of the developments in the world by listening to the radio. See Bhat, *Prison Dairy*.

[73] This is the buzz one hears in Swami, *India, Pakistan and the Secret Jihad*, 56–7. This is also supported by local sources who were part of the resistance movement. 'Pakistan paid money to Plebiscite Front,' according to Ishaq, *Nida-i-Haq*, 160.

[74] Swami, *India, Pakistan and the Secret Jihad*, 57.

different towns of the valley as well as in the district of Poonch,[75] to bring pressure upon the government, by organizing protests, sit-in dharnas, and marches to the United Nations Military Observers Group headquarters, to hold a plebiscite for the resolution of the Kashmir problem. On 29 September 1965, thousands of students, after adopting the following resolution, marched in a procession to present it to United Nations Headquarters in Srinagar:

> We shall fight in the schools, we shall fight in the colleges, we shall fight in the streets, we shall fight in the villages, we shall fight in the towns, but we shall never submit before the might of Indian imperialism. Either we shall perish or just we will triumph.[76]

The whole of September and October 1965 saw massive student protests in which both male and female students participated, demanding the holding of a plebiscite. A large number of students were killed in police firing, and hundreds were arrested and put behind bars, attracting streams of press people both Indian and foreign to Srinagar.[77] The student uprisings forced even the moderate Plebiscite leadership to join them and deliver full-throated revolutionary speeches followed by the arrest spree. The military entered into the precincts of Hazratbal, which made the situation even more explosive. Almost all the office bearers of the Plebiscite Front up to the halqa level were arrested. The spree of arrests continued for many months.[78]

The Students and Youth League furnished recruits to a hierarchical string of cells established under the superordinate control of what is known as the Master Cell to launch an organized covert campaign against Indian rule in Kashmir. Of the many subsidiary cells, mention may be made of the Students' Cell, whose job was to organize demonstrations and strikes in colleges; the Poster Cells that printed and issued posters; the Narwara Cell instructed the cadre in the use of weapons; the Buchwara Cell which ferried

---

[75] Swami, *India, Pakistan and the Secret Jihad*, 57.

[76] Saraf, *Kashmiris Fight for Freedom*, vol. II, 1263.

[77] For details see Saraf, *Kashmiris Fight for Freedom*, vol. II, 1263–6; Bhat, *Prison Dairy*, 21–32.

[78] Ishaq, *Nida-i-Haq*, 323.

weapons; and the Infiltrator Liaison Cell that facilitated the working of infiltrators from the Ceasefire Line.[79]

Detailed information about the activities of the Master Cell and other underground organizations of the time including Alfatah (busted in 1971) has been provided by Yousuf Saraf, A. M. Watali and Praveen Swami.[80] Suffice it to say that though the Indian rule in Kashmir was resisted right from the installation of the 'popular government' in 1948, the Sadiq period is underlined by an unprecedented challenge by underground forces which, along with the Plebiscite Front, were no doubt supported by Pakistan. The challenge was such that, in the words of Praveen Swami, 'if some had abandoned the struggle to throw India out of Kashmir, though, another generation was readying itself to take up the baton.'[81] This is the infallible truth about the Kashmir problem.

Farce Elections

Sadiq's claim that he would fight his opponents (the Plebiscite Front and other 'disloyal' groups) on the political plane proved hollow on the election front. The Gajendragadkar Commission appointed by the Sadiq government to enquire into the causes of 'irritations and tensions' in the state and to suggest remedial measures, made the following observations about elections in its report submitted in December 1968:

> Some representatives who appeared before us have cast doubts on the manner in which elections have been conducted in the state in the past, particularly about the fairness of the last General Election. They have pointed out that at the last General Election, 141 nomination papers were rejected and 26 members were returned unopposed to the State Legislative Assembly and that the whole of Anantnag District failed to get an opportunity to go to the polls for electing its

[79] For details see Swami, *India, Pakistan and the Secret Jihad*, 58–9.

[80] Swami, *India, Pakistan and the Secret Jihad*, 58–103; Watali, *Kashmir Intifada*, 313–14; Saraf, *Kashmiris Fight for Freedom*, vol. II, 1271–4.

[81] Swami, *India, Pakistan and the Secret Jihad*, 77.

representatives to Parliament or to the State Legislature. These persons feel that all the elections held so far were systematically interfered with by the State authorities and that this has undermined the faith of the common man in democracy.[82]

The commission made a significant observation, which, if heeded, might have given a different direction to the history of Kashmir:

This [manipulation of elections] is a cause of irritation and tension and a note of this feeling has to be taken. We should, therefore, like to add that as the Jammu and Kashmir State occupies a strategic area, it is necessary to nurture the faith of the common man in democracy and democratic institutions in the state. We hope that the State Government will do all they can to build and sustain that faith.[83]

In 1966, Jayaprakash Narayan wrote to Prime Minister Indira Gandhi, 'We profess democracy, but rule by force in Kashmir . . . we profess secularism, but let Hindu nationalism stampede us into ... establishing it by repression. ... Kashmir has distorted India's image in the world as nothing else has done. . . . the problem exists not because Pakistan wants to grab Kashmir, but because there is deep and widespread discontent among the people.'[84] Commenting on this sad state of affairs, Balraj Puri says, 'As part of Sarvodaya observers team deputed by Jayaprakash Narayan, we met several officers of the state who told us that Bakhshi had to be defeated in the national interest. Prime Minister Indira Gandhi, publically stated in her election tour that there was no need for an opposition party in Kashmir.'[85] He further states,

When I showed the chief election commission, K. Sundram, a bundle of duplicate ballot papers, he argued that Bakhshi also used to do the same. To this I retorted that I was not representing Bakhshi's

---

[82] Gajendragadkar Commission Report, 79.

[83] Gajendragadkar Commission Report, 79.

[84] Jayaprakash Narayan, in a private communication to Prime Minister Indira Gandhi, cited in Akbar, *India: The Siege Within*, 267.

[85] Puri, *Kashmir towards Insurgency*, 49.

case but rather that of citizens of the state who had been deprived of their democratic rights by Bakhshi as well as by Sadiq. Instead of taking cognizance of my complaint, Sundram threatened to take action against me. He said that it was illegal to possess ballot paper. Obviously he too believed that 'the national interest' was more important than the demands of democracy and his office.[86]

The rejection of nominations, which facilitated many unopposed returns election after election, was carried out on flimsy grounds.[87] The victims of rejections were, however, the opposition candidates. Having observed the election campaigns of different political parties in full bloom in Delhi on the eve of the 1970 parliamentary elections, the editor of Kashmir-based weekly *Aīna* and an emerging political personality of Kashmir, Shamim Ahmad Shamim, rued the bizarre situation prevailing in Kashmir, which he delineated in his editorial under the rubric 'The Distance between Delhi and Srinagar'. After narrating the electioneering scenes in Delhi, he wrote:

In Kashmir too the elections for three Parliamentary seats are going to be held. But [alas!] here the practice of elections is quite different. Here the election results are being announced before the votes are polled. Here there is no 'need' for election rallies, campaigns, posters, speeches and other electioneering activities. Here only the writ of the ruling party prevails. And when the ruling party feels that the presence of a party, candidate or individual is harmful to their monopolistic project, they clear their way by exiling, imprisoning and disempowering them. Presently, well before the commencement of electioneering, all the opponents were put behind the bars so that the favourities of the ruling party can go around without fear and fight the election by misusing their position. In all the three districts of Kashmir the

---

[86] Puri, *Kashmir towards Insurgency*, 49.

[87] The grounds of rejection were: candidates were government contractors, they held positions of profit, they failed to enclose papers regarding their inclusion in the electoral rolls, they had not taken an oath of allegiance to the constitution in the manner required by the electoral law, etc. Nevertheless, it is intriguing why only the opposition cadidates faced the rejections to the complete exclusion of the ruling party candidates.

Article 144 has been imposed, and the anti-ruling party candidates are not permitted to hold election rallies and mobilize the people. The Congress candidates say it openly that 'we have no need of votes; we can win elections without votes'. How can one win without votes? This cannot be understood by anyone in any part of India. But since in Kashmir it is a commonplace that generally the candidates win without having got the votes, it is, therefore, easily understandable to every Kashmiri. From this you could have well imagined that there is not just seventy minutes distance (by air) between Srinagar and Delhi; but it is [actually] the distance of seventy years.[88]

The Gajendragadkar Commission also found that elections to local bodies were also not held for ulterior motives, and they had lost their representative character:

We were told that the elections to local bodies such as panchayats, town area committees, notified area committees and municipalities, are long overdue and that these bodies are no longer representative in character. Some parties have alleged that besides depriving the public of their fundamental right of franchise, this has resulted in the Government managing to have their own candidates elected to the Legislative Council from the local bodies constituencies. We think that it is necessary that the State Government should immediately take all steps to arrange for the holding of the elections to the local bodies as early as possible. We should also like to add that these elections should be held regularly in future.[89]

It is worthwhile to mention that even the ruling Congress president did not consider these elections as democratic. In an interview with the editor of the *Āina* he rhetorically asked: 'If Sadiq is such a democrat and regards liberation as the basis of his government, where was this democratic temperament during the panchayat elections? You [editor] yourself know what happened during these elections [of August 1969].'[90]

[88] *Āina*, 7 February 1970.

[89] Gajendragadkar Commission Report, 79.

[90] R. K. Bhatt, 'Kashmir: Politics of Integration', in *Fourth General Election in India*, eds S. P. Verma and Iqbal Narain (New Delhi: Orient Longman, 1968), 158.

A Degree of Change

All said and done, Sadiq sincerely believed that the 'liberalization' policy was the best instrument to neutralize the opposition. He would sometimes take issue with the centre and refuse to be persuaded by its advice to deal firmly with disloyal people. When once the home minister of India Gulzari Lal Nanda pressurized him to be tough with the disgruntled youth, Sadiq asked him, 'Do you want me to use tanks against my people?'[91] A. M. Watali also relates that when once the home minister, Y. B. Chavan, asked him under what circumstances 'a dangerous doctor was released', 'Sadiq shot back that the decision was taken by the state government in public interest. . . . it was none of his business as the matter fell within the purview of the state government's powers and he should mind his own business as Home Minister of India.'[92]

The students who were arrested in connection with subversive activities were not treated as enemies under the instructions of the chief minister. Instead, they were considered 'misguided' youth who could be won over by empathy and by helping them build their careers. In the prisons, facilities were provided to them to appear in the examinations. According to the contemporary evidence, all these young men were provided with government jobs by the Sadiq government. As a result, 'some of them retired at very high positions in the state service hierarchy. Some others established themselves as prominent leaders in mainstream parties.'[93] Mohammad Yusuf Bhat, an active member of the Master Cell, says that the Counter Intelligence Department (CID) had prepared a case of treason against some of the arrested members of the Master Cell, including himself. But the then chief minister Sadiq did not agree; instead he directed the CID to slap the Defence of India Rules.[94]

Sadiq exhibited a similar policy towards those people of Poonch, Rajouri, Uri, Karnah, and Keran who to save their lives had crossed

[91] B. L. Kak, *Kashmir Problems and Politics* (New Delhi: Seema, 1981), 79–80.

[92] Watali, *Kashmir Intifada*, 291–2.

[93] Watali, *Kashmir Intifada*, 294.

[94] Bhat, *Prison Dairy*, 31.

the Line of Control during the Indo-Pak war. Desirous as they were to come back and resettle in their native places, it was difficult for them to trek back in the absence of any government policy. The home minister, a hardliner, was opposed to the idea of allowing them to come back, but Sadiq prevailed upon the prime minister and the migrants were allowed to return under a well-formulated government policy. It is heart-warming to quote the police officer who was assigned the job of overseeing and coordinating the execution of this policy:

> The Home Ministry, headed by Gulzari Lal Nanda, a hard line Congress leader, was not favorably inclined to allow the return of these migrants to their home and hearth. But G M Sadiq, the Chief Minister, put his foot down and impressed upon the Prime Minister, Indira Gandhi, that under the circumstances in which almost the whole Muslim population had to migrate during war to save their lives, they were within their rights, as the citizens of India, to return.[95]

He further says:

> While this process was on, a letter was received from army in the Home Department, requesting to allow their representative also to participate in the meetings of the screening committees. This letter was put up by me to the then Home Secretary, Ghulam Rasool Renzu, for orders of the Chief Minister. Sadiq in his own hand wrote a note on the file that the request of the Army is not tenable as in a democratic setup they do not have any *locus standi* to participate in the deliberations of such meetings, which were purely within the domain of civil administration. He further noted that in case army had any information about any matter relating to the defense of the state, they are welcome to convey the same to the concerned District Magistrate. This principled stand taken by Sadiq resulted in smooth return and re-settlement of all the families who had migrated to Pakistan in the wake of 1965 war.[96]

---

[95] Watali, *Kashmir Intifada*, 308.
[96] Watali, *Kashmir Intifada*, 309.

On 9 October, the Central Reserve Police Force (CRPF) resorted to unprovoked firing resulting in many deaths, besides wounding many. Sadiq was so disgusted that he demanded the eviction of the CRPF from Kashmir. As a result, the CRPF was recalled from different places and stationed in the barracks.[97]

Though law and order is a state subject, nevertheless, in case any activity is deemed to be militating against maintaining the integrity of the country, it immediately becomes a security issue directly dealt with by the central government. This explains the contradictions we find occasionally between the 'liberalism' of Sadiq and the violations of human rights or curbing of civil liberties when the activities crossed the limit fixed by the state. The burning down of Bemina, excesses committed by the army in Poonch and Rajouri, repressive policies against Plebiscite leaders, and temporary restrictions on the print media have, thus, to be understood in the context of security-related subjects dealt with directly by the central government, though formal consent, wherever required, was also sought from the chief minister. As Sadiq was by conviction a believer in 'liberalism', the restrictions on the press or civil liberties used to be temporary. The ban on some newspapers was lifted without much delay, and all the detainees belonging to the Plebiscite Front, the Political Conference, Awami Action Committee, and Hindu Action Committee were released.[98]

The ban on the Plebiscite Front was lifted and the party was persuaded to fight elections, against much opposition from adversaries within and without the state. The first panchayat elections on the basis of adult suffrage were held in J&K in August 1969. The Plebiscite Front directly took part in the elections and used the opportunity to mobilize public opinion. According to Karan Singh, they played a part in these elections 'with a view to wrecking the constitution and disrupting the present relationship of the state as an integral part of India'.[99] Karan Singh, who was dead set against Sadiq's policy of 'liberalization', says that the involvement of the

---

[97] Butt, *Kashmir in Flames*, 127.

[98] Butt, *Kashmir in Flames*, 126.

[99] Alam, *Kashmir and Beyond*, 128–9.

Front in these elections led to the spread of the Front's 'tentacles into the rural areas of the valley'.[100] 'The policy of liberalization', says Karan Singh, 'can obviously not mean allowing secessionist parties to hold the valley to ransom whenever they decide to escalate their opposition.'[101] Indeed, Sadiq's policy of giving space to the Plebiscite Front in the politics of Kashmir as a safety valve and as a way of mainstreaming the secessionist voice, was looked at askance by Jammu Hindu Dogras regardless of their secular or communal credentials.

Sadiq's policy of fighting the Plebiscite on the political front succeeded to a large extent. It was during this period that many Plebiscite leaders left the Front, participated in the elections, and won with the patronage of the Congress.[102] We also learn from contemporary sources of a huge trust deficit among the Plebiscite Front ranks, and allegations of some important leaders having been 'purchased' by the government.[103] And it was also during this period that Sheikh Abdullah came much closer to patching up with the Government of India, with Sadiq cooperating with the centre. By conviction, Sadiq was so ardent a believer in persuasion rather than persecution that he adopted a sympathetic and conciliatory attitude towards a known anti-Indian and pro-Pak Plebiscite leader, Munshi Mohammad Ishaq, and the latter was so moulded by the attitude of Sadiq that he acknowledged in writing the gratitude he owed to the chief minister.[104]

Although Sadiq was subjected to untold harassment and character assassination by the Pandit agitationists, and he complained that such matters could have been resolved without taking to the streets, nevertheless he hastened to add that 'the freedom of conscience is one of the fundamental rights enshrined in the constitution and its exercise within the constitutional framework has to be upheld.'[105] The Gajendragadkar Commission also 'noted

[100] Alam, *Kashmir and Beyond*, 128–9.

[101] Alam, *Kashmir and Beyond*,131.

[102] Ishaq, *Nida-i-Haq*, 343–4.

[103] Ishaq, *Nida-i-Haq*, 343–4.

[104] Ishaq, *Nida-i-Haq*, 349.

[105] Statement of the Chief Minister in the Legislative Assembly on 17 August 1967, *Roshni*, Srinagar, 21 August 1978.

with satisfaction' that 'Notwithstanding the delicate nature of the subject, the present state government have allowed ample scope to the freedom of expression to the citizen and have deliberately decided to deal with the challenge posed by the subversive trends in the state on a political plane.'[106]

In 1968, Sheikh Abdullah decided to hold a Kashmir State People's Convention in Srinagar with the objective of allowing the different shades of opinion to have a free and frank discussion on the Kashmir question. Invitations were extended to stakeholders across the Line of Control. Although, according to Mir Qasim, 'implicit in the declared objective of the convention was the rejection of the finality of Kashmir's accession to India', Sadiq did not show any reservation in facilitating the convention. The Gajendragadkar Commission's elaborate explanation of the meaning of democracy in the context of the Sadiq government's policy of liberalization is actually an endorsement of the course of action adopted by Sadiq. Appreciating the policy of Sadiq, who had allowed the Plebiscite Front to convene the Kashmir State People's Convention, the commission observed:

It seems appropriate at this stage to make a passing reference to the Kashmir State People's Convention held at Srinagar in October last. At this Convention divergent opinions were freely and fully expressed. Some of the views which were ardently advocated appear to be plainly inconsistent with the correct constitutional position under the Constitution of India and the Constitution of Jammu and Kashmir. We feel that it is a tribute to the democratic way of life adopted by India, and by the State of Jammu and Kashmir as an integral part of the Union of India, that a Convention of this kind should have been called and such views openly, freely and fearlessly canvassed.[107]

The liberalization policy of Sadiq was strongly detested by a section of the ruling Congress, who either out of their own convictions or informed by the strong Kashmiri Pandit lobby did not favour the continuation of Sadiq as chief minister. In Sadiq's own

[106] Gajendragadkar Commission Report, 76.
[107] Gajendragadkar Commission Report, 81.

cabinet and party, some, like D. P. Dhar, disliked him out of conviction for being soft towards the counter-discourse, and some
made the policy of liberalization an excuse to see Sadiq out, as his
policies restrained the habitual misusers of power from misusing
their power. Though seemingly a strong lobby at the centre had
given the 'green signal' to create a 1953-like situation, and even
if Mir Qasim had also given the nod to the dissident group, Prime
Minister Indira Gandhi's 'harsh dressing-down' to the group saved
the situation.[108] Perhaps this conspiracy had been hatched without her knowledge, or Sadiq's illness prevented her from changing
the leadership.

Pandit Agitation

In July 1967, the marriage of a Hindu girl, Parmeshwari, with a
Muslim boy put a matchstick to the tinderbox. The Pandit community of Kashmir was seething with discontentment ever since
the adoption of the policy of land to the tiller without compensation. This disaffection was further fuelled by the reservation
policy of the state which, aimed at ameliorating the conditions of
the socially, economically, and educationally backward communities by reserving a substantial quota for them in state services
and admission to higher education institutions.[109] Both policies
affected the Pandit community adversely. The inter-faith marriage
became an excuse for pent-up sentiments to burst forth. The 'con-
tentious politics' marked by street protests, provocative speeches,
counter-demonstrations, lathi charges, firing, deaths, curfew, and
hartal kept Kashmir on the boil for months.[110]

---

[108] Qasim, *My Life and Times*, 119–21.

[109] Gajendragadkar Commission Report, 74–5.

[110] Bhat, *Prison Dairy*, 161–78; Butt, *Kashmir in Flames*, 122–7;
Saraf, *Kashmiris Fight for Freedom*, vol. II, 1269–70; Qasim, *My Life and
Times*, 117–8. A wealth of information on this agitation is contained in
Khalid Bashir's book, *Kashmir: Exposing the Myth*, chapter 6, 'Agitation',
167–224.

Grievances of the Muslims

While the basic cause of the Pandit agitation was their grievance against the reservation policy which had harmed their monopoly over state services, the Muslims were also not happy with their proportionate representation in government service. In 1967, they submitted a memorandum to the Indian home minister Y. B. Chavan at Srinagar complaining against their underrepresentation in different state departments. It was brought to Chavan's notice that out of 2,252 gazetted posts, Muslims held 924 while Kashmiri Pandits occupied 638, though according to 1961 census figures Muslims constituted 68 per cent and Kashmiri Pandits only 1.5 per cent of the total population.[111] Many memorandums were also submitted to the chief minister at Srinagar and Jammu. In one memorandum it was stated that since March 1966, only 45 Muslims had been appointed to the gazetted cadre as against 91 non-Muslims. The other memorandum presented at Jammu in September 1967 reproduced the detailed statistics about the underrepresentation of Muslims in medical and engineering colleges. It was recalled that in 1966, out of the total number of 235 seats available in medical and engineering, only 104 were allotted to Muslims, and that in the engineering colleges, Muslims got 77 seats, Pandits 66, and Jammu non-Muslims 67, while in 1967 Muslims got 45 seats, Pandits 32, and Jammu non-Muslims 39 seats. The memorandum also complained about the absence of Muslims in the central services, which were being exclusively manned by Pandits.[112]

To make a detailed enquiry into the grievances of the different regions, communities, and sections, Sadiq appointed a commission of enquiry on 6 November 1967. Though appointed in response to the Pandit agitation, the commission stated,

> Community wise the position of Muslims [in state services] has shown an improvement though their total share in the services on 1 April 1967 was appreciably less than what it should be on the basis

[111] *Hindustan Times*, 20 July 1967.

[112] Saraf, *Kashmiris Fight for Freedom*, vol. II, 1269; see also Bhat, *Prison Dairy*, 188.

of population. The position of Hindus has shown a decline though their share in the state services as on 1 April 1967 was substantially higher than what was due to them according to their population.[113]

Gajendragadkar Commission of Enquiry and the Follow-Up Action

Ever since the transfer of power from the Jammu-based princely order to the Kashmir-based 'popular' rule, the Hindu-dominated Jammu districts and Buddhist-dominated Leh had been complaining of discrimination against them with regard to development matters and appointments to state services. From the Kashmir region the Pandit community was unhappy with the government not only because of the abolition of landlordism, but also because of the modified recruitment rules and rules for admission to higher education institutions, which sought to bridge the wide gap between the Muslims and non-Muslims in state services and higher educational institutions. The Scheduled Castes and backward classes were also complaining about lack of reservations for employment in government services and admission in educational institutions. To address these issues, the government appointed a commission of enquiry in November 1967 under the chairmanship of the former chief justice of India, Dr P. B. Gajendragadkar. The commission had to make a detailed enquiry into the grievances of different regions, communities, and classes and to suggest appropriate measures for the distribution of an equitable share of resources among different regions and equitable participation in the integrated development of the state, and to examine the recruitment policies and the policies of the government regarding admission to institutions of higher education having special regard to the claim of the Scheduled Castes and other economically, educationally, and socially backward communities, classes, and groups of the state.[114] After thoroughly examining the issues which led to tensions in the state, the commission submitted its

[113] Gajendragadkar Commission Report, 54.

[114] Gajendragadkar Commission Report, 1.

report in December 1968, pursuant to which the government took appropriate measures, which are briefly described in what follows.

Balanced Regional Development

The Gajendragadkar Commission, while admitting that much of the feeling of discrimination against Jammu is psychological, however emphasized that the perceptions of neglect cannot be ignored. It, therefore, stressed that the people of each region should have a feeling of equal participation in the integrated development of the state.[115] It was also found that the hilly and inaccessible parts of Jammu and Kashmir regions were much more backward than other parts. Therefore, the commission emphasized that these areas needed the special attention of the government.[116]

The development of Ladakh during the period of the first three plans was inadequate, observed the commission. Though a number of roads had been constructed, transport remained unsatisfactory. For equitable development, as well as given the strategically sensitive location of the region, the commission called for the immediate attention of the centre and the state to the development of Ladakh. The commission seriously felt that there should be an arrangement whereby the regions could have the maximum opportunity to draw up their own development programmes and to implement them subject to the interest of the state as a whole.

To meet this urgency, the commission suggested that there should be a statutory state development board and statutory regional development boards for the three regions—Jammu, Kashmir, and Ladakh. Each regional board should be headed by the chief minister or the planning minister and should consist of legislators from the region, economists, experts, and concerned officials. The regions should be adequately represented on the state development board headed by the chief minister. While the state development board would scrutinize and approve the plans prepared by the regional boards, it would also plan such schemes

---

[115] Gajendragadkar Commission Report, 42.
[116] Gajendragadkar Commission Report, 43.

and projects as were of an all-state interest. The commission also recommended that at the end of each financial year, as well as at the end of each plan period, these boards would prepare a report indicating financial and physical targets and achievements.[117]

Sadiq took immediate steps to implement the recommendations by establishing the regional development boards and providing reservations to the backward regions and classes. The policy of granting reservation to the Scheduled Castes and backward classes in state services was initiated in 1968, when the government notified reservation of 5 per cent and 2 per cent in state services for Scheduled Castes and residents of Ladakh, respectively. In 1970, the policy received a major fillip when the government announced 8 per cent reservation for Scheduled Castes and 42 per cent for backward classes, including 2 per cent for Ladakh in the state services.[118]

During the period 1961–2 to 1968–9, apart from a non-plan expenditure of about Rs 27.5 million, an amount of Rs 30 million was spent on the development of agriculture, animal husbandry, health, forestry, cooperation, communication, irrigation, and electricity in Ladakh. The Fourth Five-Year Plan outlay for the district was Rs 45 million. The Sadiq government also appointed a 10-member high-powered committee known as the Ladakh Development Committee to advise the government with regard to policies and issues affecting good governance and speedy development of Ladakh.[119]

Towards Transparency, Accountability, and Efficiency

Ghulam Mohammad Sadiq is called aina saz 123 (law maker or institution builder). He is rightly so described for having institutionalized governance to minimize discretionary powers. A series of measures taken in this regard in different sectors are mentioned in the following:[120]

---

[117] Gajendragadkar Commission Report, 43–4.

[118] Government of J&K, *Jammu and Kashmir: 50 Years*, 223.

[119] Kaul and Kaul, *Ladakh through the Ages*, 224.

[120] *Shiraza* (Sadiq Number), Jammu and Kashmir Academy of Art, Culture, and Languages, Srinagar, 1973, 112–13.

Recruitment Policy

Section 10 of the constitution of J&K, which came into force on 17 November 1956, guarantees to the permanent residents of the state all the rights guaranteed to Indian citizens by the constitution of India, including the provisions of Articles 15, 16, and 29. In Section 23, the constitution of J&K lays down the directive principle that the state shall guarantee to the socially and educationally backward sections of the people special care in the promotion of their educational, material, and cultural interests and protection against social injustice.

In June 1956, the state government issued the Jammu and Kashmir Civil Services Rules, in which a clause was included to enable the state government to make reservations in government services in favour of any backward class which, in the opinion of the government, was not adequately represented in the service.[121] Nonetheless, the Gajendragadkar Commission did not find the state government to have issued any clear-cut orders, statutory or otherwise, 'specifying which are the backward classes for the purpose of recruitment/promotions in state services'.[122] There had been some working rule which was subjected to modifications from time to time to suit particular situations. The latest working rule, as per Writ Petition no. 107 of 1965 (*Triloki Nath and Others v. State of Jammu and Kashmir*), was that 50 per cent of the vacancies were reserved for the Muslims of the state, 40 per cent for the Jammu Hindus, and 10 per cent for Kashmiri Pandits. The state justified this reservation policy on the ground that both the Muslims of the state and Hindus of Jammu formed a backward class, and 'none of them is adequately represented in the state services'. As this policy was flawed, the aggrieved Pandit community approached the Supreme Court in *Triloki Nath and Others v. State of Jammu and Kashmir* (Writ Petition no. 107 of 1965). The Supreme Court quashed the promotions made on the basis of the aforementioned working rule.[123]

---

[121] Gajendragadkar Commission Report, 56.

[122] Gajendragadkar Commission Report.

[123] Supreme Court of India, *Triloki Nath and ANR v. State of Jammu and Kashmir and Ors* (April 1968), http://indiankanoon.org/doc/1006376 (accessed 20 November 2017).

Since it was constitutional obligation of the state to make reservations for Scheduled Castes, Scheduled Tribes, and economically and socially backward classes, the Gajendragadkar Commission recommended drawing up afresh the list of backward classes in the state on the basis of multiple criteria—economic backwardness, occupation, place of habitation, average of student participation per 1,000 in the class and caste in relation to Hindus. For determining the criteria for identifying the backward classes, the government on 3 February 1969 appointed the Backward Classes Committee under Justice Wazir. The committee submitted its report on 29 November 1969. Accepting its recommendations, the government on 1 May 1970 announced 8 per cent reservation for Scheduled Castes, and 42 per cent for backward classes, including 2 per cent for Ladakh in the state services.

Sadiq also set up a State Subordinate Services Selection Board for making appointments in the State Subordinate Services, as up to this point only the gazetted appointments had been institutionalized by the establishment of the Public Service Commission during the time of Bakhshi. For recruitment to district services, the job was assigned to district selection boards.[124]

Admissions Policy for Professional and Technical Institutions

By 1965–6, a sufficient number of primary, middle, high, and higher secondary schools had been opened; the number of colleges had risen to 17, and two divisions of Jammu and Kashmir University were running postgraduate classes in many subjects. At the school level or in the arts faculties at the university level, there was no problem with getting admission for studies. Problem were faced in getting admission to professional and technical institutions, as the number of aspirants for these courses greatly outnumbered the availability of seats. To address this problem, the government granted liberal loans to those who got admission in professional and technical institutions outside the state. And

---

[124] Government of J&K, *Ghulam Mohammad Sadiq*, 3–4.

those who belonged to the Scheduled Castes and backward classes were granted scholarships.[125]

Regarding admissions to professional and technical institutions in the state or outside the state (in case of selection under the nomination quota), the government devised a policy in 1963 according to which the principals of the concerned colleges would receive applications, the selection committee framed by the principal framed the list in order of merit, and the merit list was submitted to the chief minister for final selection. However, for vacancies reserved in institutions outside the state, the State Public Service Commission invited applications and submitted a merit list for final selection by the chief minister.[126]

The procedure was modified in June 1966, and a committee of two members was constituted for the selection of candidates for all professional and technical courses within and outside the state. The committee comprised the principal of the concerned college, the secretary of the concerned department of government, and an expert related to the course. The committee scrutinized the applications for each programme as per the set guidelines. Thereafter, the committee would conduct an interview or a written test for which 100 marks were assigned.[127] The list prepared was sent to the chief minister for approval.

The matter was reviewed again in July 1967, and instead of one selection committee for all courses, a separate selection committee was appointed for each of the technical courses. The marks for the interview were reduced from 100 to 50, and 100 marks were allotted for academic merit. Admissions to the Regional Engineering College, being an autonomous body, were made by the admission board constituted by the governing body of the college. The Gajendragadkar Commission described the revised procedure as 'unexceptionable'.[128]

With special regard to the claims of the Scheduled Castes and backward communities or classes for equal opportunity to improve

[125] Gajendragadkar Commission Report, 66.
[126] Gajendragadkar Commission Report, 67.
[127] Gajendragadkar Commission Report, 67–8.
[128] Gajendragadkar Commission Report, 68.

their position, the government had framed a reservation policy for admission to higher educational institutions on the same pattern as was followed for recruitment and promotion purposes. Yet, as the classification of people as 'backward' on the basis of region and religion alone was repugnant in light of the fundamental rights guaranteed by the Indian constitution as made applicable to J&K, the High Court of J&K state, while delivering its judgment in the case of *Lalit Suri Tikku v. State of Jammu and Kashmir and Others* related to admissions to Srinagar Medical College, pointed out that 'the selection has been made on a purely discriminatory basis'.[129] The remedy, therefore, was to draw up a list of backward classes on the basis of a multiple test, as suggested by the Gajendragadkar Commission in the case of reservation policy for appointments and promotions. Having articulated the same policy with regard to admissions to higher educational institutions, the commission recommended the following:

> Reservation of places in educational and professional institutions should be made for the backward classes and the Scheduled Castes in proportion to their respective population, subject to the condition that the total reservation does not ordinarily exceed fifty per cent. The balance of the places in such institutions should be filled strictly and solely on the basis of merit. In filling the places reserved for the backward classes and those reserved for the Scheduled Castes, merit should be the criterion for selecting the persons in each of these categories.[130]

As was done in regard to the reservation policy for recruitments and promotions, the government drew up a fresh list of educationally and economically backward communities on the basis of constitutionally valid criteria and prescribed the same percentage of quotas for Scheduled Castes and socially and economically backward classes as was prescribed for the purpose of appointments and promotions.

---

[129] Quoted in Gajendragadkar Commission Report, 69.
[130] Quoted in Gajendragadkar Commission Report, 98.

Postings and Transfers

In the interest of efficiency, the Sadiq government also devised a new pattern of postings and transfers. The officers heading departments at the district and tehsil levels, including deputy commissioners, were not posted in the districts and tehsils to which they belonged. The only exceptions were teachers, professors, and clerks. The non-gazetted employees working in the Revenue Department and in the police were, however, debarred from being posted in their home villages or the halqas to which they belonged. Women employees were treated exceptionally as they had to be posted near their homes; where both husband and wife were in government service, the woman employee had to be posted at the same place or as near her husband as would be administratively practical.[131] So far as the transfer policy was concerned, it was ordered that transfers should be invariably made after five years, and in no case before two years. In cases of misconduct, postings had to be made to more distant places.[132]

Anti-Corruption Measures

To promote good governance, persons of integrity were encouraged. And to check corruption and malpractice, it was made incumbent upon government servants and ministers to submit statements of their assets. Ministers had also to submit statements of the assets of their dependents and families.[133] Also, ministers and MLAs holding recognized political offices were debarred from buying government property or selling personal property to the government. Where a deviation from this principle became necessary, the advice of the chief justice had to be sought, which was binding on the government.[134]

The seriousness with which Sadiq sought to check corruption and malpractices among government servants is also indicated by

[131] Government of J&K, *Ghulam Mohammad Sadiq*, 3–5.
[132] Government of J&K, *Ghulam Mohammad Sadiq*, 18–20.
[133] Government of J&K, *Ghulam Mohammad Sadiq*, 6–7.
[134] Government of J&K, *Ghulam Mohammad Sadiq*, 7.

the wide powers given to the State Anti-Corruption Commission and the strengthening of the Anti-Corruption Investigating Agency. To ensure speedy action against corrupt officers, the chief secretary was empowered to issue orders of suspension of gazetted officers against whom prima facie cases were established by the Anti-Corruption Commission.[135] Significantly, the government adopted the convention that any recommendation that would be made by the commission in respect of any official, whose case came up before it for probe, would be accepted by it as a rule.[136]

A Central Vigilance Committee comprising the chief secretary, home secretary, forest secretary, secretary works and power, inspector general of police, deputy inspector general of police (Anti-Corruption Organization), and deputy inspector general of police (CID) was formed to coordinate measures taken by various government departments to fight corruption. It had also to ensure compliance with the code of conduct prescribed for government servants, besides reviewing the working of the Anti-Corruption Organization and making sure that the complaints relating to corruption were disposed of promptly.[137]

The government also took steps to eradicate the evil of corruption from public life. The most glaring examples of malpractices related to the misuse of road transport permits and import licences for raw material. While the route permits were given to influential people and to kith and kin of the ruling class, those who actually operated the transport had to pay huge rents to the owners and permit holders. The Sadiq government immediately issued route permits to all such persons who operated vehicles but possessed no permits.[138] Similarly, the issuing of import licences was rationalized to afford adequate opportunities to small industrialists to obtain raw materials for genuine and fair use.[139]

[135] Government of J&K, *Ghulam Mohammad Sadiq*, 7–8.
[136] Government of J&K, *Ghulam Mohammad Sadiq*, 7–8.
[137] Government of J&K, *Ghulam Mohammad Sadiq*, 10.
[138] Government of J&K, *Ghulam Mohammad Sadiq*, 8–9.
[139] Government of J&K, *Ghulam Mohammad Sadiq*, 9.

Updating of the Rationing System

The government also updated the food rationing system, which had been based on obsolete population figures. Further, in smaller towns, there was either no arrangement for rationing or the number of fair price shops for foodgrains was too small to meet the needs of the consumers. In order to tackle this problem, the government took necessary steps. As a result, the ration strength of Srinagar city was increased by about 60,000 and that of Jammu city by about 7,000.[140] Besides, a number of suburban villages with a population of about 6,000 which had not so far been covered under rationing arrangements were also included for the purpose of food rationing. The scale of rations in the mufassils was also increased from 6 kilograms to 12 kilograms per head per month. The number of foodgrain depots was raised from 67 to 143, which covered almost all the areas of the state.[141]

Price Control

Steps were also taken to hold the price line of other commodities, especially mutton, milk, and edible oils. A price stabilization committee was set up to revive the trends of prices from time to time, and advise the government on various measures for the purpose of controlling the prices. A special officer for stabilization and control of prices was appointed. And to ensure a regular flow of supplies of mutton at reasonable prices, 73 fair price shops were opened by the Cooperative Department. The number of milk depots was increased from 54 to 76.[142] While in the rural areas there was a network of cooperative societies, for the urban population wholesale stores and super bazaars were established in most of the districts.

Employment

With a view to increasing employment opportunities in the rural areas of the state, the government launched a works programme

[140] Government of J&K, *Ghulam Mohammad Sadiq*, 12.
[141] Government of J&K, *Ghulam Mohammad Sadiq*, 12.
[142] Government of J&K, *Ghulam Mohammad Sadiq*, 13.

scheme, sponsored and financed by the central government, to provide work for 240,000 man days.[143] This was besides the several ongoing construction projects affording employment to large sections of population. For providing employment to educated youth, employment exchanges were set up in all the districts of the state in addition to the two employment exchanges already functioning in Srinagar and Jammu cities.[144]

Judicial Reforms

Besides the establishment of regional boards and district recruitment boards, the Criminal Procedure Code was amended and sessions courts were set up in all districts of the state to provide speedy and easily accessible justice to the common people.[145]

Agrarian Reforms

True to his socialist leanings, Sadiq adopted the policy of progressive taxation by exempting small landholders assessed below Rs 10 from payment of land revenue, reducing the annual land revenue to the state by about Rs 3 million.[146] The existing land laws of the state had imposed constraints on the commercial banks in financing industrial and agricultural schemes, as the tenants could not mortgage their right of protected tenancy with the banks; therefore, they could not take advantage of the existing credit facility. The Tenancy Amendment Act, 1965, removed this disability, enabling the tenant to transfer land in the form of a simple mortgage in favour of the Jammu and Kashmir Bank or a bank for the time being included in the Second Schedule to the Reserve Bank of India Act, 1934.[147]

---

[143] Government of J&K, *Ghulam Mohammad Sadiq*, 13–14.

[144] Government of J&K, *Ghulam Mohammad Sadiq*, 13–14.

[145] Government of J&K, *Ghulam Mohammad Sadiq*, 16.

[146] Government of J&K, *Report of the Land Commission* (1968), 10.

[147] Jammu and Kashmir Alienation of Land Act, Samvat 1995, Section 4(A).

Economic Development

Consistent with his ideology and the needs of the state and society, Sadiq focused on achieving self-sufficiency in agricultural production, export-oriented production, and generation of internal resources to meet the growing needs. To achieve these targets he gave priority to technological and infrastructural development. The government quickly responded to the new technologies and economic strategies and disseminated them in the society under state patronage. Having an ideological commitment to the nationalization of big industries and a mixed economy, he also extended liberal state patronage to small-scale industries and handicrafts as per the Naya Kashmir Programme. His concern for the poor is reflected in his agrarian policy, schemes launched to provide employment, subsidies, and subsidized loans to the poor. Educated unemployment in Srinagar was posing a challenge, and Sadiq explored all means to respond to it. The developmental and economic measures taken during the period in different sectors of the economy illustrate the efforts made by Sadiq to implement the Naya Kashmir Programme, in the drafting of which he had played an important role.[148]

Agriculture and Allied Activities

The late 1960s saw what is called the Green Revolution in Kashmir, involving the adoption of intensive farming, high-yielding varieties, balanced manuring and fertilization, multiple cropping, improved agricultural implements, bringing new lands under irrigation, huge subsidies on fertilizers, and liberal loans for the sinking of wells and installation of pump sets. Since agriculture still provided a livelihood to the bulk of the people, a substantial part of the state income accrued from this sector. Given the severe financial burden on state resources due to the import of foodgrains and its rationing at subsidized rates, it is no wonder that the Fourth

---

[148] For a detailed report on development in different sectors during Sadiq's period, see Saraf, *Jammu & Kashmir Trade Guide*.

Five-Year Plan launched in 1969–70 put agriculture among the top priority areas. Out of the total outlay of Rs 1,584 million, Rs 301.5 million was earmarked for agriculture-related schemes to ensure an adequate rate of growth in this sector.[149]

The IADP, which was launched in the state in 1961–2 to step up agricultural production through intensive use of all the important farm inputs, was extended to all 13 blocks of Anantnag district and 6 more blocks of Jammu district during 1966.[150] The most important development of the period was the adoption of the New Agricultural Strategy in 1968–9. The New Agricultural Strategy adopted a three-dimensional approach to increase production: (a) use of high-yielding varieties of seeds; (b) adoption of modern chemical technology; and (c) foodgrain support policy. As the high-yielding varieties and chemical fertilizers had brought radical changes in agricultural output elsewhere, the state adopted a pro-active role in disseminating the use of high-yielding seed varieties among the farmers.[151]

For this purpose, the state launched the High Yielding Variety Programme during kharif 1966, and selected two districts (Anantnag and Jammu) previously under IADP. The total area covered in acres during 1966–7 under paddy, hybrid maize, and Mexican wheat was 22,525, 162, and 1,600, respectively. Distribution of farm inputs during 1966–7 in quintals of improved seeds was: paddy (8,429), hybrid maize (45,793), wheat (13,955), fertilizers (167,670), and pesticides (284). Due to the implementation of the High Yielding Variety Programme during 1966–7, 12,000 tons of additional foodgrains were produced in the state.[152]

The Crash Agricultural Production Programme was initiated during kharif 1967 in three blocks of Jammu and six blocks of Anantnag. The area covered in acres by this programme during

---

[149] Government of J&K, *Fifth Five Year Plan* (Srinagar: Planning and Development Department), 9.

[150] Aziz, 'Economic History of Modern Kashmir', 100–1.

[151] Aziz, 'Economic History of Modern Kashmir', 100–1; H. S. Mann, 'Agriculture: An Analysis', in Saraf, *Jammu & Kashmir Trade Guide*, 148–9.

[152] Mann, 'Agriculture: An Analysis', 149.

1967–8 was: paddy (125,100 acres), hybrid maize (5,000 acres), hybrid bajra (1,300 acres), and Mexican wheat (70,116 acres). The target of 85,000 tons of additional foodgrains was achieved during the period. For this purpose, farmers were provided various farm inputs as well as loans for minor irrigation works. The programme was extended to 50 blocks during 1968–9.[153]

In order to assist small and marginal farmers as well as to improve agricultural activities in areas facing irrigation problems, special programmes were sanctioned by the government. Composite projects called Small Farmers Development Agencies were launched in Anantnag and Jammu-Kathua to help small and marginal farmers having up to 5 acres of irrigated land or 6 acres of unirrigated land. During 1970–71, two similar projects known as Marginal Farmers and Agricultural Labourers Development Agencies were started for Baramulla and Poonch-Rajouri districts. The aim was to provide subsidies and loans on a liberal basis to small and marginal farmers and to encourage them to take up activities like development of minor private irrigation facilities, soil conservation, poultry farming, sheep breeding, and dairy farming.[154]

The government placed greater emphasis on the output of foodgrains from existing cultivated lands. It was realized that intensity of cropping can be raised by growing two or more crops a year on the same piece of land. In Kashmir, there were some areas where two crops were raised, but the second crop invariably used to be oilseed crop alone. The experiments however changed the cropping pattern, as Wheat Sonora-64 and Barley after China-1039 were grown to get two cereal crops from 1967–8. This was further improved with the introduction of some more early wheat varieties like Sherbat-i-Sonora, S-308, and so on. Multiple cropping was the next step taken to increase the intensity of cropping. While more than two crops a year is not possible in the Kashmir Valley, Jammu was promising in this direction. For this purpose, the introduction of short-duration varieties coupled with assured irrigation facilities, adequate fertilizers, and facilities for combating

---

[153] Mann, 'Agriculture: An Analysis', 149.

[154] Government of J&K, *Jammu and Kashmir: On Road to Progress*, 9, Archive Reference Library, Srinagar, Document No., 28807 (nd).

pests and diseases was necessary. The scheme was initiated during 1966–7. Wheat and oilseeds were sown on the same lands after the harvesting of paddy for the diffusion of multiple cropping. A 50 per cent subsidy was provided during 1968–9 on the cost of fertilizers, etc., for introducing and implementing the programme.[155]

Apart from introducing the high-yielding crops just mentioned, many research schemes were started to improve the production of sugar beet, oilseeds, saffron, groundnut, and soya beans. Research schemes were introduced for the growing of sugar beet (in 1965), improvement of oilseeds, determining the optimum dose of fertilizers for groundnut cultivation in the Jammu region, development of soya bean in 1967, adaptability and acclimatization of saffron in different parts of the valley other than Pampore in 1968, and examining the adaptability and acclimatization of the hops plant in 1969–70.[156]

Another important step taken to increase the output of foodgrains was to bring more land under irrigation. During the Fourth Plan period, it was proposed to bring 100,000 acres more under irrigation, and cover the entire irrigated area by high-yielding varieties of food. In the said plan, a substantial increase was made for different irrigation schemes. While in Kashmir the focus was on stabilization of existing works, the converse was the case vis-à-vis Jammu where more attention was paid to creating additional irrigation facilities, including the major project on Ravi-Tawi.[157]

Under a centrally sponsored scheme, the government established the Agro-Industries Development Corporation whose main aim was to bring tractors within the reach of the poor strata of the agricultural community. The corporation started with 40 tractors in 1970, which rose to 113 in 1972. The raising of commercial crops also saw a great spurt during the period. While in Jammu sugar beet and soya bean cultivation was encouraged by state

---

[155] Mann, 'Agriculture: An Analysis', 146–7.

[156] Mann, 'Agriculture: An Analysis', 150–51.

[157] 'Development of Irrigation and Flood Protection Measures Undertaken in Jammu and Kashmir State', Department of Irrigation and Flood Control, Government of J&K, cited in Saraf, *Jammu & Kashmir Trade Guide*, 295.

patronage, plantation of fruit trees, especially apple, was the priority of state and society.[158]

Horticulture also received the special attention of the government. During 1967–8, the State Horticulture Department started the Area Development Programme in collaboration with the Agricultural Refinance Corporation, under which potential areas other than irrigated belts were brought under fruit culture. Under this programme, the Agricultural Refinance Corporation bore 75 per cent of the total share of the loan advanced, and the state government provided the remaining 25 per cent. From 1967–8 to 1968–9, an amount of Rs 4,662,827 was advanced under the scheme, and 9,817 acres of land were covered under the programme.[159] Due to these efforts, an area of about 6,000 acres was brought under fruit cultivation during 1969–71, while the target fixed for the year 1971–2 was 8,000 acres.[160] In the Fourth Plan, the ban imposed on conversion of paddy land into orchards or for growing other crops was removed essentially for promoting horticulture. About 8,000 to 9,000 acres were brought under orchards every year. In order to meet the growing demands of orchardists for plant material, a number of plant nurseries were set up. The number rose from 17 in the pre-plan period to 72 by the end of 1968–9.[161]

One of the most important measures taken by the government was the opening of training schools to train young farmers in improved horticultural practices. This process was started during the time of Bakhshi Ghulam Mohammad and carried forward by the Sadiq government. Under this initiative, 633 farmers were trained during the period 1956–7 to 1968–9. During the Fourth Plan period, 1,000 young farmers were trained in different horticultural development programmes. In this regard, one school at Sopore and three in Jammu division were opened during 1969–70.[162] Exhibitions

---

[158] Aziz, 'Economic History of Modern Kashmir', 118–19.

[159] Government of J&K, *Jammu and Kashmir: On Road to Progress*, 11.

[160] Government of J&K, *Jammu and Kashmir: On Road to Progress*, 11.

[161] Government of J&K, *Jammu and Kashmir: On Road to Progress*, 11.

[162] Government of J&K, *Jammu and Kashmir: On Road to Progress*, 12.

were held throughout the state with the aim of educating orchard-ists in scientific methods of production. Awards were given to dis-tinguished orchardists. An intensive Fruit Production Programme was started in 1969 in order to persuade farmers to adopt improved practices of fruit cultivation, such as fertilization, pruning, protec-tion of orchards from pest and diseases, and so on. For preservation of fruits, a Community Canning and Fruit Preservation Centre was established at Lalmandi, Srinagar, in 1966–7.[163] As a result of these measures, fruit exports showed remarkable improvement. Exports rose from around 23,000 tons in 1960–61 to approximately 73,500 in 1969. During 1970, the exports reached an all-time high of 100,000 tons.[164]

Animal Husbandry

A new chapter was opened in the sphere of animal husbandry in the state with the launching of the Fourth Five-Year Plan. The inadequacy in providing veterinary aid was made up for during the period, with the implementation of Intensive Cattle Development Projects (ICDPs) in both Jammu and Kashmir regions to meet the standards recommended by the Royal Commission of Agriculture appointed by the Government of India.[165] To save the farmer from incurring heavy losses of livestock, an institute for the manufac-ture of biological products to protect livestock and poultry against major diseases was established at Srinagar. This institute assumed considerable significance with the launching of ambitious poultry and cattle development schemes under Crash Programme in the Fourth Plan.

During the Fourth Plan, two ICDPs and one medium project were established covering 250,000 cows and buffaloes of breeding age. Each ICDP had 100 sub-centres covering a cow population of

[163] Government of J&K, *Jammu and Kashmir: On Road to Progress*, 12.

[164] Government of J&K, *Jammu and Kashmir: On Road to Progress*, 12.

[165] B. A. Shirazi, 'Animal Husbandry', in Saraf, *Jammu & Kashmir Trade Guide*, 154–5.

100,000 with a view to bringing about quicker transformation of cattle wealth towards increased milk production. Cross-breeding operations were propagated through the adoption of artificial insemination technique using pedigreed Jersey bulls. The livestock experts recommended the use of milch breeds such as Sindhis in the valley and Sahiwals, Therparkars, and Murrahs in Jammu. To facilitate the production of pedigreed bulls of these breeds, Jersey farms were established at Manasbal and Sindhis at Rambirbagh in the valley, besides Sahiwals, Therparkars, and Murrahs in Jammu.[166]

Industrial Sector

The Naya Kashmir Programme *inter alia* emphasized the nationalization of important industries, introduction of new industries, and promotion of small-scale and traditional crafts to realize its dream of shaping an economically prosperous Kashmir. The growth of population, pressure on land, and educated unemployment particularly in Srinagar and other major towns due to easily accessible education, now made it urgent to create new and productive avenues for non-agricultural employment. For many reasons, especially political, geographical, and infrastructural, the position of the industrial sector was not satisfactory. It was therefore necessary to give a great push to the sector without, however, compromising the basic developmental ideology along the same socialist lines as obtained in the rest of India. Carrying forward the policy pursued by his two predecessors, Sadiq gave a further impetus to the small-scale industrial sector and handicrafts, besides strengthening the policy of nationalization of important industries and forests. The steps taken in this regard are outlined in what follows.

Small-Scale Industrial Sector

Notwithstanding the steps taken by the Bakhshi government to provide facilities and concessions to offset local disadvantages

---

[166] B. A. Shirazi, 'Animal Husbandry', in Saraf, *Jammu & Kashmir Trade Guide*, 154–5.

so as to create desired interest both among local and outside entrepreneurs, the condition of the small-scale industrial sector was still dismal. The contribution of factory establishments to state income was only 2 per cent as compared to 13.85 at the national level.[167] To overcome this problem, the Sadiq government announced a package of incentives to provide facilities and concessions for entrepreneurs to establish units in the state in accordance with an approved set of schemes. A vital concession was granted to the entrepreneur in regard to the lease of industrial plots. Industrial areas were developed at suitable places and leased to entrepreneurs for the construction of work sheds. These were leased for a period of 75 years in all, with two optional renewals after 25 years. The lessee entrepreneur had to pay only a nominal premium of Rs 1,000 per kanal and ground rent of Rs 100 per kanal per annum. This concession was applicable to both local and outside entrepreneurs.[168]

To overcome the deficiency of power supply, the entrepreneurs were helped with free-of-interest loan assistance to acquire diesel-generated sets as a substitute for power. Seventy-five per cent of the loan was made recoverable in easy instalments over a period of seven years, and 25 per cent of the loan was treated as subsidy.[169] Further, the government provided 50 per cent reduction in the transport costs of finished goods, provided they were booked through the government transport undertaking. Similar concessions were also granted on the import of raw materials. Concessions by way of exemption of surcharge on road toll and payment of central sales tax were also announced.[170]

In the preparation of feasibility and project reports of identified new industries, the government extended all support to the

---

[167] S. K. Sehgal, 'Strategies for the IV Plan', in Saraf, *Jammu & Kashmir Trade Guide*, 113–14.

[168] W. S. Tambe, 'Small Scale Industrial Sector', in Saraf, *Jammu & Kashmir Trade Guide*, 125–7.

[169] W. S. Tambe, 'Small Scale Industrial Sector', in Saraf, *Jammu & Kashmir Trade Guide*, 125–7.

[170] W. S. Tambe, 'Small Scale Industrial Sector', in Saraf, *Jammu & Kashmir Trade Guide*, 125–7.

entrepreneur and bore the cost to the tune of Rs 10,000 as a token of initial help to them. Facilities were also extended for training managerial and skilled labour required for an industry. In order to provide ready capital, the state government floated the Industrial Development Corporation with an authorized capital of Rs 15 million.[171] This was in addition to the facilities already existing from the State Financial Corporation and scheduled banks. With regard to technically qualified personnel, the government considered additional liberalized facilities and concessions so that they were able to actively participate in the development of small-scale industries without being called upon to share the burden entirely on their own in the initial period of setting up of new ventures. In 1968, the government made a concerted drive for the development of this sector by launching what was called the Intensive Campaign in the valley as well as in Jammu in collaboration with the Development Commissioner, Small-Scale Industries, Government of India, and the National Small Industries Corporation of India. This move resulted in a big response, as around 700 applications were received from local and outside entrepreneurs to establish new ventures. These applications were processed and sanctioned on the spot and machinery worth over Rs 40 million was approved to be supplied on hire-purchase bases.[172]

Handicrafts

A number of steps were taken for the promotion of handicrafts which, according to the Naya Kashmir Programme, did not only constitute an important source of income, but formed a part of Kashmir's precious heritage, an expression of Kashmir's love for beauty, and an identity marker of its civilizational background.

It goes without saying that the progress of the handicrafts depended, as it does today, on their export demand. The maintenance of the export demand *inter alia* depended on the awareness of craftsmen and traders about the changing tastes of customers

---

[171] W. S. Tambe, 'Small Scale Industrial Sector', in Saraf, *Jammu & Kashmir Trade Guide*, 125–7.

[172] W. S. Tambe, 'Small Scale Industrial Sector'.

at home and abroad, and their readiness to provide new designs to satisfy these changing passions. The previous government had established a school of design which, by the beginning of the Fourth Plan period, had developed over 1,700 new designs.[173]

In the Fourth Plan period, additional sections were covered by the school in tune with new tastes and demands in fashions, motifs, and patterns. Besides, an experimental services centre for handlooms was organized to form a part of the activity of the school for providing new designs to the handloom industry, particularly in the manufacturing of tweeds, for which the state had its own place in the market.[174]

It was felt that some of the crafts had practically died out for want of skilled workmen. An extensive training programme was launched to meet this challenge. More than half a dozen training centres were established in various parts of the state and over 200 new hands were imparted training through these centres annually.[175] The large number of trainees was aimed at decentralizing handicraft industries in rural areas, then mainly confined to Srinagar and to some extent Jammu. The government also helped the trained hands to organize their own production centres by giving financial assistance, for which separate provision was made in the Fourth Plan.

With the growing demand in the post-independence period, the quality of handicrafts had deteriorated in some areas in the misplaced quest among dishonest craftsmen, especially the traders, to kill the goose that laid golden eggs. Fortunately, the state immediately realized its role and worked out a scheme to introduce quality in the sales of handicraft goods. The scheme incorporated specifications of materials used in production, frequency of inspection, sampling details, and level of control requirement for testing products at various steps.[176] The namda rug, which was

---

[173] J. M. Mengi, 'Development of Handicrafts and Export Promotion', in Saraf, *Jammu & Kashmir Trade Guide*, 175.

[174] J. M. Mengi, 'Development of Handicrafts and Export Promotion'.

[175] J. M. Mengi, 'Development of Handicrafts and Export Promotion', 176.

[176] J. M. Mengi, 'Development of Handicrafts and Export Promotion', 176.

one of the important foreign exchange earners, came under quality control, after which its export went up considerably.[177]

An important development of the period was the establishment of the State Industrial Development Corporation, with an export promotion wing to render overall assistance to potential exporters in the field of foreign trade.[178] Experts with specialization in foreign trade were always available in the office of the corporation for guidance regarding the potential demands of overseas markets, names of importers, and the nature of the competition. The corporation also functioned as a link between the exporter and buyer, and rendered services to the exporters right from the preparation of documents to the pre-shipment inspection stage and other formalities. It also used to keep in touch with the handicrafts and Handloom Export Corporation of India and various export promotion councils in the interest of boosting exports of products from the state. The corporation had also on its agenda the setting up of an export house in the state to eliminate delays and unhealthy competition and to improve selling efficiency. It also established showrooms of standardized Kashmiri goods in some important foreign trade countries and participated in industrial exhibitions and fairs both within and outside the country.[179]

The government also encouraged the establishment of the Export Corporation in the private sector to help bring the state on the export map of India. Also, seminars were organized in which central agencies dealing with the export trade were persuaded to participate to explain the different types of assistance available to traders and provide expert guidance regarding the solution of problems.[180] An exhibition of exportable items was also held to provide first-hand knowledge to the visiting party about exportable products.

---

[177] J. M. Mengi, 'Development of Handicrafts and Export Promotion'; Habibullah Mir, 'Kashmir's Export Potential', in Saraf, *Jammu & Kashmir Trade Guide*, 173–4.

[178] Mir, 'Kashmir's Export Potential', 174.

[179] Mir, 'Kashmir's Export Potential', 174.

[180] Mir, 'Kashmir's Export Potential', 174.

Public Sector

There was a fair amount of industrial activity in the state sector. The J&K Industries and J&K Minerals were running several units, including manufacturing woollens, silk, cement, briquettes, rosin, turpentine, sports goods, matches, and furniture. Programmes for modernizing and expanding many of these units were undertaken during the period. Sadiq brought more economic sectors under the state sector or reinforced their position as vital organs of the economy and social service. Of the additions, mention may be made of the Jammu and Kashmir Projects Construction Corporation Ltd (JKPCC), for undertaking the construction of major government projects at an accelerated pace and at reasonable costs.[181] Switching over to the public contract undertaking, the JKPCC was established in May 1965. Although the lumbering project—a step towards the nationalization of the forest industry—had been established in 1960–61, it started its meaningful work of extraction of timber from the time Sadiq assumed power. As against 129,603 cft in 1962–3, 710,544 cft of timber were extracted in 1968.[182] From 1967, the work of supplying firewood ration was also entrusted to the lumbering project.[183] This spectacular increase was the result of two factors: mechanization of extraction of timber, and the abolition of forest leases and making the timber trade a state enterprise.

The Government Transport Undertaking was one the major industries in the public sector both from the point of view of revenue as well as from its service as a public utility concern. Significantly in 1965, it yielded a profit of 6.4 million against an annual loss of up to Rs 2.3 million in the past.[184] Ever since its inception, the undertaking had depended for its extension entirely on its own sources. It was only from 1964–5 that funds were allotted to the undertaking under the plan scheme to enable it to implement schemes like the Nationalization of Passenger Services

---

[181] Saraf, *Jammu & Kashmir Trade Guide*, 298.

[182] Saraf, *Jammu & Kashmir Trade Guide*, 13.

[183] Saraf, *Jammu & Kashmir Trade Guide*, 13.

[184] J. S. Jamwal,'Transport Undertaking', in Saraf, *Jammu & Kashmir Trade Guide*, 206.

on the highway, expansion of its fleet strength, construction of a central workshop for the maintenance of the fleet on scientific lines, construction of depot offices, and also the development of city/district and long route services.[185] A marked expansion was also made in regional workshops. Steps were taken to start, with immediate effect, inter-state services between Jammu and Delhi and between Jammu and important cities of Punjab. This was followed by a direct service between Srinagar and Delhi.[186] This connectivity afforded a vast scope to the traders in their businesses. A direct goods service to various important business centres outside the state particularly facilitated the fruit industry of the state with a cheap and fast transport system.

Electricity

One of the serious bottlenecks in the development of industrial sector in Kashmir has always been deficiency of adequate availability of electricity. While referring to the problems faced by industrial sector the Commissioner Planning and Development of Sadiq government, S.K. Sehgal, very confidently stated, "all these problems can be overcome if electricity could be made available in abundance and comparatively at low rates" (Sehgal, 'Strategy for IV Plan', 114–15.). That is why it was agreed that in the IV Plan the highest priority had to be given to building up the generation, transmission and distribution capacity. It is true that the outlay on the development of power generation had been increasing from plan to plan. Between the III Five Year Plan and the Beginning of IV Five Year Plan (including the plan holiday of 3 years) about 25 crores were allotted for power generation. For the IV Plan the outlay for the purpose was about Rs 37 crores. During the period of Sadiq hydel projects at Chenani and on upper Sind were completed and executed. This increased the installed capacity by 46 MW. A thermal station at Kalakote was also built and completed. Moreover a 96 MW station on the Jehlum river at Baramulla was

---

[185] J. S. Jamwal, 'Transport Undertaking'.
[186] J. S. Jamwal, 'Transport Undertaking'.

under execution. It was in 1970 that the work on 360 MW Salal project was started. At the same time great attention was paid to the provision of an adequate transmission and distribution system. A 132 KV line was constructed between Chenani and Srinagar. A 220 line was added to the existing 132 KV line between Pathankot and Jammu. Jammu was already connected with Chenani. As such the Valley was connected with Punjab which considerably improved the electricity needs of the state.

Roads

Along with the need for developing electricity, the development of the communication system was considered by the policy planners of the time as equally important. And in the peculiar geophysical conditions of Kashmir, the road system had to meet the entire transportation needs, and it had to be more intensive than in the plains. As in the case of power generation and transmission, so in laying the road network, the Sadiq government showed a remarkable interest. This is evident from the plan outlays for the purpose. In the Third Plan 88.4 million was spent on road construction; in the subsequent three years the expenditure was over 70 million, and in the Fourth Plan there was an outlay of 180 million for the purpose. Significantly, a major share of the amount earmarked for the purpose of construction of roads was spent in connecting the far-flung areas of Jammu, Ladakh, and Kashmir with nearby urban centres, as the objective behind road building was to develop the backward areas and give a push to the full exploitation of forest resources, and develop tourism, industries, horticulture, and agriculture.[187]

Social Services Sector: Education, Health, and Drinking Water

Massive expansion at the lower level without losing sight of qualitative changes, and minor expansion at the higher level but with

---

[187] R. C. Bhargava, 'Economic Background', in Saraf, *Jammu & Kashmir Trade Guide*, 121.

greater focus on consolidation, guided the educational policy of the Sadiq government. The broad objectives of educational policy, namely, cultivating a scientific temper, tolerance, harmony, patriotism, proper channelization of the energies of the youth, employability, and equitable development of all sections, especially the economically and socially weaker sections, had maintained continuity ever since the transfer of power to 'popular' rule. It may be recalled that Sadiq was the education minister during the Bakhshi period.

At the primary level, the target was to achieve 100 per cent enrolment of boys and at least 70 per cent enrolment of girls. After the efforts made to achieve this objective, we see a considerable increase in the percentages. Thus, whereas the number of primary schools for boys and girls was 940 and 1,755, respectively, in 1960, the numbers increased to 3,648 and 1,453 respectively in 1968–9, that is, before the implementation of the Fourth Five-Year Plan. The number of boys' secondary schools and girls' secondary schools increased from 204 and 46 respectively in 1961 to 443 and 118 respectively in 1970–71.[188] Priority was given to schemes designed to bring about qualitative improvement in the schools and the standards of teaching.[189] Science teaching was encouraged, the backlog of untrained teachers was completely wiped out, and adequate equipment was provided in the schools. Radio Schooling Broadcasts on various subjects was introduced in 1969–70 in order to impart quality and updated knowledge.[190] For the first time in the state, a Sainik school was started at Nagrota in Jammu at the cost of Rs 6.8 million. Work on a similar residential school at Manasbal was also started. One of the important objectives of education was to produce and promote the emotional integration of Kashmir with the Indian Union. In his report, the director of education says:

National integration is now an aim of education and various activities and techniques are being introduced and adopted in the schools

[188] Rasool and Chopra, *Education in Jammu and Kashmir*, 38.

[189] Sehgal, 'Strategies for the IV Plan', 116.

[190] G. R. Azad, 'Education', in Saraf, *Jammu & Kashmir Trade Guide*, 289–92.

to achieve this aim. Writing of uniform textbooks has been adopted on an all India level pattern and textbooks have been nationalized from 1st to 8th class. Ideals of one Nationhood and Secularism in the country are stressed in the context of the textbooks.[191]

While the focus of the technical education programmes was mainly on consolidation and improvement, two non-technical colleges—one in Srinagar and another at Jammu—were opened in view of the lack of scope for expansion of the existing colleges. The number of colleges grew from 4 in 1947–8 to 18 in 1969–70, including 5 for women.[192] The enrolment in colleges went up from 10,511 in 1965–6 to 18,055 in 1969–70.[193] Special attention was also given to training the teachers. By 1969, there were 19 teachers' training schools, besides two teachers' training colleges and a State Institute of Education.[194] Further, in 1969, the Jammu and Kashmir University was bifurcated into two separate universities—the University of Kashmir at Srinagar and the University of Jammu at Jammu.

To promote higher education, especially technical and professional education, governments prior to Sadiq's provided scholarships to the weaker sections and loans without interest. This policy was maintained by Sadiq and approvingly mentioned by the Gajendragadkar Commission Report, which includes details that one is tempted to quote verbatim:

> Besides making education free from the kindergarten to the University stage, the state government have also been granting scholarships and study loans on a liberal scale. The scholarships are available to members of the Scheduled Castes and backward classes and to those of the remaining students who belong to the 'lower income group' as defined by the government of India. Educational loans are granted to students who are selected or nominated for various training and technical courses in and outside the state in terms of the Educational

---

[191] G. R. Azad, 'Education'.

[192] Government of J&K, *Jammu and Kashmir: On Road to Progress*, 47.

[193] Government of J&K, *Jammu and Kashmir: On Road to Progress*, 47.

[194] Government of J&K, *Jammu and Kashmir: On Road to Progress*, 46.

Loans Rules promulgated by the state government in 1964. These loans are granted to hereditary state subjects, 'for courses of study, education or any other training approved by them'. They carry no interest. The rate of loan admissible is normally Rs. 1,500 per annum for diploma courses, Rs. 2,500 per annum for technical/professional degree or post-graduate courses and Rs. 1,500 per annum for other courses. The initial grant of loan is also dependent upon the income of the parents or guardians of the students and the number of children dependent on them. Students whose parents or guardians have a monthly income exceeding Rs. 1,000 are not generally eligible for study loan. Loans are also granted to students who secure admission privately in various courses of studies such as medicine, engineering and post-graduate courses in science subjects. The payment of the loanees is generally made for a period between 2 and 5 years, while the period for recovery or loan extends up to 12 years.[195]

The overall budget for education went up from Rs 3.35 million in 1947–8 to around Rs 90 million in 1970–1.[196]

The health sector was further expanded and improved, particularly by providing an adequate strength of trained nurses and paramedical staff. The bed strength was increased to make a bed available every 1,000 people. More than 100 new dispensaries were opened, besides taking up the programme to control communicable diseases in the centrally sponsored sector.

About 700,000 rural people were provided drinking water by 1968–9. The target of the Fourth Plan was to provide 700,000 more rural people this basic facility,[197] as the dominant village population continued to remain deprived of this facility till the 1970s.

* * *

In 1969, the weekly *Āina* published from Srinagar brought out a special issue on *shakhsiyat* (personalities) which *inter alia* included a sketch on G. M. Sadiq under the title 'Āina-Saz' (law

[195] Gajendragadkar Commission Report, 66.

[196] Government of J&K, *Jammu and Kashmir: On Road to Progress*, 45.

[197] Sehgal, 'Strategies for the IV Plan', 116.

maker). The author makes a comparison between Sadiq and his predecessor, Bakhshi, as rulers, throwing light on the differences in governance under the two heads of government. Besides mentioning many other differences, the author writes:

> In comparison to Bakshi Sahab he [Sadiq sahib] is more civilized [muhazab], educated and humble [shaista] person. But in comparison to him, Bakshi Sahab was more practical and realistic. Sadiq sahib is man of rules and regulations; and generally speaking he does not go against rules. Bakshi Sahab's person [zat] in itself was a rule [qaida] and regulation [zabita], and he did not care for any rules and regulations. Bakshi Sahab was a generous [farakh dil] and bountiful [sakhi] person. This cannot be said about Sadiq Sahib. During the period of Bakshi Sahab no one could dare to raise eyebrows even against any personal servant of any member of Bakshi dynasty. On the contrary, during the period of Sadiq Sahib, he himself is being criticized, reproved and rebuked. During the period of Bakshi Sahib it was not only difficult but also impossible to see Shaikh Sahab and his fellow travellers taking part in political activities. During the rule of Sadiq Sahab, Sheikh Sahab and his companions are free [to indulge in political activities].[198]

'But the special quality of Sadiq Sahab', continues the writer, 'which impressed me the most is his courage of conviction. Without caring for the consequences, he sticks to his convictions.'[199] Indeed, Sadiq was a man of principle, but in the wrong place and at the wrong time. He was the ruler of a state where his ideals and the ground realities of the place militated against each other. This fact is best captured by his well-wisher, Prem Nath Bazaz, who was outraged by his integrationist policies which gave no cognizance to the collective aspirations of the Muslims and failed to 'strike a balance between the demands and emotions of the Indians and the aspirations, urges and sentiments of the Kashmiris'.[200] Bazaz also

---

[198] Quoted in *Shiraza* (Sadiq Number), Jammu and Kashmir Academy of Art, Culture, and Languages, Srinagar, 1973, 112–13.

[199] Quoted in *Shiraza* (Sadiq Number), Jammu and Kashmir Academy of Art, Culture, and Languages, Srinagar, 1973, 112–13.

[200] Bazaz, *Kashmir in Crucible*, 233–4.

contested the charges of parochialism levelled against Kashmiri Muslims for their political demand informed by ethno-nationalism: 'It is unfair to accuse Kashmir patriots of parochialism or narrow mindedness when they display communal tendencies as long as Indian Nationalism, despite its tall presumptuous claims, itself remains based on religious beliefs and Hindu mythology.'[201]

Sadiq also ruled at a time which did not favour his idea of 'liberalization'. Neither of the two rulers of post-1947 Kashmir who preceded Sadiq had faced such a terrible challenge of political instability as he did. Anti-Indian forces—both from within and without—created a condition of war against the Indian state. Kashmir burst out like a volcano. In fact, given the conditions of Kashmir, there was wide scepticism in India about the efficacy of the policy of liberalization. But Sadiq received the full support of Nehru. However, with Nehru's death, the scenario changed. Forces ranged against liberalization came to power at the centre, and the Hindu nationalists raised a hue and cry following the mass mobilization by activists of the Plebiscite Front. And Sadiq was forced to apply a break to his policy of liberalization. The jails and interrogation centres which were emptied by Sadiq immediately after he assumed power, came to be filled up again. In fact, the mass antipathy against India has been so deep-rooted and pervasive in Kashmir that once the state tests its liberal principles in Kashmir, the valley erupts into mass protests, which the governments have never tolerated as they attract international attention. Hence the state has always preferred and continues to prefer the policy of curbing civil liberties in Kashmir. This was done ultimately by Sadiq too, who in this regard was a staunch loyalist of the Indian nation-state. The reason for not allowing free and fair elections was also the same, as in the words of Bazaz, 'if free elections are held, it may be taken for granted that the majority of the seats will be captured by those unfriendly to India.'[202]

What is, however, important about Sadiq is that he was sympathetic to his people, and instead of treating the secessionists as enemies, he tried to win them over by persuasion and patronage.

[201] Bazaz, *Kashmir in Crucible*, 233–4.
[202] Bazaz, *Kashmir in Crucible*, 157.

Apart from liberalism, Sadiq stands out for his policy of clean and transparent government and pro-poor policies.

Yet liberalization, clean government, and the developmentalism of the benevolent state did not constitute more than an attempt to manage the orderly disorder. The radicalization of youth, so profusely borne out by the underground activities during the period, clearly pointed to the direction in which Kashmir was moving more rapidly from the mid-1960s. The political discontent was fuelled by communal riots in many Indian towns, interference of communal forces in Kashmir, and the communalization of the inter-faith marriage of two grown-up individuals. Sadiq's willing cooperation in bringing about the 'forced integration' of Kashmir into the Indian Union added fuel to the fire.

Mir Qasim (Chief Minister, 12 December 1971– 25 February 1975)

On 11 December 1971, G. M. Sadiq passed away, and on 12 December 1971 Sayyid Mir Qasim, cabinet colleague of Sadiq and one who shared his ideological moorings, took over as chief minister. Immediately after assuming office, Mir Qasim issued his policy statement in which he promised that the welfare of the people would be his guiding principle and that he would try to bring back to the national mainstream all those who had been alienated in the past. He promised a policy of tolerance and conciliation towards those who had become rebels in the past and assured full freedom of expression.[203] Also, he sternly warned the heads and secretaries of state departments that he wanted a clean administration.[204] Nevertheless, the government of Mir Qasim was not ready to allow those parties to take part in elections whose stand on Kashmir's accession with India was not clear. That is why it did not permit the Plebiscite Front to fight elections. The Front was banned and its leaders were arrested.[205] Yet in the process of

---

[203] Qasim, *My Life and Times*, 131.
[204] Qasim, *My Life and Times*, 131.
[205] Qasim, *My Life and Times*, 132.

bringing Sheikh Abdullah into the mainstream, Mir Qasim played an important role; he even readily agreed to step down and leave the chair to Abdullah, though many of his colleagues in the cabinet were not in favour of this. Mir Qasim was also not ready to allow Jamat-i-Islami to join electoral politics; but the compulsions of Indian politics, where disqualifying the RSS for elections was not possible, Qasim had to reluctantly give in.

Mir Qasim adopted a conciliatory attitude towards those members of the underground organization, Al-Fatah, who had been arrested during the period of Sadiq. To quote Mir Qasim:

An exemplary punishment would do no good either to society or to these young people. I, therefore, wanted to reform them. I called the DIG, Kashmir Police, to explain my stand on this matter. We released the youths, whose crimes were minor, and sent the rest to the Central Jail in Srinagar, where special care was given to their health and education. They were given all the facilities to prepare themselves for school or university examination. Kashmir University and the Board of School Education were instructed to make arrangements for holding examination in the jail. Science students were taken to the laboratory of the S. P. College, Srinagar, for practicals. I am happy to say that this policy brought a positive revolution in the lives of the condemned youth. The Director General of Police, Pir Ghulam Hasan Shah, played a remarkable role in this mission.[206]

Like his predecessors, Mir Qasim took special interest in agrarian reforms. He abolished land revenue on agricultural produce and passed the Jammu and Kashmir Agrarian Reforms Act of 1972 with the objective of rationalizing the existing ceiling and conferring ownership rights to the tillers of land.[207] The act extinguished the right to ownership of any person in land not held by him for personal cultivation on 1 September 1971. Ownership rights were

[206] Qasim, *My Life and Times*, 129.

[207] In this regard, the J&K legislature adopted a bill in October 1972 known as State of Jammu and Kashmir Agrarian Reforms Act of 1972. Government of J&K, 'New Agrarian Reform in Jammu and Kashmir: Salient Features' (Srinagar: Department of Information), 2.

conferred on tillers on payment of levy in easy instalments and at rates lower than market rates. A ceiling of 12½ standard acres was prescribed for all those who were owners of land.

Many long-term projects initiated during Bakhshi and Sadiq's time were completed during the period of Qasim. The new projects started by him include the Tawi Lift Irrigation Project, the Nala Mar Project, and the construction of new roads. Almost all the phases of the Jhelum Hydel Project were completed besides expanding the Uppper Sind Hydal Project and the Chanani Hydel Project.[208] The Dodh Ganga Water Supply Project, started in 1969, was completed, and so was the Nisbal Water Supply Scheme.[209]

Mir Qasim also paid attention to the development of education and solving the problem of educated unemployment. Significantly, all the graduates who had obtained their degrees till 1972 were employed without any conditions.[210] To ensure a clean administration and to solve the problems of the people, Mir Qasim used to meet the people every day early in the morning. He also reconstituted the Anti-Corruption Department.[211]

The faction-ridden Congress party with no social base in Kashmir Valley, save its dependence on goons, coupled with Mir Qasim's disinterest in administration caused by the inevitability of Abdullah's eventual resumption of power (in which Qasim also played a role), promoted *goondas* (goons) in society who, under the patronage of ministers and other officers, had a field day. According to a contemporary bureaucrat, 'As they [ruffian elements] enjoyed the protection of minister concerned and divisional officers, they had become quite unruly and a pain in the neck for the government.'[212] Mir Qasim very tactfully brought them around and absorbed some in government services and some in gainful self-employment enterprises.[213]

[208] Qasim, *My Life and Times*, 144–5.

[209] Qasim, *My Life and Times*, 145.

[210] Qasim, *My Life and Times*, 145.

[211] Qasim, *My Life and Times*, 145.

[212] Ahmad, *My Years with Sheikh Abdullah*, 43.

[213] Ahmad, *My Years with Sheikh Abdullah*, 43.

5

Kashmir Summer (1975–89)

After a long struggle of more than two decades for the right to self-determination under the inspiring leadership of Sheikh Mohammad Abdullah, the 'lion' finally agreed to climb down, and a peace deal was signed between the prime minister of India, Indira Gandhi, and Sheikh Abdullah, known as the Indira–Abdullah Accord of 1975. The Accord, however, did not concede any demand of Abdullah. The prime minister vehemently stated that the 'clock cannot be put back'. Finally, Abdullah had to be contented with holding state power as chief minister. There was an apparent peace for around a decade, and with it development works also received a great fillip. However, the deep-seated collective political sentiment promoted for a long period of 22 years did not allow the 'peace' to be sustained for long. India's refusal to respond creatively and Abdullah's appetite for power without considering the march of Kashmir history—the march in whose characteristic momentum he had played the role of an engine—created an unimaginable crisis—unimaginable because the 'peace' attained during Sheikh's time was so deceptive that the past seemed to have been completely buried beneath existential pursuits, and any talk of the 'political rights' of Kashmir apparently had hardly any takers among the general public, except a tiny minority.[1] True, the collective political sentiment of any given society is the 'conscience of conscience'. This is Hegel's

---

[1] Based on my interviews with contemporaries of the time.

'Collective Will of the Society'. But this will remains only a sentiment till it is represented by someone. The one who asks the society what its will is and represents this will, he becomes the 'great man'.[2] As long as Sheikh Abdullah represented the collective sentiment, he was the Sher-i-Kashmir (lion of Kashmir). Now, the sentiment came to be represented by other varieties of leadership. And to the surprise of Kashmir watchers, the new leadership won a mass response immediately after the demise of the Sheikh, though the dissent was born in 1975 itself.[3]

After studying the rise and fall of the major civilizations of the world, Arnold J. Toynbee, the famous historian-philosopher of the twentieth century, concluded that breakdown occurs when a civilization fails to respond creatively to its challenges. Instead of changing with the times, the declining civilization idolizes, say, an outworn technique, or institution, a form of government, a particular type of marketing, and so on, which was effective for one challenge but not for the next one. 'A society may disintegrate because it adopts a "time machine" with which to escape into a future or with which to bury itself in a past.'[4] The efforts to peg a given civilization in its present stage ultimately give birth to a 'time of troubles'—rebellions and wars break out all over the place.[5] To overcome this problem, a 'universal state' is established to put down these wars. The 'universal state' appears to be

---

[2] Hegel's classic description of a great man is as follows: 'the great man of the age is the one who can put into words the will of his age, tell its age what its will is, and accomplishes it. What he does is the heart and essence of his age; he actualizes his age.' *Hegel's Philosophy of Right* (English translation) S.W. Dyde, 1896, G. Bell, London, p. 307; Avaliable at https://www.marxists.org/reference/archive/hegel/works/pr/philosophy-of-right.pdf Accessed on 12 September 2018.

[3] It is interesting that while on the one hand, Sheikh Abdullah got a rousing reception from the people on his arrival in Kashmir two days after taking over as chief minister, Kashmir observed a complete hartal on the day he took the oath of office in Jammu, in response to the call given by the prime minister of Pakistan Z. A. Bhutto, against the Indira–Abdullah Accord. Ishaq, *Nida-i-Haq*, 382; Butt, *Kashmir in Flames*, 194.

[4] Kenneth Winetrout, *Arnold Toynbee* (Boston: Twayne, 1975), 31.

[5] Kenneth Winetrout, *Arnold Toynbee*.

a blessing. There is peace which is welcomed. The universal state also performs important services for the people and they are happy with this renewed prosperity. But these 'are not summers but Indian summers', masking autumn and presaging winter, because they have been created through force and deception, not through the process of creative energies; they represent a holding action. Therefore, winter is round the corner.[6]

Almost the same situation applies to the post-1953 history of Kashmir. Instead of creatively responding to the Kashmir problem, the Indian state used its coercive methods and forced the sole spokesperson of Kashmir, Sheikh Abdullah, to sign a peace deal without accommodating the demand around which he had mobilized the people, especially the Muslims of the state, for not less than 22 years. The Accord, which made Sheikh Abdullah the chief minister of the state under the domineering Indian government led by Indira Gandhi, no doubt, like Toynbee's 'universal state', brought 'peace' and some prosperity to Kashmir, but like the 'universal state' this promised land was just a prelude to disaster, because it represented a holding action. It was not a response to the challenge. It was an attempt to secure the status quo, ignoring the ground reality that the status quo was outworn in the fast-changing circumstances, especially those created by the intensive mobilization of the Plebiscite Front and the great shift in the conditions and outlook of the people in the wake of developmental measures taken by the post-colonial state. As the 'peace' created by the Accord was sustained just for a few years, less than a decade, instead of naming it 'Indian Summer' after Toynbee, I prefer to call it the 'Kashmir Summer', because in Kashmir the summer is very brief, not more than three months' duration, while the rest of the year, like the enduring instability of Kashmir, is cold in the Vale.

Indeed, the circumstances that obtained in Kashmir were explosive, and neither New Delhi nor Abdullah could fathom the implications of a humiliating deal. While for the Indian government it was a great triumph, for the collective Kashmir psyche it was a burning shame. Right from 1947, and especially from 1953, the society of Kashmir had not been indoctrinated into creating the

---

[6] Kenneth Winetrout, *Arnold Toynbee*, 30–31.

circumstances to make Sheikh Abdullah the chief minister of the state. Remember, he had sacrificed a far better deal in 1953.

The Plebiscite Front established in 1955 mobilized the people around one demand, namely, conducting a referendum for deciding the finality of Kashmir's political future. The slogans of the Plebiscite Front were *rai shumari fauran karau* (hold plebiscite immediately), *ye muluk hamara hai, iska faisla hum karain gai* (this country [Kashmir] is ours, we [only we] will decide its future); *jis Kashmir ko khoon se seencha woo Kashmir hamara hai* (the Kashmir that we have irrigated with our blood belongs to us), *Sher-i-Kashmir ka kay irshad? Hindu-Muslim-Sikah itihad* (what is the wish of the Sheikh? [His wish is] the unity between Hindus-Muslims and Sikhs), long live Abdullah, and the like. To take this message to every nook and corner of Kashmir and to every individual, the Plebiscite Front established a well-knit organizational structure, about which a participant observer of the movement, Zahid G. Mohammad, says:

> As a student of contemporary history, I see the organizational setup of the Plebiscite Front and commitment of its cadres as forte of the post 1947 political struggle in the state. It was a grassroots level organization with a Mohalla committee in almost every town, village committee in remotest villages all over the state of Jammu and Kashmir. Every second year the party carried out membership drive throughout the state. Basic members of the party were in millions. Through ballot or other democratic methods these members elected delegates for the general council that in turn elected office bearers for district, provincial, and central offices. The organization had very elaborate working committee with members from not only Kashmir, Jammu, and Ladakh but all districts in the state. Only a few top office bearers were supported by the organization and rest of the cadres had to fend for themselves—in true sense it was an organization of plebeians.[7]

In the first annual session held at Sopore in 1964, around 1,500 delegates representing different areas of the state participated. And

---

[7] Zahid Mohammad, *Kashmir: Changing Shades* (Srinagar: Gulshan Books, 2013), 167.

it was announced that the organization had nearly about 700,000 basic members.[8] The Plebiscite cadres inculcated the ideology of self-determination deep in the Kashmiri mass psyche. The propaganda machinery was supplemented by its two organs, namely *Front* and *Mahaz*. For the most part privately, and sometimes openly, the Front propagated a pro-Pak stance.[9] This is why during the Plebiscite Movement (1955–75), for the Muslims of Kashmir freedom meant accession to Pakistan.[10] The internationalization of the issue; discussions on Kashmir at international fora; promises made by the Indian prime minister on the floor of the Indian Parliament and through All India Radio as well as by Indian representatives during UN Security Council debates;[11] the publicity given to these statements by the Pakistani media; the propaganda/counter-progananda programmes broadcast by Azad Kashmir Radio and listened to with much interest by the Muslims of Kashmir;[12]

[8] *Mahaz* (Srinagar), 10 October 1964, 5, cited in Altaf Para, 'Emergence of Modern Kashmir: A Study of Sheikh Mohammad Abdullah's Role', PhD thesis, Department of Political Science, University of Kashmir, 2008, 233.

[9] Ishaq, *Nida-i-Haq*, 329–33; Balraj Puri, 'Jammu and Kashmir', in *State Politics in India*, ed. Myron Weiner (Princeton: Princeton University Press, 1968), 237.

[10] This is also substantiated by the following verse of *wanwun* (a genre of song sung by women during marriage ceremonies and other festive occasions) sung by Kashmiri women in every nook and corner of the valley in praise of the bridegroom on the occasion of marriage. The verse goes: *Sabaz dastaras Nabi chhuai razi; Pakistanuk gazi aaw* (The Prophet is happy with your green turban; You look as if the Gazi [one who fights a religious war] of Pakistan has come). This song was also sung in praise of Prime Minister Morarji Desai when he visited Mirwaiz Manzil, the headquarters of Mirwaiz Maulvi Farooq's party, the Awami Action Committee, during the 1977 assembly elections.

[11] For details see Lakhanpal, *Essential Documents and Notes on Kashmir Dispute*.

[12] During my interactions with some politically conscious contemporaries of the time, I was told that the programmes broadcast from Azad Kashmir Radio like *booni naar* (blazing chinar) and *manet dbhab* (a heavy blow) were very popular and were listened to enthusiastically by the people of Kashmir.

and above all the spirited public speeches the Sheikh gave to the mammoth public gatherings which received him after his release from jail and during his tours to different parts of the state,[13] especially in the valley, constantly nourished the deep-seated plebiscite-mindedness among Kashmiris. The repressive measures of the government against the Plebiscite leaders and workers further alienated the people and reinforced the plebiscite ranks.

Quite interestingly, while the different agencies, especially the Plebiscite Front, were cultivating anti-Indian and pro-Pakistan political sentiment among the Kashmiris, Abdullah stealthily started changing his stance.[14] With her finger on the pulse of Sheikh, who was in 'search of power with dignity', Indira Gandhi with inputs from D. P. Dhar, the Machiavelli of Kashmir politics, set forth on a long process of negotiations—or rather the negotiation of attrition[15]—to exhaust the Sheikh and coerce him into dropping 'with dignity' from his demand. The leadership in Delhi laboured under the misunderstanding that Kashmir was Sheikh-made, rather than Sheikh being Kashmir-made. The ailing Sheikh Abdullah, who was in the twilight of his life and under the tremendous pressure from his family to accept office once again,[16]

[13] Ishaq, *Nida-i-Haq*, 283–5; Butt, *Kashmir in Flames*, 72. See also Abdullah, *The Blazing Chinar*, 451–3.

[14] The change in the Sheikh's stance began from late 1965 during his stay in Delhi, where he was shifted after his incarceration in Ootacamund in the south. Ishaq, *Nida-i-Haq*, 358.

[15] The negotiations began in June 1972 and continued up to November 1974.

[16] Writing on the pressure that was being exerted on Sheikh Abdullah by his family to accept power, Ghulam Ahmad, the principal secretary to Sheikh Abdullah from 1975 till Sheikh's death, writes: 'Syed Mir Qasim spent many sleepless nights with the Sheikh in marathon sessions to win over the almost tired Sheikh to his way of thinking. Meanwhile Sheikh Abdullah came under tremendous pressure from the family to accept the office once again. For too long, they argued, he had remained in wilderness, sans creature comforts, sans power and sans authority. The allurements of office were too attractive to resist. While negotiations were going on ... the family members and Sheikh's close relations were itching for the authority that they would wield on acceptance of the Gaddi (throne) by the Sheikh.

climbed down from the demand of plebiscite to the demand for power. It may be mentioned that for about 11 years (1953–64), the Sheikh had demanded plebiscite for the resolution of the Kashmir problem. From 1964 to 1968–9, he insisted on the resolution of the Kashmir issue through negotiations between India, Pakistan, and the Kashmiri people. And from 1969 to 1974, Sheikh Abdullah, while making concessions to the circumstances, demanded that Kashmir's constitutional position be restored to its pre-1953 status.[17] And finally he had to satiate himself with merely state power, for which he had so much hunger that he wrote a letter to Mir Qasim saying that 'he would go back to the old position of demanding Plebiscite in Jammu and Kashmir if settled matters were not implemented immediately.'[18] To add insult to injury, the Plebiscite Movement was, in the words of Sheikh Abdullah's point man Mirza Afzal Beg, 'wandering in wilderness' (*siyasi awaragardi*). According to Beg, 'they roamed the desert, forlorn, forgotten, neglected, unwept, and unsung despite the turnabout of his party National Conference and formation of Plebiscite Front.'[19]

One important insight regarding Sheikh Abdullah is that he would either change his statements as per the changing audiences, or he would use ambiguous language to confuse or satisfy mutually opposing groups,[20] or he would hide his agenda from the

---

Most of them set up their camp in Delhi and were eagerly looking forward to the day when the Sheikh would assume office and open way for them to secure their hold on the government as also on the people.' Ahmad, *My Years with Sheikh Abdullah*, 41–2.

[17] For the shifts in Sheikh's stated position vis-à-vis the solution of the Kashmir problem, see Gundevia, 'On Sheikh Abdullah'; Butt, *Kashmir in Flames*, chapters 13–15; Ishaq, *Nida-i-Haq*, 355–82; Abdullah, *The Blazing Chinar*, 498ff.

[18] Butt, *Kashmir in Flames*, 190.

[19] Ahmad, *My Years with Sheikh Abdullah*, 39.

[20] 'For example, after a near total victory of his nominees in the civic polls held in Kashmir in September 1972, Sheikh Abdullah addressed a public meeting in Srinagar on 17 December. *The Times of India* reported the Sheikh's statement the next day under a two-column headline which read *NO Confrontation, Says Sheikh*. While the *Indian Express*, which has

people and provoke anti-India sentiments among them to wring a better deal for himself without caring about the explosive society he was creating. From 1968, Sheikh adopted a more rhetorical, deceptive politics. He kept his political agenda a secret from the public, restricted knowledge of it to his close coterie, and played with demagoguery to create 'favourable' circumstances to bargain a better deal. Interestingly, Sheikh himself makes an admission of this fact. Commenting on Jayaprakash Narayan's displeasure over a fiery speech Abdullah had made at Hazuribagh in 1968, in which he had exhorted the youth that 'freedom is not given, but it is to be snatched', Sheikh said, 'but he [Narayan] was not aware of the conditions obtaining here [Kashmir]. Our most urgent need is to keep up the spirit of the people.'[21]

Between 1969 and 1975, when the Accord was signed, a number of protests emerged in Kashmir against the news that Sheikh Abdullah was striking a deal with the centre having sacrificed at the altar the right to self-determination.[22] At each such occasion, either Sheikh himself or his lieutenant came before the people and vehemently denied such reports, and forcefully reiterated their commitment with the basic objective of the Plebiscite Movement.[23]

---

sometimes boldly taken up the cudgels on his behalf, surprisingly enough published a report of the same meeting—also a two-column headline, which read Sheikh Talks of Confrontation.' Abdullah, *The Testament of Sheikh Abdullah*, 'Publisher's Preface' (1974), 7–8. The former R&AW chief A. S. Dulat also says, 'Sheikh Saheb (was) adept at wording political matters in such a way that satisfied both Kashmiris and Delhi.' A. S. Dulat, *Kashmir: The Vajpayee Years* (New Delhi: Harper Collins, 2015), 190.

[21] Abdullah, *Ātash-i-Chinar*, 594. Also see Butt, *Kashmir in Flames*, 133–4.

[22] Of these demonstrations, mention may made of the ones held on 29 September 1965, 12 October 1965, 8 May 1968, 7 August 1969, 23 August 1969, May 1970, 24 January 1973, 21 May 1973, 7 November 1973, 11 April 1974, 20 April 1974, 24 February 1975, and 28 February 1975. Butt, *Kashmir in Flames*; Ishaq, *Nida-i-Haq*.

[23] *Kay Samjota Huwa Hai?* (Has Agreement Been Reached?), *Aftab*, February 1975; Butt, *Kashmir in Flames*, chapters 13–15; Ishaq, *Nida-i-Haq*, 355–82.

Significantly, during the Parthasarthy–Beg talks which stretched from June 1972 to November 1974, Abdullah reiterated not less than 17 times his commitment to the people that no compromise would be made with the demand for referendum, and whatever agreement was reached, it would be placed before the people for their final approval.[24] The leadership also assured the people that the final decision on Kashmir would be taken in consultation with Pakistan.[25] Interestingly, even after having signed the Accord, which did not offer anything except the chief ministership to Sheikh Abdullah, and that too on crutches,[26] Sheikh told a large public gathering: 'A few days back, the top leaders of Pakistan, on a visit to "Azad Kashmir", said the UN Resolutions on Kashmir should be consigned to the wastepaper basket. And I made a strong protest. I said the Kashmiris had got those resolutions passed after great sacrifices. We do not think they are useless.'[27]

That at the time the Accord was signed Kashmir was on the boil is also evident from the frequent demonstrations, hartals, fires, ubiquitous presence of forces, and underground activities as discussed in the preceding chapters. The National Conference leadership was aware of the pro-plebiscite mentality of the Kashmiri Muslims. That is why they did not fulfil their repeated promise that 'whatever decision would be taken, that would be placed before the public for their final approval.' They made the people

[24] The commitments were made on 10 May 1972, 14 June 1972, 19 June 1972, 20 June 1972, 4 July 1972, 30 January 1973, 1 February 1973, 2 April 1973, 13 July 1973, 19 July 1973, 11 November 1973, 30 November 1973, 28 March 1974, 1 April 1974, 4 April 1974, 21 April 1974, 17 June 1974, 8 July 1974, 18 September 1974, 20 October 1974, 13 November 1974, and 15 November 1974 at Khanyan, Hazratbal, Srinagar Airport, Hazuribagh, Sopore, Pulwama, Srinagar, Srinagar (Mujahid Manzil), Marty's Graveyard Srinagar, Tral, Ganderbal, Badgam, Lal Chowk, New Delhi, Idgah Srinagar, Kupwara, and Charar-i-Sharief, respectively. See Butt, *Kashmir in Flames*; Ishaq, *Nida-i-Haq*.

[25] *Aftab*, 13 February 1975.

[26] To make him a dependent chief minister, Sheikh Abdullah was forced to lead the existing Congress-dominated assembly, rather than assuming power through proper elections, which he had requested.

[27] Abdullah, *The Blazing Chinar*, 611.

believe that the Accord was nothing more than a change of strategy, and that the stand of National Conference was uncompromising. Sheikh even told the mammoth gathering on his reception after taking over as chief minister that, 'for us UN Resolutions, for which we have given our blood, are still relevant.' And no less a lie sold to the public was that the Rawalpindi road would be opened, which to the people bordered almost on the fulfilment of their long-held aspiration.[28] Because of the unflinching faith the people had in Abdullah, they did not mind giving him adequate time to let the promises come to fruition. But the creative minority, which was aware of the nature of the Accord, withdrew from the newly created environment. According to the survey conducted by Balraj Puri, the Accord created frustration among the 'secessionists'.[29] In other words, those sections of people who were aware of the actual outcome of the Accord were disappointed. The disappointment led to frustration, and threw up a section of people who withdrew from the society only to return with a new strategy for achieving what the society had been prepared for since 1947 by the 'lion of Kashmir'. That is why we had, in the words of John Ray, two Kashmirs after 1975—a 'world of choice, the hope of education, development and even of imperfectly functioning democracy' on the one hand, and on the other, 'lathi charges, teargas' and even 'shooting in Maisuma and Gawkadal'.[30] The world of protests was enlarged with every passing day, as we shall discuss elsewhere in this chapter.

Why Abdullah Continued to Remain Popular

Even after giving a send-off to the 22-year-old mass movement, which had entailed huge sacrifices for the people, and with which were deeply wedded the emotions of the Muslims of Kashmir, Sheikh Abdullah continued to have a mass following. It is true that a section of people immediately withdrew, only to return to

---

[28] Nehru, *Nice Guys Finish Second*, 597.

[29] Puri, *Jammu and Kashmir*, 181.

[30] Ray, 'Kashmir 1962 to 1986', 203.

revive, with new energy, the struggle abandoned halfway by the Sheikh. However, the dominant majority still clung to the Sheikh, as evidenced not only by the rousing receptions accorded to him on his entry into the valley after taking over as chief minister,[31] but also from the absolute majority the National Conference got in the 1977 assembly elections,[32] and, even more so, by the epic outpouring of grief among millions of mourners on the day of his death in September 1982.

Why did the Sheikh retain mass support till his death even after his political turnabout in 1974–5? Did the people march with their leader in burying the past and in opening a new chapter in the chequered history of Kashmir—the chapter of 'strengthening the bonds between the Union and the state', in which 'Jammu and Kashmir continues to make its contribution to the sovereignty, integrity, and progress of the Nation',[33] to quote the letter written by Sheikh to Indira Gandhi expressing his concurrence with the conclusions reached between Beg and Parthasarthy. To be sure, this was not the case. There were some objective reasons for Abdullah's continuous mass appeal which are outlined below.

Memories of Sheikh's Contribution and Sacrifices

The most important factor behind the enduring popularity of Sheikh was that he was the leader of those people who were either his contemporaries or near contemporaries. And minus some exceptions, they were deeply beholden to him for his contribution to securing their freedom from oppression and exploitation. The emotional attachment of the masses with the Sheikh was too deep to be easily effaced from their minds. To be sure, from the side of the masses, it was not less than a lover–beloved kind of relationship, where the lover either ignores the limitations of his beloved or construes them in such a way that vice becomes virtue or, at

---

[31] Abdullah, *The Blazing Chinar*, 541–2.

[32] In the 1977 assembly elections, the National Conference got 47 seats.

[33] Text of the letter written by Sheikh Abdullah to Indira Gandhi dated 2 February 1975, cited in Butt, *Kashmir in Flames*, 203–5.

least, its bitterness evaporates. And if the National Conference still commanded some support, it was the lingering legacy of the generation which had lived with the Sheikh and seen a transition from darkness to light owing to his sacrifices and role.

It is true, as we have seen, that the brief rule of Abdullah (1948–53) was not all rosy. But the long struggle—longer than the earlier one—which he fought subsequently rejuvenated Abdullah in a new avatar in the popular imagination. For this whole period he was unblemished, only making sacrifices for the people, not even allowed to participate in the wedding ceremony of his daughter.[34] He was involved in false cases,[35] and placed in jail in far-away Ootacamund, which to the popular imagination was situated in a deserted place in some corner of the world with no humans. Even the place 'Kud', which is situated on the way from Srinagar to Jammu, was to the common people an unknown place.[36] The term 'jail' and, more so, 'interrogation centre', would send shivers down one's spine. The stories of the horrors of jail life were circulated among the people by the well-knit National Conference cadre. The local leadership took every care that the reality of jail life of the Sheikh and his close associates was not leaked to the public.[37] Indeed, at the time when the compromise between the Sheikh and Indira Gandhi was arrived at, Sheikh was the beloved of the people for all the sacrifices he had made, first to get freedom from the oppressive and discriminatory princely order, and then in actualizing the political sentiment of the people.

Sheikh as Wali

Abdullah's charisma was no doubt mainly based on the tangible contributions he had made in general and for Muslim society in

[34] Abdullah, *The Blazing Chinar*, 531.

[35] Some of the cases slapped against Sheikh Abdullah were the Kund conspiracy case, Hazratbal conspiracy case, and the Kashmir conspiracy case.

[36] Based on my interviews with common people who were in their youth in the 1960s and early 1970s.

[37] Based on my interviews with common people who were in their youth in the 1960s and early 1970s.

particular, but he and his councillors did not remain oblivious of existing beliefs about the attributes of a 'perfect man', as it was the fulfilment of these conditions which qualified one to be followed blindly without demur. It goes without saying that Sufism is the keynote of Islam in Kashmir. And in Sufism two conditions are indispensable for recognizing a person as Wali (the Friend of God). These are control of the *nafs* (carnal self) and working *karamat* (miracles). Though popular belief in Kashmir gives overwhelming importance to performing karamat over control of the carnal self in acknowledging someone as Wali-e-Khuda, Abdullah had a good record in the fulfilment of this condition too. By 1975, he had fought for about 39 years, languished in jails for years together, and suffered physically as well as mentally in the interest of the larger cause of the people, whereas his junior colleagues who collaborated with Indian government were enjoying power. To appreciate this did not need a philosophical mind. But given the most popular imagery of a Wali, the Machiavellis of Abdullah's team wove a halo of spirituality around him. The basis for this was laid by Abdullah himself through his power of holding his audience spellbound by reciting verses from the Holy Quran in his sweet and bewitching voice before delivering the political speech.[38] The use of symbolism and imagery by Sheikh in his speeches conveyed, at least to the common masses, that he was in communion with God.

> On one side people were beckoning to me to live a life of comfort, luxury, affluence and authority at the cost of my conscience. They were asking me to forget my ideals of self-determination and the rights of the people of Kashmir to govern themselves and to barter away the rights of Kashmiris. . . . On the other side the Holy Quran was warning me in God's own voice not to reject God's path and not to fall prey to the comforts of life.[39]

Though the stories about the miraculous powers of the Sheikh prevalent at the local level have not yet received the same scholarly

---

[38] Abdullah, *The Blazing Chinar*, 66–7.
[39] Gundevia, 'On Sheikh Abdullah', 30–31.

attention as the stories about Mahatma Gandhi have,[40] a few stories and popular beliefs regarding the deification of Sheikh were commonly known throughout the length and breadth of the Kashmir Valley, and were believed by the general masses without any iota of doubt. These stories and beliefs have survived through oral history and not so much in writing. Indeed, the process of shaping the personality of Sheikh in the crucible of spirituality started during the heyday of the freedom struggle. One of the fantastic rumours projecting the Sheikh as a spiritual personality was the rumour that the leaves of the chinar tree carried on them the words 'Sher-e-Kashmir!'[41] The consciously coined rumour was spread by the National Conference cadres and believed by the people without question in their collective quest for proof of the invincibility of their leader. The rumour had become so widespread that the maharaja had also desired to see the leaves.[42] Another hot rumour on the tip of everyone's tongue—old, young, and children—was that when the maharaja was about to throw Sheikh Abdullah into a *kadai* (frying pan) of optimally boiling oil, Sheikh Abdullah put his little finger into it, and to the utter surprise of everyone, he said with a smile, 'the oil is very cold.' And seeing this miraculous power of Sheikh, the maharaja set him free.

Believing him to be a *pir* (saint) with supernatural healing powers, there used to be large gatherings waiting for his reception and aspiring to get their ailments cured by him. Sheikh would walk through these huge crowds waving his stick gently. Anyone who was touched by the stick would yell, 'my ailment has been cured.'[43] According to Sheikh, people had such blind faith in his spirituality that 'they pulled strands of hair from my head and made souvenirs of them'.[44] Again, Sheikh Nooruddin Rishi, the popular indigenous Sufi saint of fifteenth-century Kashmir, is said to have made

[40] See Shahid Amin, 'Gandhi as Mahatma: Gorakhpur District, Eastern UP, 1921–2', in *Selected Subaltern Studies*, eds Ranajit Guha and Gayatri Chakravorty Spivak (New York: Oxford University Press, 1988), 288–348.

[41] Abdullah, *The Blazing Chinar*, 88.

[42] Abdullah, *The Blazing Chinar*, 88.

[43] Agha Ashraf Ali, *Kuch to Likhye ki Loag Kehte Hain* (Srinagar: Shalimar Art Press, 2010), 86.

[44] Abdullah, *The Blazing Chinar*, 69; Ali, *Kuch to Likhye*, 86.

a prophesy saying, *Zainagire āb faireh, Soure manz lal naireh*
(Zainagir area will be irrigated; and a jewel will be thrown up by
Soura [the birthplace of Sheikh Abdullah]). It would be in place to
mention that the mass popularity of Sheikh Nooruddin Rishi was
exploited by a variety of interest groups at different periods of history. Verses for serving an agenda were composed and attributed
to Sheikh Nooruddin to establish their authenticity/legitimacy. In
this regard, it is worth quoting Ashraf Wani:

> Sheikh Nooruddin had won so much faith and following among
> his contemporaries and subsequent generations of Kashmir, and
> his fame as a preacher-poet of the people was so well known that
> he became a sanctifier, a legitimizer. If anyone desired his/her dis-
> course to earn mass acceptability, appropriating the Sheikh by con-
> structing the pre-conceived *shrukhs* (verses) and attributing them to
> him became the convenient way.[45]

Such was done consciously by the well-wishers of Sheikh Abdullah.

All in all, therefore, Abdullah was deified. After being released
from jail, Sheikh routinely used to make tours of the valley. In
the long queues of receptions, the women folk sang *wanwun* (folk
songs) in praise of the Sheikh and people struggled to get access to
him to touch, at least, his clothes with their hands with the same
veneration and belief as one sees a Kashmiri Muslim smearing his
throat and body with the holy dust of the sacred precincts of a
shrine. The principal secretary to Sheikh Abdullah also observed
the masses' reverence for the Sheikh, and the latter acting like a
pir to whom the people turned for miraculous solution of the crisis of their lives: 'Large crowds of gullible and credulous women
would wait at his residence for hours on end to receive his bene-
dictions. He would invariably dole out pieces and crumbs of bread
and bottles of water which, they believed, had healing powers.'[46]

---

45 Mohammad Ashraf Wani, 'In Search of an Authentic Text of Sheikh
ul Alam's Poetry', *Alamdar* (Centre for Shaikh ul Alam Studies, University
of Kashmir, 2012), 5, no. 5, 39–42.

46 Ahmad, *My Years with Sheikh Abdullah*, 30. See also the cover photo
of Nyla Ali Khan's book, *Sheikh Mohammad Abdullah's Reflections on
Kashmir* (New York: Palgrave Macmillan, 2018).

Politics of Deception

Despite all this public appeal of Abdullah woven out of mundane and spiritual warp and weft, the policy makers of the National Conference had no doubt that the root cause of the mass appeal of Sheikh among the Muslims was the sense that he represented their collective political sentiment. Clearly, the rumours about the Sheikh's spiritual position and occult powers were believed by the people because Abdullah's name was associated with them; and it was for this reason that the public on their part also added to this piquant propaganda because they wished for him an unconquerable power. Therefore, alongside hammering out political issues with Indira Gandhi, a strategy to maintain the Sheikh's appeal among the people was also devised. Thus it was decided not to tell the people that the Plebiscite Movement they had engaged in since the mid-1950s had been rolled back. The people were told that for achieving 'our ideals and objectives', power was a necessary instrument.[47] It was advertised that instead of fighting for rights while remaining in the opposition, acceptance of power was a 'change of strategy', not a sell-out. In his autobiography, Ātash-i-Chinar, Sheikh captioned the chapter on Accord as *Kashmir Accord: Hikmat-i-'Amli ki Tabdili* (Kashmir Accord: Change of Strategy). Not only this, Sheikh made the people believe that with regard to representing the collective sentiment of Kashmir, he was more sincere to the people than Pakistan.[48] The Plebiscite Front leaders and office bearers canvassed a 'wait and watch'[49] policy among the people. In other words, they kept the people in suspense by propagating the idea that something was in the offing, especially the opening of Rawalpindi road. At the same time, the brain behind the plot coined a pungent slogan, *ala karay ga wangan karay ga, bab karay ga, bab karay ga*. Loosely translated, this would mean, whatever good or bad would be done; only Sheikh Abdullah is entitled to do it. It may be mentioned that the term

[47] Abdullah, *The Blazing Chinar*, 541.

[48] Abdullah, *The Blazing Chinar*, 542.

[49] Butt, *Kashmir in Flames*, 207; also based on my interaction with people.

bab, by which Sheikh is called here, is synonymous with Wali/ pir. This was a deliberate attempt on the part of the leadership, with its guilty conscience, to reinforce the position of Abdullah as sole spokesperson of Kashmir; and to exact acceptability from the innocent masses about the deal by making them shout, *bab karay ga, bab karay ga*, so that when the reality unfolded with the passage of time, the sloganeering public would feel morally obliged to defend the new meaning of the slogan, though they had been innocent about the actual game plan behind the coining of the slogan.

It must be mentioned that the Plebiscite Front had thousands of workers, and a good number of them were either, like the workers of any other political party, blind followers of their leader or they were bereft of any ideological moorings, and therefore could be, according to a contemporary observer, 'jubilant because they felt they had secured power'.[50] It was this regiment of workers which was used by the Sheikh and his colleagues to spread such canard to maintain his mass support; and to be in the front rows of public gatherings to second loudly and with applause every statement of the leadership, or to be in the vanguard of public processions to raise pro-leadership slogans. It was the category of young workers called the Youth Federation which was used to revive the old *sher–bakra* (lion–goat) conflict between the followers of Sheikh and Maulvi Farooq to consolidate the traditional mass following of the Sheikh.[51]

The politics of deception reached its acme during the 1977 assembly elections. Indeed, none knew the heartbeat of Kashmiris better than the National Conference leadership. Taking (undue) advantage of the popular faith enjoyed by the National Conference, the leaders again cashed in on the Pakistan sentiment to win the popular vote. Mirza Afzal Beg carried rock salt (called in Kashmir 'Pakistani salt' for its source in the Khewra salt mines of Pakistan) and a green handkerchief—both symbolizing Pakistan—to each of his election rallies and showed these two 'symbols' of Pakistan to people in order to beguile them into believing that the party

---

[50] Butt, *Kashmir in Flames*, 194.
[51] Butt, *Kashmir in Flames*, 233.

continued to maintain its historic stand.[52] Moreover, the National Conference leadership made the opening of the Rawalpindi road an important election plank, promising the people that if the National Conference came to power, the Rawalpindi road would be opened.[53] Afzal Beg would also sarcastically call Kashmir's accession to India, not *ilhaq* (meaning accession in Urdu), but *alla haq*, meaning a dish consisting of a combination of pumpkin and knol khol—a combination not preferred by Kashmiris.[54]

Sheikh Abdullah's doublespeak is proverbial. It was not only his political rival and rabid communalist Pandit Prem Nath Dogra who denounced his doublespeak by saying that Sheikh was 'communalist in Kashmir, a communist in Jammu and a nationalist in India';[55] but it was common knowledge to all Kashmir watchers of the time that the Sheikh was a past master in playing many games simultaneously, and he played them as per the demands of time and space. Ghulam Ahmad, who watched the Sheikh closely, being his principal secretary, says:

> Sheikh Abdullah has been variously called an enigmatic as well as magnetic personality, a crowd puller, a volatile public figure who used different languages at different places and on different occasions depending upon the mood and response of his audience; a demagogue who swayed crowds in whatever direction he wished. In one breath, he would beguile gullible crowds by his polemics and indulge in the other breath in pontificating and shibboleths moving them into tears.[56]

Though Sheikh was 'mortally afraid of Mrs Gandhi',[57] and even compromised on his towering personality to placate her,[58] he

---

[52] Ahmad, *My Years with Sheikh Abdullah*, 60; Ishaq, *Nida-i-Haq*, 3789.

[53] Ahmad, *My Years with Sheikh Abdullah*, 60; Nehru, *Nice Guys Finish Second*, 597.

[54] Ahmad, *My Years with Sheikh Abdullah*, 61.

[55] *Hindu* (Madras), 15 October 1952, cited in Korbel, *Danger in Kashmir*, 207.

[56] Ahmad, *My Years with Sheikh Abdullah*, 8–9.

[57] Ahmad, *My Years with Sheikh Abdullah*, 75.

[58] Sheikh Abdullah became so mortally afraid of Indira Gandhi that 'a day [he was humiliated by Mrs Gandhi] later he sent a bouquet

would occasionally make rhetorical public speeches with a self-seeking purpose. But to the masses, the language projected him as the Sheikh of the plebiscite days. To quote B. K. Nehru:

> The Sheikh is convinced at the moment that the Centre wishes to remove him from the Chief Ministership of the state, whether by totally unconstitutional means or by destabilizing him to such an extent as to give the Centre an excuse to pretend that the Constitution has broken down. His recent occasional very wild utterances are explicable by his desire to defend himself against such a Central onslaught by creating public opinion in his favour through stressing the special status and situation in Kashmir, by recalling what happened in 1953 and threatening that a repetition of those events will be accompanied by an even worse reaction, and occasionally going beyond this and mentioning Pakistan in the same breath as India. I do not believe that he is genuinely pro-Pakistan or anti-India but he does wish to remain in power and to pass that power on to his son.[59]

Sympathy Wave

During the decisive electioneering days of the 1977 assembly elections, a few developments took place which created a sympathy wave for Sheikh and his party. Sheikh Abdullah suffered a heart attack; the news about his precarious condition spread far and wide. The governor and other top functionaries visited him. The home minister of India sent two leading heart specialists to treat Abdullah. The prime minister, external affairs minister, and Indira Gandhi sent messages of sympathy to him. These were broadcast and published by the newspapers. All India Radio also broadcast the news that the condition of Abdullah was precarious. Farooq Abdullah also announced in a public gathering in Srinagar, '*Sheikh sahib zindagi aur moat ki kashmakash main hain*' (Butt, *Kashmir*

---

of saffron flowers and Diwali greeting to her. And on 2 November 1976, Mr. Beg was sent to New Delhi to assure the Centre that the National Conference leadership was fully loyal to India.' The National Conference also accepted seat adjustment for Lok Sabha polls with the Congress party. Butt, *Kashmir in Flames*, 217.

[59] Nehru, *Nice Guys Finish Second*, 593.

in Flames, 242). (Sheikh sahib is fighting between life and death). On the last day of the campaign, a taped message of the bedridden Abdullah was relayed in a big public gathering in which Abdullah appealed to the public to vote for the National Conference 'for the protection of their interests'. The daily *Aftab* also published two photographs of the ailing Sheikh which deeply moved the public.[60] D. D. Thakur, cabinet minister in the Sheikh government, corroborates that 'it was his [Sheikh Sahib's] illness which united the whole Valley in his support and made his success at hustings a certainty.'[61]

Simultaneously, the Indian prime minister, home minister, and defence minister visited Kashmir. Instead of visiting the ailing Sheikh, they made many irresponsible statements against him: 'Sheikh Sahib is old and sick, he should now retire.'[62] They vilified the Sheikh. The way the central leaders conducted themselves, it appeared that the Indian government (which in Kashmir means India) was determined to control Kashmir and kick the local party out by any means. For the first time in the history of Kashmir, the Indian prime minister visited the headquarters of the Awami Action Committee party, Mirwaiz Manzil. The home minister openly told the administrative machinery to help the Janata Party to win, and at his instance a large-scale transfer of officers took place. This was an affront to the people of Kashmir. As a result, the election wave heavily gravitated in favour of the Sheikh. The impression of the people that the ruling party at the centre was determined to capture power in Kashmir by hook or by crook—the impression gathered by media is represented here by the editor Aftab, Sanaullah Butt—is based on the following facts:

a) Back-to-back visits of the top leadership of Janta Party including the Prime Minister, Home Minister, Defense Minister etc. to Kashmir.
b) Vilification of the NC leadership by them even though Sheikh Abdullah was seriously ill and bed-ridden.
c) Transfers of the officers at the instance of the central government.

[60] Butt, *Kashmir in Flames*, 238–42.
[61] Thakur, *My Life and Years in Kashmir Politics*, 283.
[62] Butt, *Kashmir in Flames*, 239; Qasim, *My Life and Times*, 155; Singh, *Kashmir: A Tragedy of Errors*, 9.

d) Alliance of the Janta Party with the known pro-Pakistan *Awami Majlis-e-Amal* party led by Mairwaiz Maulana Farooq who wielded considerable following in the politically sensitive downtown area of Srinagar.

Since the credentials of the family of Maulana Farooq and the *Awami Majlis-e-Amal* were widely known, no Indian leader prior to Morarji Desai had visited the family of Maulana or the headquarters of his party. Since Morarji Desai made a sharp departure from the political position of his predecessors by entering into an alliance with Mirwaiz to seek his help in winning elections, it was quite clear that the Janta Party was determined to capture power in Kashmir and kick out NC by using all conceivable means.[63]

Sheikh's Autonomy Murmurings

After winning a landslide victory in the 1977 assembly elections, the Sheikh continued to play the politics of manipulating popular political sentiment. To be fair to Abdullah, he never ceased to nurse the dream of an 'independent' Kashmir, but he was caught between his two mutually opposing ambitions—the yearning for independence, and the quest for power which destroyed him as well as Kashmir. In the early 1950s, Josef Korbel, who was intimately associated with the politics of Kashmir, made an objective assessment of Abdullah—a characterization that remained true of him till his death. Korbel said, 'The story of Sheikh Abdullah is a sad and sorry one. It is a story of a patriot, once passionately devoted to his people's welfare, but one whose patriotism was too shallow to reject the temptations of power.'[64]

Immediately after assuming power in 1975, Sheikh Abdullah constituted a four-member committee headed by Mirza Afzal Beg to draft a proposal for the restoration of autonomy to Kashmir. As Beg knew that there was no scope for restoring the autonomous position of Kashmir, he avoided working on the proposal. After his dismissal from the cabinet, Sheikh Abdullah appointed two committees

---

[63] For details see Butt, *Kashmir in Flames*, 239–41.

[64] Korbel, *Danger in Kashmir*, 207.

to submit separate reports on the issue. One was headed by D. D. Thakur and the other by G. M. Shah. Thakur in his report opined that this demand would only impair relations with the centre. The Sheikh observed silence thereafter.[65] It is also important to mention that in 1975–6, the National Conference published a pamphlet, *Why Autonomy to J&K State?* A delegation of the National Conference, which was invited to attend the All India Congress Committee meeting at Chandīgarh, attempted to distribute the pamphlet among the members. Indira Gandhi took serious note of this and had the move stopped, and did not provide Sheikh Abdullah the opportunity to address the meeting. The Sheikh felt so humiliated that he emotionally asked his cabinet members to be ready to resign.[66]

Sheikh Abdullah continued speaking from his heart, but could do nothing practically. On the day of the National Conference victory in the 1977 elections, he gave a statement to Voice of America pledging to work for the restoration of autonomy.[67] Governor B. K. Nehru, with whom Sheikh frequently used to spend his leisure time discussing freely and frankly the affairs of J&K, besides 'the whole universe', writes:

> it became clear from our conversations from time to time as well as from the difficulties he was creating in accepting wholly the Constitutional demands of the Centre that his objective was ... the eventual creation of a separate independent state consisting of at least the Valley together with such Muslim areas of Jammu division as could be tagged on to it.[68]

Ahmad, principal secretary to Abdullah, corroborates Nehru's account with thick detail. He observes that the ambition to carve an independent position for Kashmir stayed with the Sheikh permanently, till his last breath:

> He was till his end toying with and working for the idea of a Greater Kashmir comprising the valley including Kargil District,

[65] Thakur, *My Life and Years in Kashmir Politics*, 288–90.
[66] Ishaq, *Nida-i-Haq*, 391.
[67] Ishaq, *Nida-i-Haq*, 380.
[68] Nehru, *Nice Guys Finish Second*, 593–4.

Muslim majority areas of Doda, Bhaderwah and Kishtwar and parts
of Udhampur District namely Ramban, Batote, Gool Gulab Garh,
Banihal, and Poonch and Rajouri Districts of Jammu province with
minor adjustments for reasons of topography. It was with this objective in view that a special Division of R&B Department was created reviving and converting the old Mughal Road linking Rajouri
District with Shopian of Pulwama District in the Valley into a permanent and all weather road. Similarly, work on construction of
Daksum (in district Anantnag) Kishtwar road was started in right
earnest. This road was to provide a link with Kishtwar, Doda and
Bhaderwah. The Central Government was also asked to upgrade,
improve, and realign the Batote-Doda road, which was subject to frequent landslides and resultant closure of the road for days on end.[69]

Meanwhile, Abdullah, at least, demonstrated that he was
a chief minister with a difference. He was reticent in accepting
Indian Administrative Service officers from the centre to form a
part of the state cadre; nor did he permit the chief justice of the
high court who was a Kashmiri to be transferred. Abdullah also
resurrected the Resettlement Bill.[70] Besides other things, the bill
signalled to the people a move towards the reunification of people
divided by the Line of Control—and the possibility of returning to
the time when the state was undivided.[71] Writing about the challenges which the authority of the centre faced on Kashmir when
he took over as governor, B. K. Nehru says:

> The challenge to the authority of the Centre which the Sheikh was
> making when I took over charge were three. The first was his refusal
> to permit the Centre to send IAS officers to form part of the state
> cadre as was custom in every other state of the Union. The sec-
> ond challenge consisted in the refusal of the Sheikh to permit the
> transfer of the acting Chief Justice of the state and his replacement
> by a nominee of the centre from outside the state as was, and is,
> the practice in all the other states of the Union. The third problem

[69] Ahmad, *My Years with Sheikh Abdullah*, 123–4.

[70] Aditya Sinha, *Farooq Abdullah: Kashmir's Prodigal Son (A
Biography)* (New Delhi: UBS, 1996), 133–4.

[71] Aditya Sinha, *Farooq Abdullah: Kashmir's Prodigal Son (A
Biography)*, 133–4.

was much more serious. Some considerable time earlier a bill had been introduced in the Assembly by a private member, entitled the Resettlement Bill. It provided that residents of the state who had left it between 1947 and 1954 to go to Pakistan could be allowed to come back and resume all their rights as Indian citizens.[72]

Indeed, in the contemporary political history of Kashmir, Abdullah is the last example of the head of government trying to assert his authority. He differed, though politely, with the governor who refused to sign the Resettlement Bill. What transpired between Abdullah and the governor is interesting to hear from the then governor himself, as it alludes to an assertive role of the chief minister as the representative of the people vis-à-vis the head of the state as representative of the central government:

> I warned Sheikh that I would not sign the bill as it was so clearly unconstitutional. His argument was that it was none of my business whether the bill was constitutional or not; that was the business of the Supreme Court. I should sign the bill when it was passed, if anybody had any objection to its constitutionality they could go to court.[73]

Congress–National Conference Confrontation

No sooner did the Sheikh take up the reins of government than the Accord was converted into discord, following the confrontation between Sheikh and the Congress party over the politics of hegemony, which continued throughout the Congress-backed rule of Abdullah[74] and after. Leaving the details to the following section, it suffices to say here that the people construed this embittered relation between Congress and the National Conference as a fight between New Delhi and Srinagar to dominate Kashmir in a struggle among unequals. Abdullah issued a press statement saying, 'since the Congress had withdrawn its support to me, therefore,

[72] Nehru, *Nice Guys Finish Second*, 595–7.

[73] Nehru, *Nice Guys Finish Second*, 597.

[74] Butt, *Kashmir in Flames*, 210–24; Thakur, *My Life and Years in Kashmir Politics*, 224–9; Abdullah, *The Blazing Chinar*, 546–54.

the Accord between me and Mrs Gandhi should be treated as cancelled.'[75]

All the factors mentioned in the preceding discussion help us fathom the sustained popularity of the Sheikh till his death. Soon after his demise, the 'rebels' who had 'withdrawn' during the hustle and bustle of the Abdullah government, to conceive of a new strategy to actualize the will of the people left half-done by the mass leader, returned to revive the suspended course of history, which found considerable favour with the young generation, more than ever, in the changed circumstances.

Reform and Development (25 February 1975– 8 September 1982)

Having received no response to his letter dated 29 December 1974, written as his last attempt to persuade Indira Gandhi to concede to his demand of granting autonomy as per the accession agreement, the tired Sheikh Abdullah, under pressure from his family and many of his colleagues, ultimately agreed to be satisfied with accepting state power to strengthen the 'bond between the Union and the state' and 'to afford to the people of the state full scope for undertaking social welfare and developmental measures'.[76] With his sole aim now 'to initiate measures for the wellbeing of the people of the state',[77] Sheikh Abdullah took over the office of the chief minister on 25 February 1975, and focused on 'peace, development, administration, and economic reforms'. With a brief interregnum of three months and 13 days (27 March 1977–8 July 1977), Sheikh Abdullah in his second innings ruled Kashmir from 25 February 1975 to 8 September 1982, that is, seven years and three months. On account of his long career in political struggle, memories of his role as a harbinger of change from a medieval to a

---

[75] Butt, *Kashmir in Flames*, 222–3; Qasim, *My Life and Times*, 153.

[76] Letter dated 11 February 1975 from Sheikh Mohammad Abdullah to Mrs Indira Gandhi, quoted in Noorani, *Article 370*, 412–3.

[77] Letter dated 11 February 1975 from Sheikh Mohammad Abdullah to Mrs Indira Gandhi, quoted in Noorani, *Article 370*, 413.

modern polity and economy, his mass support, age, courage, bold-ness, and authoritarian style, Abdullah's colleagues and associates held him 'in awe as well as grudging esteem'.[78] As a result, he took some bold steps, making it abundantly clear that a strong leader-ship with a political will to perform can do wonders.

On the eve of Abdullah's assumption of power in 1975, law and order had become a casualty even in the Secretariat and the universities. Both Sheikh Abdullah and D. D. Thakur begin the narrative of their achievements with the observation that the Secretariat had become the hub of hooliganism,[79] and the situ-ation had deteriorated to such a pass that the hooligans used to force their entry even into the chamber of the chief minister. The indiscipline had not only affected the normal functioning of the Secretariat, which had become free for all rogue elements, but it had adversely affected the conduct of the lower staff too. Therefore, the first thing that the new government did was to send a signal of zero tolerance to indiscipline. Rules and regulations were framed and strictly adhered to, which changed the environment for good. Abdullah speaks with pride about this:

> After taking stock of the administrative set-up we set to reform its functioning. The secretariat is the core of a government. Ours had become a veritable fish-market. Anyone could barge in any time. There were some infamous men seen constantly lounging along its corridors. Sometimes they even went on the rampage. The officials took their cue from them. The atmosphere prevailing in the offices was one of non-seriousness. To remedy the situation, we streamlined rules for entry of visitors into the secretariat. Within few days there was a visible improvement in the things and the functionaries began to attend to their work with regularity. Cafes that were abuzz with the government officials during day time became more or less desolate.[80]

If one talks to the teachers of the University of Jammu and the University of Kashmir who were in service at the time Abdullah

[78] Ahmad, *My Years with Sheikh Abdullah*, 9.

[79] Abdullah, *The Blazing Chinar*, 542–3; Thakur, *My Life and Years in Kashmir Politics*, 202–3.

[80] Abdullah, *The Blazing Chinar*, 542–3.

took over, everyone would say the same thing: the professors and vice chancellors were abused and assaulted. D. D. Thakur says, 'no respectable dignitary could be invited to the University without jeopardizing his honour and prestige.'[81] 'Copying', or cheating, in the universities and colleges had become a routine affair,[82] none having the courage to check the rogue elements promoted by puppet regimes. As education minister, Sheikh Abdullah assured all help to the university authorities in restoring law and order. Strict instructions were issued to the police to deal with the criminal elements sternly. And for checking the menace of copying, he deployed police at the examination centres to ensure discipline. Sheikh also issued orders to the officers and the staff associated with the conduct of examinations against showing any lenience in checking the menace of copying. Abdullah himself visited the examination centres to demonstrate the seriousness of his government in wiping out copying root and branch. The result was that a new era of fair and clean examinations was ushered in in the state. 'What seemed impossible', says Abdullah, 'just a week before has been almost uprooted.'[83] This successful measure made it abundantly clear that no problem, howsoever difficult it might appear, could escape a solution provided there was political will to do it.

To restructure the economy on healthy lines, the Sheikh government in 1975 appointed a Development Review Committee of eminent economists of India under the chairmanship of L. K. Jha, the governor of the state and an economist of repute. In its report (1976), among other things, the committee articulated a stance against the policy of providing heavy subsidies on food.[84] Heavy subsidization of certain basic needs—food and fuel—in Kashmir was justified for a certain period of time to give poor people a

[81] Thakur, *My Life and Years in Kashmir Politics*, 35.

[82] Abdullah, *The Blazing Chinar*, 612.

[83] Abdullah, *The Blazing Chinar*, 612.

[84] With regard to food subsidy, the Jha Committee reported, 'the state government was pursuing a mistaken price policy with regard to food grains. Grain cannot be sold cheaply if it is not bought cheaply, and as subsidy is kept low, the price offered to farmers is also very low.' Quoted in Prakash, 'The Political Economy of Kashmir since 1947', 2055.

breather for improving their economic position and to enable them to take advantage of the status-improving routes provided by modernization. And once the conditions of the people had significantly improved, the once progressive measure became retrograde in that subsidization unnecessarily became a major cause of budgetary crisis and resultantly a bottleneck in the process of development. Having realized its baneful implications for the economy of the state and for the collective mentality and honour of Kashmiris, Abdullah took the bold step of abolishing it, though this was a risky move considering that Bakhshi and his successors had used subsidized rice to turn the tide. Clearly, living on subsidized food had become such a deep-seated habit of Kashmiris that none other than a mass leader like Abdullah could take such a daring step. Appreciating the boldness of this act, Mir Qasim, the predecessor of Abdullah, noted:

> right from the days when Mr. Sadiq was the Chief Minister, the Centre had been telling us to stop the food subsidy because it put a heavy burden on the Exchequer. At every meeting of the Planning Commission we were told to stop the subsidy as it had become a luxury for the people of Kashmir. Neither Mr. Sadiq nor I as the Chief Minister had the guts to do that; only a leader of the Sheikh's stature could perform that feat. Once Mr. Sadiq's move to only slightly reduce the subsidy provoked protests. He gave up. When I became the Chief Minister, I invited political leaders of all hues to discuss this matter. They warned against it pointing to the fate of Mr. Sadiq's move.[85]

According to Abdullah, the state exchequer was subjected to a burden of more than Rs 200 million annually on account of the subsidy on food, impinging on the infrastructural development and employment generation capacity of the state.[86] The money thus saved was spent on development and employment. Strangely enough, though the Sheikh took the Congress party on board before taking the decision on withdrawing subsidy, they, for the

---

[85] Qasim, *My Life and Times*, 150.
[86] Abdullah, *Ātash-i-Chinar*, 613.

sake of unseating Abdullah through popular uprising, subsequently opposed it.[87]

In far-flung areas like the hilly districts of Ramban, dealers in foodgrains used to misappropriate the foodgrain supplies provided by the government, and thus 'made money out of the flesh of the hungry',[88] says D. D. Thakur who belonged to the area. For this reason it was called the 'dealer raj'. To check this menace, the Abdullah government abolished the dealer system, and instead established government depots which proved a significant improvement over the old system.

Another area in which the Development Review Committee recommended stringent reforms was the transport sector. It noted with regret that huge losses were incurred by the Government Transport Undertaking (GTU), 'though the private sector was doing well'. The committee, therefore, recommended that the undertaking should be turned into a corporation and managed and run on commercial lines.[89] Considering the enormous burden of GTU on the state's resources, the government accepted the recommendation and converted the GTU into the Road Transport Corporation. The Development Review Committee was also not satisfied with the existing system of permits. Instead of allowing free competition among desirous aspirants to become transport operators, permits were issued selectively, and only a favoured few had the privilege of getting these route permits. The result was that over the period, monopolistic trends developed in the private sector. Although permits were normally given on the basis of not more than two per family, in actual fact there was a good deal of cartelization.[90] According to the then transport minister, 'the black market value of route permit ranged between half a million

[87] Abdullah, *Ātash-i-Chinar*, 613; Qasim, *My Life and Times*, 150.

[88] Thakur, *My Life and Years in Kashmir Politics*, 204.

[89] Government of J&K, *Report of the Development Review Committee*, Part IV, Industrial Development (Srinagar: Industries & Commerce Department), 45–64.

[90] Government of J&K, *Report of the Development Review Committee*, 45–64.

and one million rupees.'[91] After a great deal of fieldwork and expert advice, the Sheikh government abolished the ban on issuing of route permits. Alongside this, the Pathankot–Srinagar route was nationalized. As a result of this reform, a competitive transport industry developed to the great relief of those who wanted to enter the field for livelihood. It also checked the rise of transport charges and provided relief to commuters and dealers in goods.[92] The Road Transport Corporation also benefited financially as a result of its exclusive privilege of plying vehicles on the Jammu–Pathankot route.

In view of the Darbar Move (the shifting of offices between the summer and winter capitals every six months), and the geographical hazards separating the three regions, coupled with the vast expansion of each department and inconvenience caused to the people, the Gajendragadkar Committee had recommended that there should be at least one deputy head of each department and one high court judge stationed at Jammu when the offices were shifted to Srinagar, and vice versa.[93] During the Sheikh government, it was felt that in the interest of efficiency the two divisions of Kashmir and Jammu should have separate heads for each department. The memorandum prepared by the concerned minister, D. D. Thakur, in this regard was approved by the cabinet, and almost all departments were placed under regional heads for Jammu and Kashmir regions.[94]

Single-Line Administration

One of the most significant contributions of the Abdullah government was the introduction of what is known as 'single-line administration', by making the district a unit of planning and development with the objective of equal development of all regions and sub-regions, especially with a view to covering the entire countryside by involving local representatives, officials, as

---

[91] Thakur, *My Life and Years in Kashmir Politics*, 207.

[92] Thakur, *My Life and Years in Kashmir Politics*, 208–9.

[93] Gajendragadkar Commission Report, 99.

[94] Thakur, *My Life and Years in Kashmir Politics*, 212.

well as the people in the process of development and its speedy implementation. Sheikh Mohammad Abdullah and his cabinet ministers toured different parts of the state and found that the people were not satisfied with the pace of development, and there was a widespread demand for administrative reforms across the state.[95] By this time, the Planning Commission had also recast the guidelines for district planning. While it was being debated at the national level, the J&K government introduced single-line administration in 1976 through the constitution of district development boards.[96] Thus, the J&K has the distinction of being among the pioneering states in the country where the process of decentralization was carried out at the district level by way of constitution of the district development boards.[97]

The district development board consisted of the district development commissioner (chairman), the district superintending engineer, Public Works Department/Roads & Buildings (PWD/R&B) (vice chairman), other district S.Es (member), all members of Parliament, MLAs, and Members of Legislative Council (MLCs) of the district (member), chairman of TACs-NACs and president of municipalities in the district (member), and nominated members representing weaker sections and women—not more than three (member).

The board was given the powers to formulate long- and short-term plans and oversee their implementation. The deputy commissioner of the concerned district was appointed as ex-officio district development commissioner with wide financial and administrative powers. The district development commissioner was also appointed the chairman of the district development board

[95] Veerana Aivalli, 'Single Line Administration: An Administrative Experiment in Jammu & Kashmir', *Indian Journal of Public Administration* 43, no. 3 (1998), 353.

[96] For further details on the subject see Veerana Aivalli, 'Single Line Administration: An Administrative Experiment in Jammu & Kashmir', 280–81.

[97] Shushma Choudhary, 'Does the Bill Give Power to the People?', in *Panchayat Raj in Jammu and Kashmir*, ed. Mathew George (New Delhi: Institute of Social Sciences & Concept Publishing House, 1990), 63.

of the concerned district. The district development commissioner was given overall charge of the development of the district with the powers of a major head of the department for all development departments functioning in the district. Further, each district was provided with a district superintending engineer. He was provided overall and unified charge of the public works programme and staff of the district in different branches, namely, roads and buildings, irrigation, and public health engineering.[98] The superintending engineer, however, had to work under the overall control and direction of the district development commissioner.

The intention behind conferring the powers of a major head of department to the district development commissioner, and overall charge of engineering departments to the district superintending engineer of each district, was to enable these officers to issue the bulk of administrative approvals and technical as well as financial sanctions with respect to the district plan at the district level itself. A related objective was to secure an integrated approach to development at the level of the district.[99] The district development boards were to hold meetings once every quarter or whenever the district development commissioner found it to be necessary. Finalization of the annual plan for the concerned district and undertaking a mid-term review of its implementation were the priority areas.[100] The board meetings took the shape of cabinet meetings whenever the chief minister chaired meetings along with the council of ministers accompanied by the chief secretary, secretaries, and heads of the departments concerned.[101]

Up to 1996, the board was chaired by the district development commissioner. From 1996 the minister of the cabinet has been made the chairperson of the board; and representation was also

[98] Government of J&K, 'Administrative Re-organization', Order no. 2380-GD of 1976, General Administration Department, Srinagar/Jammu, 14 October 1976.

[99] Government of J&K, 'Re-organization of District Administration', Order no. 2973-GD of 1976, General Administration Department, Srinagar/Jammu, 28 December 1976.

[100] Government of J&K, 'Re-organization of District Administration'.

[101] Government of J&K, 'Re-organization of District Administration'.

given to the panchayats, town areas, weaker sections of the society, and women.

The concept of decentralized planning cannot prove effective unless a larger freedom is given to the district development boards for fixing priorities, inclusion of projects having local area relevance, and making reappropriations keeping in view public needs. This was done during the Ninth Five-Year Plan, and a beginning was made during the year 1997–8 by effecting functional and financial decentralization together with establishment of appropriate budgeting and reappropriation procedures at the decentralized levels.

Agrarian Reforms

On representations from various sections of society, the Agrarian Reforms Act of 1972, passed during the period of Mir Qasim, was repealed by a new act called the Jammu and Kashmir Agrarian Reforms Act of 1976, passed on 27 July 1976.[102] Some of the most important features of the Act were as follows.[103]

1. It abolished absentee landlordism in all forms. All those owners of land who were not cultivating their lands on the first day of September 1971, lost rights over their lands with effect from 1 May 1973. However, in the case of small landowners, some concession was made. They could retain up to 5 acres of land provided they were land revenue payers, but not among income tax payers. However, custodial land was left beyond the purview of this Act.
2. The ceiling was fixed at 12.5 acres in case of *abi* land (technically meaning paddy land). Unlike the previous acts, orchards were also brought under the ceiling. However, their ceiling was

[102] 'Agrarian Reforms Act Bill Passed with Some Modifications', *Srinagar Times*, 28 July 1976.

[103] 'State of Jammu and Kashmir Number One in Land Reforms', *Srinagar Times*, 19 August 1976; Government of J&K, Jammu and Kashmir Agrarian Reforms Act, 1976, Revenue Department, Srinagar, 1976. Act is also available at *http://jkfcr.nic.in/acts_rules.html*.

increased to between 10 and 20 per cent of the standard ceiling (12.5 acres) depending on the variation in the fertility of the land subject to the overall upper limit of ceiling at 200 kanals.

3. The act provided for the distribution of surplus land among the landless.

4. The ceiling area related to the family, and the family under the law meant husband, wife, and their children excluding their married daughters.

5. Landowners were not entitled to receive any compensation for the area left after resuming the permitted limit on rent basis and tenants of such land were vested with ownership rights, free from any levy. However, in case the landholding of the owner was small (not more than 5 acres) and he was not able to resume the land which he was otherwise entitled to, the tiller had to pay. The levy was, however, much lower than the market rate.

6. The act prohibited the alienation of land by way of sale, gift, and mortgage with possession, bequest, and exchange.

7. To facilitate the implementation of the act quickly and without any hazards, there was a significant provision in the act leaving the cultivator and the owner to their discretion if they could parcel the land among themselves as per the division of the produce of the land between the owner and the cultivator. In simple terms, if the division of the produce was 50:50, they could also distribute the ownership of the land accordingly. In such cases, the peasant could immediately become the owner of the land without any undue hassle from revenue officials.

Besides the institutional reforms, the Sheikh government also made efforts to bring the dry land under cultivation. And to realize this objective a comprehensive and ambitious programme for an irrigation network was chalked out during the Sixth Five-Year Plan (1980–85) to create an additional potential of about 338,000 hectares, for which an outlay of Rs 445 million was proposed for minor irrigation and Rs 609.4 million for medium and major irrigation. During the plan, an additional irrigation potential of 46,320 hectares was created under medium and major irrigation schemes. Also, the Command Area Development Programme was initiated in Kashmir in 1981–2 as a state sector scheme to minimize the time lag between the creation of irrigation potential and its optimal utilization.[104] The programme was entrusted with the responsibility of managing the command area of Maraval Lift Irrigation, Lethpora Lift Irrigation, Yusmarg Niu Karewa, Banimulla, and Manal Zora Storage Projects.[105]

Uplift of Weaker Sections

In order to improve the conditions of the weaker and marginalized sections, emphasis was laid on improving the educational status of Scheduled Castes and Other Backward Classes by enhancing the amount of pre-matric scholarships. In 1978, a post-matric scholarship scheme was sanctioned for providing scholarships to Scheduled Caste, Scheduled Tribe, and Other Backward Class students for pursuing studies from higher secondary to postgraduation level. Arrangement was also made to provide them coaching in science, maths and English subjects. Reimbursement of examination fee for students from these categories pursuing professional courses was introduced in 1976. To empower women, 10 per cent jobs were reserved for women in government service in 1976. Subsequently, 50 per cent seats were earmarked for women in government medical college alongside opening a women's polytechnic college.

Development

Like its predecessors, the thrust of the state development plans during the Sheikh government (1975–82) was on creating adequate infrastructure, namely, power and transport, and provision of social and community services, together with promoting agriculture, industry, trade, tourism, and above all human resources to overcome the growing problem of unemployment and to increase per capita income.

[104] For details, see Government of J&K, *Sixth Five Year Plan*, 1985–90 (Srinagar: Planning and Development Department), 48–9.

[105] For details, see Government of J&K, *Sixth Five Year Plan*, 1985–90.

The state domestic product (at constant prices of 1970–71) doubled from Rs 2,750 million in 1973–4 to Rs 4,170 million in 1981–2. The per capita income at the same prices which had shown sluggish movement from Rs 548 in 1970–71 showed quick growth from Rs 559 in 1973–4 to Rs 673 in 1982–3.[106] The annual plan allocation which was Rs 480 million in 1974–5 rose to Rs 1,850 million during 1983–4. Revenue receipts moved from Rs 1,230 million to Rs 3,530 million during the same period.[107] The export of fruits doubled from 1.689 million quintals in 1973–4 to 4.461 million quintals in 1983–4. Road transport was taken to all villages. The installed generation capacity moved up by 126 MW from 82.87 MW in 1973–4 to 209 MW in 1982–3 covering 81 per cent of the villages.[108]

The number of small-scale units registered with District Industries Centers (DICs) multiplied five times from 2,203 to 12,902 during 1973–4 to 1982–3. The number of khadi and village units too doubled from 489 to 874. The production of handicrafts, the traditional industry of the state, increased to Rs 813.7 million against Rs 200 million, and employment went up to 165,000 against 80,000 in 1973–4. The number of tourists increased from 200,000 in 1973 to 600,000 in 1982.

Literacy rate moved from 18.58 per cent in 1971 to 26.67 per cent in 1981.[109]

Crisis in Governance

Though the Sheikh's rapprochement with the centre was a great victory for India, the opportunity it threw up was not fully realized because of the struggle for hegemony between the Congress and

---

[106] Government of J&K, *Economic Review of Jammu and Kashmir, 1973–84* (Srinagar: Directorate of Economics & Statistics, Planning and Development Department), 1.

[107] Government of J&K, *Economic Review of Jammu and Kashmir, 1973–84*.

[108] Government of J&K, *Economic Review of Jammu and Kashmir, 1973–84*.

[109] Government of J&K, *Economic Review of Jammu and Kashmir, 1973–84*, 2.

National Conference. While during the period of Indira Gandhi, the Indian state pursued the policy of one party, one rule, Abdullah replicated this policy in J&K. True, in his heart of hearts, he had never forgotten the constitutional position he had enjoyed in 1953 when he was removed. After the Accord, he, at least, wanted to satisfy his 'totalitarian' ego to be allowed to rule without any challenge from any side, much less from the centre/party ruling at the centre. After the Accord, the National Conference leadership reduced the meaning of autonomy to *izat wa ābru ka muqam* (place of honour and dignity). So did it reduce its sponsored concept of *kashmiriyat*[110] to a guarantee from the centre not to disturb the National Conference monopoly of state power. But given the political project of Indira Gandhi, it was unimaginable to expect this from her. She even eyed the possibility of Sheikh joining the Congress. Minus Mir Qasim, the state Congress party obediently followed their leader, Mrs Indira Gandhi. Indeed, she was not unoften provoked by them against the Sheikh. The result was that during the entire period of Congress rule at the centre, Kashmir politics reeled under a tussle between the centre and the state, the former represented by Congress and the latter by National Conference. Contemporary sources present the period like a cold war, in which the Congress party used all its resources to harass Sheikh.[111] The then governor captures the hostile relations, saying:

> The relations between the Prime Minister and Chief Minister were certainly not good. Apart from her basic mistrust of the Sheikh's

[110] The term *kashmiriyat* was coined by the National Conference elite after 1947. It signified a distinct Kashmiri identity, and therefore a distinctive political status for Kashmir. Kashmiriyat, the concept conceived and illustrated by the ideologues of NC, has both narrative and discursive aspects. Its narrative aspect was a function of the concept to demonstrate 'unique' history of Kashmir, whereas its discursive aspect was its function to justify its clamouring for the political autonomy of Kashmir which, the party however reduced to respecting the hegemonic power of NC after 1975.

[111] For details see Abdullah, *The Blazing Chinar*, 544–54; Butt, *Kashmir in Flames*, 207–24.

loyalty to India—a mistrust that was not unjustified—her complaint was that his people deliberately put obstacles in the work of the Congress Party and subjected its members to discrimination of all kinds.... The Sheikh's complaints were different. He said that he was quite willing to cooperate with the Centre—a statement not wholly correct—but he was not willing, as Chief Minister of Jammu and Kashmir, to be a slave of the Prime Minister. What she wanted was not cooperation but unquestioning obedience. If she said stand, he was to stand up; if she said sit, he was to sit down. This he was not prepared to accept.[112]

That the Congress-led central government as well as the local Congress party were desperate to snatch power from the National Conference by bringing the Sheikh down through machinations and rigging elections is also mentioned in the assessment note of the then governor of Kashmir, a copy of which he sent to the prime minister:

> In any case, there seems, as long as he [Sheikh Abdullah] is alive, no option but to accept his rule. All that one can endeavor to ensure is to try and get him off his present track and back to normal and orderly behavior. There are neither men nor political parties who have the slightest chance of winning an election if the Sheikh were to oppose them, even if every possible effort were made to rig the election in their favour. In the conditions of today, this rigging is simply not possible and the present leadership of the Congress (I) which is hoping to come to power in his way is grossly mistaken in their belief that even if it were done it would lead to the result they desire.[113]

Balraj Puri aptly remarks: 'the Indira-Abdullah Accord was evaluated (by Congress leadership and its sponsored Indian commentators) ... by the degree of Centre's control over the state.'[114] Besides criticizing his every policy, they launched a campaign alleging misuse of power by Sheikh and his party, and even provoked revolts at different places of the state.[115] A number of times, the

---

[112] Nehru, *Nice Guys Finish Second*, 591.

[113] Nehru, *Nice Guys Finish Second*, 593.

[114] Puri, *Kashmir: Insurgency and After*, 37.

[115] Ahmad, *My Years with Sheikh Abdullah*, 75.

lion of Kashmir was humiliated publicly by Indira.[116] She was constantly told by the local Congress leaders that Sheikh continued to nurse his separative political ambitions and that he was bent upon finishing the Congress in Kashmir. Therefore, she became more aggressive and authoritarian than she was routinely, producing almost the same outbursts in the Sheikh as were generated in 1953, though this time, the objective was to perpetuate the monopoly of the National Conference and establish dynastic rule.

The principal secretary to Sheikh Abdullah also says that Abdullah was so persecuted by the Congress and the Hindu nationalist parties that had he not died, the state would have witnessed a repetition of 1953:

> The euphoria and the *déjà vu* promoted by power grabbed by Sheikh in the wake of the Delhi Accord, gradually wore off, particularly when the centre began restricting and restraining his freedom of movement. Agitations stage-managed in the three regions of the state, Kashmir Pundit agitation in the Valley, Kishtwar agitation and Rajouri agitation in Jammu province, clamour of Bhartiya Janta Party to accord status of District to Reasi, the fallout of recommendations of the Boundary Commission, and frequent calls given by the Bhartiya Janta Party in Jammu for abrogation of Art. 370 of the Indian Constitution, and the Ladakh agitation were constant irritants, which frustrated the Sheikh to an extent where he could only indulge in impotent rage. Had he lived longer, he would have most probably risen in revolt against Delhi and would doubtless have been in prison again. The final *coup de grace* was dealt when a planeload of income tax officers descended upon the Valley and made raids in businesses known to be his financiers and supporters. He was enmeshed in a web where he could only fume and fret but was incapable of doing anything. This in fact was responsible for hastening the deterioration of his health from which he was not able to recover.[117]

True, the semblance of competitive politics began in Kashmir for the first time after the Plebiscite Front was converted into the

---

[116] Ahmad, *My Years with Sheikh Abdullah*, 75; Butt, *Kashmir in Flames*, 207–24.

[117] Ahmad, *My Years with Sheikh Abdullah*, 75–6.

National Conference, when the Janata Party was ruling at the centre, and elections were held under governor's rule in 1977. On the basis of first-hand information, the editor of *Aftab* says that it was for the first time since 1947 that the practice of rejecting the nomination papers of opponents was not adhered to; and it was also for the first time that the people really participated in the elections. For the conduct of this fair election in which 'every candidate had freedom, voters enjoyed freedom, newspapers had freedom and all political parties had it,' the Kashmir watchers give credit to governor's rule.[118] It was for the first time that the centre did not decide beforehand which party was to win, as had been the case till then. The election of 1977 was a 'great improvement over the record of farcical elections between 1951 and 1972'.[119] It is also important that 'in this election the winning party was not being supported by central government and it was not the case of pre-determined winner as used to be. In fact the national and the regional parties bitterly fought against each other. The rout of the ruling party by a regional party was a unique and thrilling experience for the people.'[120]

This election, however, suffered from two limitations which vitiated its fairness: one was the hoodwinking of voters by the National Conference leadership, giving them false hope that it was fighting the same battle it had fought since 1953; second, some complaints of booth capturing and bogus voting by National Conference workers, not unoften with the support of polling staff.[121] As a participant observer, Prem Nath Bazaz gives a vivid picture of polling day. He says that polling in the Jammu region was by and large peaceful on 30 June 1977. However, this was not the case in the valley, where polling was held on 31 June.[122]

Even after the victory of the National Conference, its workers resorted to violent methods. Under the cover of merrymaking and victory parades, National Conference hoodlums went on

---

[118] Butt, *Kashmir in Flames*, 236–7.

[119] Bose, *Kashmir: Roots of Conflict*, 90.

[120] Puri, *Kashmir: Insurgency and After*, 55.

[121] For a detailed report, see Bazaz, *Democracy through Intimidation and Terror*, 124–39; Gauhar, *Elections in Jammu and Kashmir*, 86–92.

[122] Bazaz, *Democracy through Intimidation and Terror*, 124–39.

a rampage in almost all mohallas of Srinagar. In its editorial on 7 July 1977, the *Khidmat* wrote, 'after the announcement of the election results the lawlessness and anarchy are witnessed on a very large scale.'[123] The pro-Abdullah correspondent of *Hindustan Times* also corroborated the occurrence of hooliganism by National Conference workers: 'Admittedly there were cases of hooliganism in the wake of the National Conference election victory.'[124]

The vested interest in hegemonizing power had not only afflicted the Congress and its client governments; even the National Conference under Sheikh after 1975 followed suit. Besides declaring emergency in J&K in line with the rest of India on 29 June 1975, the Sheikh Abdullah government promulgated the Public Safety Ordinance on 6 November 1977[125] to suppress the opposition, especially the Jamat-i-Islami, and to muzzle the freedom of the press. The declaration of emergency was followed by the closure of Jamat schools and mass arrests of its workers.[126] And when the emergency was lifted in February 1977 under external and internal pressure in India, the Sheikh Abdullah government still insisted on silencing the opposition, prompting him to promulgate a Public Safety Ordinance which armed the state to perpetuate its hegemony by misusing its power against the institutionalized opposition.

The Jammu and Kashmir Public Safety Ordinance, as it was called, enabled the government to detain any person without communicating the grounds of detention. It also put curbs on newspapers and other publications by declaring any report prejudicial to the interest of State security and maintenance of public order.[127]

---

[123] Bazaz, *Democracy through Intimidation and Terror*, 135.

[124] Bazaz, *Democracy through Intimidation and Terror*, 135. Also, see *India Today*, vide, Tavleen Singh, *Kashmir: A Tragedy of Errors*, 9

[125] Bazaz, *Democracy through Intimidation and Terror*, 151.

[126] For details, see Qari, *Vadi Ya Purkhar*.

[127] Under the ordinance, a person could be detained for two years without communicating the grounds of detention to him, if the same was considered necessary in the public interest. The person could be detained without the case even going to the Advisory Board, which was to be set up under the ordinance, and there was also no bar on issuing a second detention order after the expiration of the first. The section of the ordinance applicable to the press gave power to the government to prohibit

Commenting on the ordinance the *Statesman* maintained: 'Whatever the truth, the scope of the present ordinance goes far beyond all democratically acceptable norms of official control. It vests the Government with draconian power of arrest without giving grounds and of virtually indefinite imprisonment.'[128] Warning Abdullah that he should ponder what he was doing, *Times of India* wrote:[129] 'the issue is not one of constitutional validity but also one of political morality.' Calling it a retrograde step, the *Hindustan Times* said editorially:[130]

> What the State administration has done through the draconian ordinance is to resurrect all the dreaded features of the emergency under which the life and property, as much as freedom of speech and movement, were restricted or altogether denied at the sweet will and pleasure of the executive.... The ordinance, therefore, comes as a total surprise and will be condemned by the people of India who have paid very dearly to win back their lost freedom.

In its biting critique, the *Kashmir Times* wrote under the caption 'Shackles on Liberty': 'By hustling through the black law called Public Safety Bill, ignoring strong opposition both inside and outside the legislative Assembly, Sheikh Abdullah's 9-month-old government has only smeared its face. It is the blackest piece of treachery against the people who elected the government on the categorical assurance that their civil liberties will be restored and fundamental rights fully protected.'[131]

Interestingly, on behalf of the people of Kashmir, the central government led by the Janata Party and the Indian press raised strong objections against the ordinance, forcing the Abdullah

---

the circulation of newspapers published in the state as well as ban the entry of newspapers published outside the state if the government felt that the same contained prejudicial reports. The ordinance also regulated the entry of persons into prohibited areas.

[128] *Statesmen*, 6 November 1977.

[129] *Times of India*, 8 November 1977.

[130] Quoted in Bazaz, *Democracy through Intimidation and Terror*, 153.

[131] See Bazaz, *Democracy through Intimidation and Terror*, 189–90.

government to send its cabinet minister D. D. Thakur to Delhi and other parts of India to neutralize the opposition.[132] And 'necessary alterations in the period of detention in accordance with the assurance given to the Central Government were accordingly introduced in the bill which was passed by the legislature on 1 April 1978.'[133]

Abdullah's ire was also driven by the rising power of the Jamat-i-Islami as a politico-socio-religious party, which Abdullah saw as a challenge to his hegemonic ambitions. No wonder the Jamat became the main target of the Sheikh's misuse of power. Not only were its schools banned but the announcement of every election would usher in mass arrests of Jamat-i-Islami members and sympathizers.[134] The acme of the National Conference policy of perpetuating its hegemony can be seen in the complete silence the government observed in 1979 when the property of Jamat workers and sympathizers was looted, plundered, and set ablaze, not even sparing their animals and plants—the former being burnt and the latter uprooted. 'After Zulju's invasion of 1320 AD', says a historically conscious eyewitness of lawlessness,

> this was the second ugly event of arson, loot, and plunder in the history of Kashmir, the National Conference government did not only give its consent to actions of the law breakers by remaining silent during the chaos created by the unruly mob, but even after the event the guilty were not brought to book making it patently clear that the riot was a handiwork of National Conference which wanted to completely uproot its rising political opponent after having largely succeeded to dent its public image by propagating that the Jamat members and sympathizers are *bad a'tiqad* (mis-believers). This compaign was made through the well organized cadre based party and their patronized imams, khatibs and mutavalis of the mosques and shrines.[135]

[132] Thakur, *My Life and Years in Kashmir Politics*, 287–8.

[133] Thakur, *My Life and Years in Kashmir Politics*, 288.

[134] Based on my field study.

[135] Interview with Profesor G.M. Shad, former Professor of History, 23 April 2016.

It should be mentioned that except for romantic slogans, namely, *Yahan kai chalay ga? Nizam-i-Mustafa* (What will work here? The code of life given by the Prophet), the Jamat has never indulged in Hindu–Muslim hatred. In fact, when, for the first time after 1947, some ugly anti-Hindu incidents took place in a part of south Kashmir in 1986, the Jamat-i-Islami demonstrated the essence of Islam in practice by protecting and sheltering the terror-stricken Hindus. This is not only what I learnt from my field study and interviews with contemporaries of the time, but was also reported by independent Indian observers of the period. The full report of an independent committee was published in the *Times of India* in its 14 April 1986 issue. *Inter alia*, it mentions, 'curiously, while accusing figures were raised against some members of secular parties, we find no evidence of the involvement of Jamat-i-Islami.'[136] The Hindu Prabandhak Committee also corroborates the findings of the said 'goodwill' team.[137] The persecutionary policy against Jamat-i-Islami also received support from the central government. Strangely enough, while B. K. Nehru, the governor, was aghast at the existence of Jamat-i-Islami in Kashmir and wanted to see it wiped out,[138] he was silent about the communal forces in Jammu or Ladakh though, unlike Jamat-i-Islami, they were openly brandishing anti-Muslim venom.

Corruption

Despite having bid adieu to the vocabulary of *raishumari* (plebiscite) around which he had mobilized the people and won mass support, Sheikh Abdullah still commanded a public following that *inter alia* expected from him a sharp departure in the system of governance. Sheikh aroused public hopes when he, immediately after assuming the reins of the government in 1975, declared that *ab waqt-i-hisab ā gaya ha* (Now the time of accountability has finally come). Life-size cut-outs and graffiti that said 'war against

---

[136] *Times of India*, 14 April 1986.

[137] Jawad Hussain Rishi, 'Jammu wa Kashmir kay Majooda Halat', *Srinagar Times*, 30 July 1986.

[138] Nehru, *Nice Guys Finish Second*, 594–5.

corruption' appeared on all major crossings in the state. Massive media hype and euphoria were built up; many officers with doubtful integrity were either shown the door or shifted to Delhi with a big bang. But within only a few months the euphoria died down, and the Sheikh regime proved as corrupt as any government that had preceded it. This is corroborated also by D. D. Thakur:

> The corrupt officers stopped making money for at least three months fearing action against them. Some senior officers whose names were recommended by the Screening Committee under the Chairmanship of the Chief Secretary were prematurely retired. This had a very healthy effect on the entire administration. The people began feeling the difference. The initial apprehension about the effectiveness and bona fides of the government were removed *but only till the creation of the second rung of Ministers in the state* (emphasis mine).[139]

During the second phase of his chief ministership, according to Ghulam Ahmad, Abdullah was

> incapable of taking firm decisions and what was worst, steadily but surely loosing grip on the administration as well as the party, National Conference. He chose a team against his better judgement and counsels, of men of dubious distinction who took full advantage of the Sheikh's jaundiced judgements and weakened state of mental facilities. Corruption became rampant and touched new heights of ingenuity as well as credulity and it became almost the order of the day. The lion had become a lamb.[140]

On investigation, the Saxena Report documented over 3,330 cases of corruption in Kashmir between 1975 and 1981.[141]

A publication which became famous by the name of *Lal Kitab* (The Red Book) was brought out by the Pradesh Congress Committee, giving details of the assets of the Sheikh family based on revenue records. According to contemporary sources, this was

---

[139] Thakur, *My Life and Years in Kashmir Politics*, 202.

[140] Ahmad, *My Years with Sheikh Abdullah*, 67.

[141] Quoted by Prakash, 'The Political Economy of Kashmir since 1947', 2057.

a devastating broadside fired by the centre, which brought the name
of the Sheikh family into the mud. Morarji Desai as prime minister
in the Janta rule forwarded the copy of the Red Book to the Sheikh
for comments. A detailed reply was sent to the 'charge sheet' refut-
ing the allegations and indicating the sources from which properties
and assets had been acquired. Whether or not Morarji Desai was
satisfied by this defence is nowhere on record. But the mud thrown
on the Sheikh in the hope that some of it may stick somewhere was
fulfilled. The Red Book became the talk of the town and tarnished
the image of the Sheikh family.[142]

Jagmohan, who was a critic of the Sheikh, finds substance in the
allegations levelled in the *Lal Kitab*. He says:

A booklet, *Lal Kitab* catalogued acts of corruption committed by the
Sheikh and his family members. In the early eighties it was being
circulated clandestinely. May be the book was politically motivated.
But the specificity of the allegations was such that only the Sheikh's
'historic greatness' prevented the public from believing the charges.
In any case, it cannot be denied that corruption thrived under the
very nose of Sheikh.[143]

There was a hot rumour that the conduit of corruption was
Begam Abdullah, as it was she who was easily accessible to corrupt elements due to her being susceptible to corruption both
during the period of Sheikh Abdullah and Farooq Abdullah. The
otherwise sympathetic Tavleen Singh also alludes to it saying,
'There was gossip, of course, about the amount of money his
[Sheikh Abdullah's] wife and her family had allegedly made but
it was something that was discussed only at dinner parties at the
Oberoi.'[144] In the forest leases, in the allotment of lands for hotels,
in making concession to industrialists, in giving contracts, in dis-
posing of government property, in the misuse of lease agreements,
in making government nominations for receiving education in
professional colleges, in the appointment/transfer of officers to

---

[142] Ahmad, *My Years with Sheikh Abdullah*, 74–5.

[143] Jagmohan, *My Frozen Turbulence in Kashmir*, 204.

[144] Singh, *Kashmir: A Tragedy of Errors*, 22.

'prize positions', corruption, nepotism, and favouritism were rampant. As a matter of fact, there was a vicious nexus between the politicians, bureaucrats, and businessmen presided over by the Sheikh family. It should be remembered that it was during this period that scores of plots were allotted at cheap rates to influential businessmen for the construction of hotels and commercial complexes on the bank of Dal Lake, regardless of the resulting irreparable loss to nature's gift on Kashmir. The existence of corruption during the period of Sheikh Abdullah was even acknowledged by Farooq Abdullah when he, at a seminar presided over by Governor Jagmohan, openly lambasted his late father for allowing the construction of hotels on the Boulevard.[145]

Discretionary powers in the hands of corrupt politicians and bureaucrats were a great impediment to the delivery of justice. A typical example of this was the selection list of teachers issued by the Selection Committee in 1978 pertaining to Poonch district. It was a case of brazen bias, favouritism, and partiality by tailoring merit through the misuse of the *viva voce* test. As a result, matriculates were appointed and candidates with postgraduate degrees were left out. This led to a widespread agitation engulfing almost all the districts of Jammu province.[146] According to D. D. Thakur, the situation was pacified when the government sanctioned 100 more vacancies for the aggrieved candidates.[147] The same misuse of discretionary power was done with regard to admissions to professional colleges. Ahmad, the principal secretary to the chief minister, says that to accommodate more aspirants over and above the sanctioned seats, for admission to the polytechnic colleges, he was asked by the chief minister to prepare three lists of candidates, recommended by ministers, officers, and legislatures; and he prepared four instead of three, the fourth one containing the meritorious

---

[145] At a seminar in early 1984, Farooq Abdullah remarked, 'A great crime was committed by permitting construction of these hotels. They should all be burnt. They have raped and pillaged the bank of the lake.' Rehmani Farooq, *Sheikh Abdullah ke Naqoosh* (Srinagar, 1988), 206.

[146] Thakur, *My Life and Years in Kashmir Politics*, 296–7; Watali, *Kashmir Intifada*, 369–72.

[147] Thakur, *My Life and Years in Kashmir Politics*, 296–7.

candidates who were not recommended by any of the above-mentioned categories of persons. All the four lists were approved. This is how the admissions were made.[148] D. D. Thakur frankly admits that the government failed to stem the rot. In a desperate situation, he even suggested automatic periodic rotation of the engineers saying that, 'if we were helpless in removing corruption, at least we could try to socialize it on the basis of an equitable principle.'[149] And he hoped against hope, 'I nurse a pious hope that someday across the sands of time some people with courage, with a flair for originality will trudge on these inhospitable areas of public administration and start a decisive crusade against the evil to get rid of the day-to-day botheration, waste of time and energy.'[150]

Financial Crisis

Notwithstanding the achievements made in certain areas of the economy, Godbole Committee found failures in the context of needs, opportunities, and promises quite disquieting.[151]

> The state's full production potential remained constrained due to policy failures and shortcomings in implementation. The output employment multipliers remained very weak. Though the relative share of the tertiary sector increased over the years, within the tertiary sector, however, government services accounted for the lion's share—more than 3 lakh families directly depended upon employment from the government in early 1990s. This put a heavy strain on the budgetary resources.[152]

Labour absorption capacity in the non-agricultural sector was circumscribed by lack of adequate investment. The average size of holdings was reduced to 3.53 acres in 1960–61, 2.31 acres in 1971–7, and 2.05 acres in 1991–2. The complementary industry

---

[148] Ahmad, *My Years with Sheikh Abdullah*, 12–13.

[149] Thakur, *My Life and Years in Kashmir Politics*, 215.

[150] Thakur, *My Life and Years in Kashmir Politics*, 216.

[151] Godbole Report, 4.

[152] Godbole Report, 6; Nisar Ali, 'Where Is Work Culture?', *Greater Kashmir*, 12 June 2007.

supporting agriculture, that is, orchards, could not come up in terms of technical progress, innovations, and development along with other agricultural allied sectors like animal husbandry, sheep breeding, apiculture, sericulture, and so on. The state domestic product of the sector registered a consistent decline over five decades. It contributed to state income, at current prices, 62 per cent in 1948–9, which came down to 47.40 per cent in 1980–81. This steep and consistent decline in its contribution had to be off-set by correspondingly feeding sectors of the economy to sustain the state, which has not happened. Along with decline in average size of holding, the net area sown per capita also registered a steep decline from 0.19 hectares in 1950–51 to 0.11 in 1984–5.[153] Highlighting the economic crisis in Kashmir during the period of Abdullah (1975–82), the then finance minister wrote:

Because of the mushroom growth of institutions of higher education, the number of educated unemployed had spiralled to an alarming degree. The employment market, on the other hand, remained stagnant for decades. The government service was the only avenue. The wage budget of the state inclusive of dearness allowance had reached an astronomical figure. The state exchequer, on the other hand, despite all efforts of the state government could not meet even the salary budget and had to wholly depend on central devolution in the form of divisible taxes and the grant-in-aid under Article 275 of the Constitution. On the basis of recommendation of the finance commission, apart from these devolutions the state owned debts to the centre which ran into thousands of crores of rupees over a period of time. These massive loans notwithstanding the financial condition remained precarious. A week earlier than the end of the month, I called for the treasury account to see whether the funds available were sufficient for our monthly salaries. At times there was a shortfall. I took an overdraft from the J&K Bank. I flew to Delhi to request the Finance Minister for release of extra funds which most often I succeeded in getting. This being the state of the finances, nothing substantial was possible to mitigate the suffering of the educated unemployed. The few schemes relating to employment included in the development plan also did not make much of an impact. Red

[153] Godbole Report, 4.

tapism, lack of dedication of those who implemented the schemes, delay in sanctioning of the loans by the financial institutions and procedural wrangles impeded the growth of industry, both small and large as also the handicraft and village industry.[154]

Farooq Abdullah

The father is senile, the son-in-law is a rogue, the son is a fool. We unfortunate Kashmiris have no other choice.

—A senior Kashmiri civil servant to B. K. Nehru, the Governor[155]

When Abdullah kicked his lifelong colleague, advisor, and point man in the Indira–Abdullah Accord, Mirza Afzal Beg, out of power, the latter told a press conference on 29 September 1978, 'I will fight for the democratic rights of the people and against the dynastic rule. In this regard I have to play an important role because I possess a treasure of information about Sheikh dynasty and their friends.'[156] While rejecting the allegation of promoting dynastic rule, Abdullah announced in a public gathering on 20 October 1978 at Pampore,

The movement [freedom movement] was not started for establishing [another] dynastic rule after the end of the dynastic rule of the Maharaja. The bogy of dynastic rule has been raised simply to cause mental agony to the people and to divert their attention from the basic issues. Lo! Would a person who jumped into the battlefield to oppose dynastic rule, even think of [establishing] a dynastic rule.[157]

Yet, exemplifying the disjunction between his words and deeds, Abdullah could not think of anyone beyond his family as worthy of succeeding him, though by his own admission neither of the two claimants[158] from his family was worthy of the position.

---

[154] Thakur, *My Life and Years in Kashmir Politics*, 218–19.

[155] Nehru, *Nice Guys Finish Second*, 600.

[156] Ishaq, *Nida-i-Haq*, 392.

[157] Ishaq, *Nida-i-Haq*, 392–3.

[158] The two claimants were G. M. Shah, his son-in-law, and Farooq Abdullah, the son of Sheikh Abdullah.

That he had a very poor opinion of his son Farooq Abdullah, who nevertheless was ultimately coronated as his successor, is borne out by two important persons of the time—D. D. Thakur, the finance minister in Sheikh's government, and Governor B. K. Nehru, who played a crucial role in the nomination of Farooq as heir-apparent to the throne, and subsequently in his appointment as chief minister amidst great difficulties from the opposing lobby. Thakur, who had taken it upon himself to see Farooq succeeding his father, says that when at the instance of Farooq, he once in 1975 sought the permission of Abdullah to allow Farooq to come along with him to Lucknow to attend a conference organized by the Avadh Bar Association, Abdullah was annoyed and said, 'I think you are doing a great dis-service to him. He is not capable of managing a small polyclinic, how do you expect him to do well in Politics?'[159] In 1981, Thakur seriously broached the subject of Farooq's nomination as successor to Abdullah. 'Farooq', he said, 'is flamboyant and non-serious in anything assigned to him. His dabbling in politics would land the state in serious confusion. The other day, even when he knew that I was seriously ill, he had got himself booked for Bangalore even when you dissuaded him from proceeding.'[160]

Yet so strong was the dynastic hold of the Sheikh that he finally coronated Farooq—the choice of his wife and of Delhi—as his successor by electing him president of the National Conference with great pomp and show. He said at the public gathering, 'what father has not been able to achieve, shall be achieved by the son.'[161] Farooq was subsequently inducted in the cabinet. Even after having chosen him as his successor, praising him before the public, and attending his oath ceremony as cabinet minister despite being ill-disposed, Abdullah confided in Thakur after returning from the ceremony, 'I wish I had a successor of my

---

[159] Thakur, *My Life and Years in Kashmir Politics*, 247. Also see Ahmad, *My Years with Sheikh Abdullah*, 68.

[160] Thakur, *My Life and Years in Kashmir Politics*, 251.

[161] Quoting Pandit Motilal Nehru, who said this about his son Pandit Jawaharlal Nehru on the occasion of his election as the Congress President. Thakur, *My Life and Years in Kashmir Politics*, 308.

choice. But I am sorry to tell you that I have no pleasure in doing what I have done today.'[162]

The then governor B. K. Nehru, as we shall see, had a national interest in backing Farooq's nomination as successor. He corroborates Thakur's account regarding Abdullah's poor opinion about the suitability of Farooq as his successor. He says:

> Being as close to Sheikh as he was, DD [Thakur] was a force trying to persuade him to make up his mind in favour of the son rather than the son-in-law. This, till virtually the last moment, the sheikh did not do. He did not like Gul Shah any more than anybody else did. But he thought so poorly of Farooq as a possible successor that he was not at all sure that the future of the state would be safe in his hands. He regarded him as a spoilt playboy, utterly irresponsible, not in the slightest interested either in politics or in administration, not capable of handling either, not particularly attached to his father and, on the whole, unfit to be given the responsibility of governing a state of the complexity of Jammu and Kashmir.[163]

The governor, who was tasked with the job of finding ways and means to overthrow the Sheikh government and bring the Congress to power, had informed the centre, especially Indira Gandhi, that though it was not possible during Abdullah's time, Farooq's accession to power would pave the way for installing a government of the centre's choice, as there would be opposition to him within the party.[164]

There were also additional reasons which weighed in Farooq's favour as compared to the other contestant: G. M. Shah, the son-in-law of Sheikh. These reasons are enumerated by the governor himself, who was all for Farooq becoming the next chief minister.

> He [Farooq Abdullah] had, like modern young people, neither interest in nor knowledge of religion, whether his own or that of others. Above all he felt himself to be an Indian. Unlike his father his loyalties went beyond the narrow confines of Kashmir to the country of

---

[162] Thakur, *My Life and Years in Kashmir Politics*, 309.

[163] Nehru, *Nice Guys Finish Second*, 600.

[164] Nehru, *Nice Guys Finish Second*, 593.

India. He was certainly no lover of Pakistan. He liked singing and dancing and eating and drinking and good company, including that of beautiful women, and the game of golf. He had shown no particular interest in politics nor in public affairs, whether of Kashmir or of India. Among the two claimants, who were the only possible candidates for the succession (so strong was the dynastic hold of the Sheikh), I naturally preferred him.[165]

Given the 'national interest' in having Farooq as the next chief minister, it is not surprising that the governor insisted the prime Minister, Indira Gandhi, broach the subject with the ailing Abdullah, though he remained non-committal to the question put by her.[166] B. K. Nehru says that, 'till the very end, I was told, he doubted his son's capacity to run the state.'[167]

D. D. Thakur, sworn enemy of G. M. Shah, was able to garner the support of all cabinet ministers in favour of Farooq.[168] And well before performing the last rites of the Sheikh, he was sworn in as the chief minister on 8 September 1982. It was a few days after that the National Conference legislative party met and unanimously elected Farooq Abdullah leader of the party.

The first blunder Farooq committed was that in a public gathering, he disgracefully dismissed the entire council of ministers he had inherited from his father, and that too when the whole cabinet was sitting with him on the dais.[169] All of them were not corrupt. According to Nehru, 'a few of them were actually honest'.[170] Nehru had advised him against doing it. In spite of having assured the governor that he would act upon his advice and first see for himself their conduct and work, Farooq did the contrary.[171] Did Farooq really replace the old *chor* (thieves) lot by a 'clean' and

[165] Nehru, *Nice Guys Finish Second*, 599–600.

[166] Nehru, *Nice Guys Finish Second*, 600–1.

[167] Nehru, *Nice Guys Finish Second*, 603.

[168] Sinha, *Farooq Abdullah*, 135.

[169] Nehru, *Nice Guys Finish Second*, 606; Ahmad, *My Years with Sheikh Abdullah*, 83.

[170] Nehru, *Nice Guys Finish Second*, 606.

[171] Nehru, *Nice Guys Finish Second*, 606; Sinha, *Farooq Abdullah*, 140.

honest council of ministers? The governor of the time did not agree with the argument Farooq advanced to justify this sweeping change: 'A whole new set of ministers was sworn in, most of them as or less competent and some of them only marginally less corrupt than the outgoing lot. There was certainly one among them who was notoriously corrupt and could not by any stretch of the imagination be regarded anywhere near honest.'[172]

B. K. Nehru, notwithstanding his support for Farooq, portrays a dismal picture of governance during Farooq's first phase (1982–3) as chief minister:

> The state continued to be administrated with the usual inefficiency and corruption. The Chief Minister had no idea of administration. His idea of improving it was to go about on his motor cycle, replace a traffic constable who, in his view, was not controlling the traffic properly and demonstrate to him by standing in his place how traffic should be controlled; or by getting on to a bus and performing the functions of a travelling ticket inspector to see that the bus conductor was not swallowing the money for the fares he had charged the passengers. These antics were very endearing but hardly the way to run a government. The complaints against him were that he was far too often away from the state, that files sent up for his orders accumulate without any attention being paid to them, that when he was on tour, his idea of correcting any mistake he might have noticed in the administration was to suspend on the spot the officers he thought were responsible for it, making him unpopular with the bureaucracy. These complaints were, unfortunately, wholly justified. The fact was that, as his father had foreseen, Farooq was not really interested in the governance of the state. He was a modern young man and was bored stiff in the company of the dull, stuffy, narrow-minded, self-serving politicians with whom he had to spend his time.... His favourite haunt was Bollywood.... The rumbles of discontent continued but there was not much improvement.[173]

After a brief spell of Farooq's rule, fresh assembly elections were due in the summer of 1983. The Congress party in no way wanted to be eclipsed by the power-wielding National Conference. Indeed,

---

[172] Nehru, *Nice Guys Finish Second*, 606–7.
[173] Nehru, *Nice Guys Finish Second*, 608–9.

Indira Gandhi supported Farooq's candidature as chief minister with an eye to the seat of power in Kashmir.[174] He would either prove a failed chief minister and thus clear the decks for Congress to capture power, as was communicated by B. K. Nehru when asked to install the Congress government by hook or by crook.[175] Or he could be persuaded to share power with the Congress. Without waiting for Farooq's failure as chief minister, Indira Gandhi opted for the second choice and desired to enter into an electoral understanding with the National Conference regarding the assembly elections of 1983. Farooq Abdullah expressed his inability, saying that his mother was not in favour of any such understanding with Congress. This was too much for Indira to stomach. To her and to her darbar, the interests of the Congress party were synonymous with the interests of India. This is the reason that Indira took exceptional interest in the election campaign of the state, which Tavleen Singh rightly calls 'A Bitter Campaign'.[176]

Further, a series of other developments took place which added fuel to the fire. Of the three which seriously hurt Mrs Gandhi's ego, one was the thin audience at the much publicized public speech of Indira Gandhi at Iqbal Park, which was attributed to the National Conference's campaign against participation in this public gathering. A second was the misbehaviour of some youth at Iqbal Park, who had flashed themselves at Mrs Gandhi, which was again attributed to the National Conference. The third was Farooq Abdullah's alliance with the opposition, and holding of an opposition conclave in Srinagar.[177] Other factors, namely, the alliance with Maulvi Farooq, rigging in the elections, meeting with Bhindranwale, raising of anti-India slogans during a cricket match in Srinagar, and the non-seriousness of Farooq[178]—all these were excuses used by Delhi to avenge the 'defiant behaviour that Mrs Gandhi was not at all accustomed to'. Tavleen Singh, who covered this crucial period as

---

[174] Sinha, *Farooq Abdullah*, 139.

[175] Nehru, *Nice Guys Finish Second*, 593.

[176] See Singh, *Kashmir: A Tragedy of Errors*, chapter 3, 'A Bitter Campaign', 34–48.

[177] For details see Farooq Abdullah, *My Dismissal* (New Delhi: Vikas, 1985).

[178] Farooq Abdullah, *My Dismissal*.

an impartial reporter without allowing herself to be swayed away by the anti-Farooq wave of patronage 'national' press, depicts the impending fall of Farooq consequent upon his 'defiant mood' in an environment when the prime minister was not worried about democratic norms: 'Farooq, everyone said, was being foolish in trying to take on Indira Gandhi. No chief minister did unless he wanted to find himself jobless. That was Mrs Gandhi's way.... After becoming prime minister she made it even more clear that disobedience and defiance were not acceptable, not just from Congress chief ministers, but even those who represented opposition parties.'[179]

Thus started the process of dismissing the Farooq government with the cooperation of the 'national' press to win Indian public opinion in favour of the plan, and also by engineering defection among National Conference legislators with the active cooperation of opponents of Farooq Abdullah, especially his brother-in-law, G. M. Shah, and D. D. Thakur, who along with 11 other MLAs had been dismissed by Farooq. The then governor has given a detailed account of the pressure exerted on him to dismiss the Farooq government,[180] but he was the last man to be dictated to by the prime minister to do something which he thought was unconstitutional. But Indira was so desperate to kick Farooq out of power that she ultimately did not shy away from taking the obnoxious step of transferring the governor and replacing him by a trusted loyalist, Jagmohan, who helped her to execute the plan of dismissing the elected government and installing in its place a government of defectors who had been purchased by vast quantities of money.[181] Jagmohan did not give Farooq an opportunity to prove his majority; he was interested in governor's rule. He knew he was engaging in an unpopular act. That is why the army, the Border Security Force, and the CRPF were put on alert, and two battalions of the Madhya Pradesh Special Armed Forces were flown to Srinagar.[182]

---

[179] Singh, *Kashmir: A Tragedy of Errors*, 49.

[180] Nehru, *Nice Guys Finish Second*, 611–41.

[181] Nehru, *Nice Guys Finish Second*.

[182] For contemporary accounts of Farooq's dismissal see Nehru, *Nice Guys Finish Second*, 609–20; Singh, *Kashmir: A Tragedy of Errors*, chapters 4 and 5, 59–76; Farooq, *My Dismissal*.

Still, however, Chief Minister Shah had to prove his majority on the floor of the house, for which Jagmohan gave him one month. As the speaker, Wali Mohammad Itoo, was a National Conference man, it was apprehended that 'he could put a spanner in the works by declaring that Shah and his twelve-man party were defectors.' Hence, by trampling democracy underfoot, the armed commandos were sent into the House on the day the assembly met, where they picked up the speaker and physically threw him out. Abdul Ghani Lone and Bhim Singh who were likely to vote for Farooq were arrested on their way to the assembly.[183]

The dismissal of the elected government by engineering defections and inventing excuses came as a rude shock to the people, who since 1977 had begun to think that governments were elected by the people, quite against the early practice of governments being imposed by Delhi. It created a deep sense of alienation among the people of Kashmir, convincing them that at least for Kashmir democracy was out of bounds. There was a reaction especially in Srinagar. As a result, the government had to impose curfew for a long period. 'There was no less than seventy-two days curfew in Srinagar during the first three months alone.'[184] That is why the chief minister came to be known as 'Gul Curfew'.

B. K. Nehru was surprised to see the zeal in Indira Gandhi and the local Congress leaders in installing Shah when she as well as the intelligence agencies and many Kashmiri Pandit leaders had 'adverse' reports against him. The subsequent events unfolded the real design. It was just within one year and about seven months (2 July 1984–6 March 1986) that the Shah government was dismissed, and that too through a dubious method: a 'secular' party engineering Hindu–Muslim discord and making it an excuse to withdraw support. It so happened that the Congress party used certain elements to vandalize some religious places in south Kashmir to destabilize the Shah government. The events in south Kashmir drew forth a reaction in Jammu, where some Muslims were manhandled either by design or to take revenge. Both oral history and reports of independent observers unanimously say

---

[183] Singh, *Kashmir: A Tragedy of Errors*, 79–80.
[184] Nehru, *Nice Guys Finish Second*, 641.

that the incidents of desecration and ill-treatment towards Hindus were the handiwork of the Congress party. Significantly, the intelligence reports also mention that the communal disturbances in south Kashmir were the creation of the Congress party: 'According to the reports received by the State Government through its intelligence agencies, it appeared that these communal disturbances had been ignited by the Congressmen by a remote control device—with a view to persuade Rajiv Gandhi to agree to the withdrawal of support.'[185]

By all accounts the political culture of the Congress party had become reprehensible. Their egos and the party had taken precedence over nation, ethics, and morality. Party men did not support any faction/party for the 'national interest', but for their self-interest followed by party interests. That is why their support was always fragile and temporary, which created a situation that reminds one of the twilight days of the great ancient and medieval empires which invariably reeled under factionalism, quick enthronement and dethronement of rulers, and a deficit of political stability. Writing about the obnoxious role of the Congressmen during the Shah government, D. D. Thakur is unsparing, though he was close to Indira Gandhi:

> Relations between the Congress and N.C. (Khaleda) started worsening primarily because Congressmen both inside and outside the legislature became more ambitious than we had anticipated in the beginning. Each of them wanted a pound of flesh for himself. Since the life of the government was coterminous with the support of the Congress Legislature party, every-one started holding the government to ransom. Demands—good, bad or indifferent were made periodically. These demands were not of a public character but all personal and individual. Despite all that we could do, they continued to be unhappy. Even in the assembly, their attitude became hostile. They did not want N.C (Khaleda) to take roots in the state. For then, they thought, Congress would be rendered irrelevant leaving the two factions of the National Conference to face each other. Finally, elections had been held in June 1983. Almost three years had already elapsed. The duration of the assembly being six years,

[185] Thakur, *My Life and Years in Kashmir Politics*, 399.

the elections were due sometime before June 1989. There would have been nothing to their credit, if the *status quo* was maintained till the date of the election.[186]

To be sure, the Congress party had no love lost for G. M. Shah in backing him for power by trampling democracy underfoot. It either wanted to weaken Farooq by splitting the National Conference party,[187] or to impress upon Farooq (and others as well) that he (or anyone) could not rule Kashmir without the blessings of or alliance with the party ruling at the centre. The Congress's purpose in both cases was to wield power in Kashmir; if not alone, at least in partnership with the local party, perhaps as a prelude to ultimately marginalizing the dominant local party or to becoming an equally powerful force. And it certainly succeeded in its design. Farooq so dreaded the ruling party at the centre that he said explicitly, 'anyone who wants to form a government in Kashmir cannot do so without sharing power with New Delhi.'[188] Thus, after the deposition of the Shah government, the central government under Rajiv Gandhi did not need to exert much in befriending Farooq to enter into an accord called the Rajiv–Farooq Accord, the purpose of which was nothing except to rule the state jointly. All political commentators believe in unison that this pre-poll alliance between National Conference and Congress was a great blunder as it closed all outlets for an anti-India/anti-centre sentiment, which was expressed by supporting the local party against the national party. Now the National Conference was as much an Indian party as the Indian National Congress or any other non-local party. The collective political sentiment therefore looked for an alternative outlet. This quest gave birth to a semi-disloyal organization, the Muslim United Front (MUF)—an umbrella organization of many religious and political parties which posed a serious challenge to the National Conference–Congress alliance in the 1986 assembly

---

[186] Thakur, *My Life and Years in Kashmir Politics*, 398.

[187] The National Conference was split into two factions—the National Conference (led by Farooq) and National Conference (Khalida)—after the alliance.

[188] *India Today* (New Delhi), 10 November 1988.

elections. The central leadership of the Congress also realized, though belatedly, that the pre-poll alliance between National Conference and Congress was politically incorrect. No less a person than Rajiv Gandhi himself admitted this in a meeting at Srinagar in 1988. A. M. Watali, the then police chief of the valley, who was present at the meeting, writes: 'Rajiv Gandhi while address-ing Congress workers in a separate room, where I was also present, admitted the mistake of forging pre poll alliance with Farooq in 1987, as it had squeezed the political space for secular and lib-eral elements, strengthening the fundamentalist forces.'[189] Since by virtue of the Rajiv–Farooq Accord, the judge and the murderer became rolled into one, the elections were rigged in the 'national interest', which sounded the death knell to the fragile peace in Kashmir. Instead of allowing the anti-establishment forces to be dissolved in the melting pot of power, the denial of democracy graduated the semi-loyal groups into full-fledged rebels.

The elections of 1987 marked a watershed in the political his-tory of J&K. It was the culmination of fraud and resulted in the beginning of a new phase in J&K—armed struggle. The large-scale rigging turned Yusuf Shah, the MUF candidate from Amira Kadal constituency, into Syed Salahud-Din, the present chief of the United Jihad Council, and united four young men to form the core of the Jammu Kashmir Liberation Front (JKLF) to start an armed struggle in J&K.[190] During the election in question

---

[189] Watali, *Kashmir Intifada*, 35.

[190] Yusuf Shah was the MUF candidate for the assembly elections of 1987 from Amirakadal Constituency. As counting started, it became clear that Yusuf Shah was winning by a landslide. His opponent, Ghulam Mohiuddin Shah, National Conference candidate, sure of a rout, left the counting centre in a dejected mood, only to be summoned back (to his surprise too) to be declared winner by presiding officials. There was widespread protest and police suppressed it ruthlessly, arresting the MUF candidate and supporters wholesale, including Yusuf Shah's election manager Yasin Malik. Both were imprisoned until the end of 1987 without any formal charge or court appearance. Yasin Malik, a young boy from downtown Srinagar, during his imprisonment met three other youth—Ashfaq Wani, Hamid Shaikh, and Javid—and decided to fight India by armed struggle; they formed the nucleus of the JKLF. Yusuf Shah who now goes by his *nom*

G. N. Gauhar, the then sessions judge, was appointed as a central election observer for three electoral constituencies—Budgam, Beerwah, and Khan Sahib. He gives a detailed account of the fraud committed by the National Conference even in those constituencies where they were in a comfortable majority. For instance, in Budgam, according to the assessment made by Gauhar after visiting it, the National Conference candidate Geelani was sure to win. But the candidate himself was not confident and did not want to take any risk. When Gauhar went to the Town Hall, where the counting was done, he saw the National Conference candidate, the deputy commissioner of Budgam, and the chief counting officer locked in a heated argument. The duo wanted the chief counting officer to do what they had been usually doing. However the officer, a Kashmiri Pandit, was not ready and furiously uttered: 'If MUF is anti-national, fundamentalist and communal organization why should they be allowed to contest? Rajiv and Farooq should have banned them to contest. But I am a trustee.'[191] Later the chief counting officer received a call from Hamidullah Khan, additional chief secretary for law and order, and the hoodwinking was done— even though there was no need for it.[192] In Charar-i-Sharief MUF had all the disadvantages, and all scales tilted towards the ruling party. But since the National Conference had a deep-seated habit of rigging elections, it could not resist the temptation here too.[193]

---

de guerre, Syed Salahud Din, was also firmly convinced by his experience that 'slaves have no vote in the so-called democratic set-up of India.' He crossed the Line of Control and since 1990 has been the commander-in-chief of the Hizb-ul-Mujahidin—the largest indigenous guerrilla force fighting India. Rigging, strong-arm tactics, ballot boxes being pre-stamped for the National Conference, massive booth capturing by ruling party gangs, voters not being allowed to cast their votes, government-nominated supervisors stopping the counting as soon as opposition candidates take a lead, and many other such tactics resulted in MUF winning just four seats and the National Conference–Congress combine taking almost all other seats, however, once again at the cost of democracy.

[191] Gauhar, *Elections in Jammu and Kashmir*, 119.

[192] Gauhar, *Elections in Jammu and Kashmir*, 119.

[193] Gauhar, *Elections in Jammu and Kashmir*, 119.

Indeed, in all constituencies in J&K, bungling became the rule and seats were deliberately conceded. Frauds were committed, including the preparation of electoral rolls and even during and after counting. Jagmohan also gives evidence of this fraud. He says:

> The manner in which the state Assembly election of 23 March 1987 were conducted caused grave misgivings about their fairness. In some constituencies, counting was suspended and the margin of victory for National conference candidates turned out to be smaller than the votes rejected. The margin of victory in Bejbehara was 100, in Wachi 122, and in Shopian 336, while in these constituencies the number of votes declared invalid was 1,172, 1,703 and 1,122 respectively.[194]

Gauhar points out that after making an assessment of every constituency he came to the conclusion that instead of the five seats which were conceded to the MUF, it would have at best captured 20 to 22 seats.[195] With such a position the MUF would not even have been able to get the status of an opposition party. But the Rajiv–Farooq combination was not prepared to allow any party to take root in J&K and pose a challenge to them. What, however, is most important is, to quote Jagmohan: 'These elections unfortunately were followed by the arrest of top-ranking leaders of the Muslim United Front (MUF). The unpalatable reality soon emerged on the surface. It found expression in the angry outburst of individuals and leaders.'[196]

While Congress and National Conference were enjoying their victory, the situation in the valley was turning very serious. In 1988, increasing numbers of youth went mysteriously missing across the valley; they crossed the Line of Control in search of weapons and training and returned with the same in large numbers, leading to a paradigm shift—mass-based armed struggle for *azadi*.

Worse, like the previous alliances between Congress and National Conference and Congress and National Conference

---

[194] Jagmohan, *My Frozen Turbulence in Kashmir*, 163.

[195] Gauhar, *Elections in Jammu and Kashmir*, 122.

[196] Jagmohan, *My Frozen Turbulence in Kashmir*, 163.

(Khalida), this alliance also repeated the old story: accord turned into discord creating a political mess during the entire period of the National Conference–Congress coalition government. Infighting between the activists and leadership of the two parties at all levels became the order of the day. There was also a scramble for power among the leaders of the Congress.[197] The unfulfilment of the promise of Rajiv Gandhi that the central government would grant a special assistance of Rs 10,000 million to J&K embittered the relations between the National Conference and the Congress. Farooq attacked the Congress for its hollow promises.[198] In an interview with Ajit Bhattacharjee, Farooq Abdullah said that he had realized that 'the National Conference would be hurt by an alliance with Congress, but there was no other way to get the funds needed for Kashmir's development from the centre.'[199] There was such a great trust deficit between these two parties that the National Conference invariably attributed the underground activities during 1988–9 to the Congress for destabilizing the Farooq government.[200] While the central government and the local Congress party accused the National Conference of complacency towards militancy, Farooq attributed it to youth discontent owing to unemployment and the apathy of the Indian government towards this basic issue: 'These [bomb blasts] showed the extent of frustration and unemployment among the Kashmiri youth, because they were suspected of being anti-India.'[201]

B. K. Nehru, who sacrificed his posting in J&K by taking issue with Indira Gandhi and who had suffered mental agony for resisting the dismissal of Farooq Abdullah, was not, however, satisfied with Farooq's working as head of the government:

> The administration was out of his control simply because he had no idea of how to administer. He travelled about the state taking

[197] Watali, *Kashmir Intifada*, 47–8.

[198] Watali, *Kashmir Intifada*, 48; Puri, *Kashmir: Insurgency and After*, 62.

[199] Ajit Bhattacharjea, *Kashmir: The Wounded Valley* (New Delhi: UBS, 1994), 253.

[200] Watali, *Kashmir Intifada*, 48.

[201] *Indian Express*, 4 November 1988.

arbitrary decisions and casually suspending officers. But he did not attend to his files; nor did he know, because he had never been told, that there is such a thing as a chain of command and that the giving of direct orders by the head of administration to its lower levels destroys in no time all administration. When I asked the Chief Secretary for an explanation of what was happening, the answer I got was, 'please don't ask me. I don't know what's happening in the state.' Farooq attended office most irregularly and his absence from the state for visiting his cronies and buddies all over the country on some excuse of dealing with matters of state was continuing to increase discontent.[202]

Ghulam Ahmad, principal secretary of Sheikh Abdullah, who had retired from service in 1987, also gives a poor picture of Farooq's functioning:

> When urgent and serious matters of State deserved his attention, he was found indulging in speeding sprees on a motorbike with film actress riding upon the pillion.... Consistency was never his forte; perhaps he believed in the axiom 'consistency is genius of mediocrity.' On several occasions, he passed different orders on the same subject and on the same file. He never applied his mind to anything important or not so important. His IQ is less than a child's IQ, and grey matter less than that of a Dolphin. That explains his indulgence in childish pranks and antics. He is no respecter of his elders. He would change his decisions like quick sand.[203]

* * *

The writings of insiders and interviews with well-placed people and commoners converge to show that the Farooq Abdullah and G. M. Shah governments maintained the long history of rampant corruption, nepotism, embezzlement, malpractice, and improper decisions. Quite interestingly, however, there is no evidence of a corrupt politician or bureaucrat having been booked by the Vigilance Department for indulging in corrupt practices. While

---

[202] Nehru, *Nice Guys Finish Second*, 629.
[203] Ahmad, *My Years with Sheikh Abdullah*, 82.

praising some good deeds of G. M. Shah, his private secretary says:

> But all these good deeds were brought to nought by the team members of his cabinet. Each member considered himself/herself as a despot in his/her domain. Favours were offered at a price; the ministers became the whipping post of political corruption, which ultimately percolated down the lowest rung of administrative hierarchy. Such was the widespread reach of corrupt practices that earned sobriquet *'Gulashah bai chodah choor'* [G.M.Shah and the fourteen thieves] for the regime. G. M. Shah felt helpless and could not stem the rot. He thought that even a slight reprimand to any member of the cabinet would bring down his government.[204]

Until 1986, the subordinate staff had been appointed by ministers and bureaucrats. It is worth citing a few illustrative examples as to how they misused their discretionary powers, rather than exposing the diseased system in its fullness. In 1984–5 the transport minister gave a list of 150 persons to the managing director of the State Road Transport Corporation with instructions to appoint them as drivers. When the managing director indicated that certain procedures for selection had to be followed, he was instructed not to bother with these procedures. And when the managing director after verifying their documents found that many persons in the minister's list did not possess even a driving licence, he was told, 'what is your use as a Managing Director if you cannot arrange even a few driving licenses?'[205]

Another example of naked corruption and politicization in recruitment to the subordinate services pertains to the selection of police inspectors and sub-inspectors in January–February 1986. Here is an eyewitness account of the duel of charges and counter-charges which the recruitment provoked at the cabinet meeting where the matter cropped up for discussion. While on the one hand it shows that regardless of the recruitment rules, 30 per cent of the officers were appointed on the recommendations of

---

[204] Ahmad, *My Years with Sheikh Abdullah*, 91.

[205] Jagmohan, *My Frozen Turbulence in Kashmir*, 190.

the leaders of the ruling party, the trading of charges also reveals favouritism and corrupt practices on the part of the chief minister and ministers in other matters for which they blamed each other, and that too by citing the persons recruited and the exact amounts swallowed by them:

> The trouble started when the Director-General of police explained the selection procedure that he had adopted. He said that, in accordance with the Chief Minister's instructions, he had decided to select 70 per cent of the candidates by 'open merit' and 30 per cent on the recommendations of the leaders of the ruling party. The police chief also disclosed that, with regard to selection from Baramulla District, he had consulted two leaders of the ruling party who were close confidents of the Chief Minister. This provoked the Agriculture Minister, who also came from Baramulla area. He taunted that the two 'close confidants' were not the only leaders from Baramulla. He, too, represented the same district. An infuriated Chief Minister interjected to say that the two leaders in question had fought for him, and he could not forget their sacrifices. The Agriculture Minister remarked that he, too, had sacrificed a lot. This further annoyed the Chief Minister. 'What sacrifices have you made? You are a thief. You have swallowed Rs. 6 lakhs. Give me account of the money.' The Agriculture Minister shot back, 'You are a bigger thief. You have swallowed Rs. 28 lakhs. You should first render account of this amount. Government is not your fiefdom. You have recruited all your relations against important posts.' The Chief Minister lost control and shouted, 'You get out from my house.' The Agriculture Minister shouted back, 'This is neither your personal property nor your family jagir. Who are you to ask me to get out?' The Chief Minister started rolling his sleeves and menacingly advanced towards the Agriculture Minister.[206]

In the statutory public undertakings, influential persons grossly misused their positions. For example, in the State Financial Corporation, four sons of political elements were appointed.[207] It was also a routine matter to appoint to the subordinate services,

---

[206] Jagmohan, *My Frozen Turbulence in Kashmir*, 189.
[207] Jagmohan, *My Frozen Turbulence in Kashmir*, 192.

especially Class IV and daily wagers, what the ruling political party called their 'workers'—those who actively supported it in winning elections—to the complete exclusion of their opponents. The ex-governor of Kashmir Jagmohan says, 'In February 1988 I found that in the preceding nine months over Rupees three crore had been spent on employing daily wagers at the instance of the Ministers.'[208]

It was during governor's rule that a major reform was carried out by promulgating the Jammu and Kashmir State Subordinate Services Recruitment Act, 1986, which provided for the establishment of an independent statutory Subordinate Service Selection Board. But the tradition of manipulation and nepotism was so strongly ingrained in Kashmir's political system that as soon as Dr Farooq Abdullah came back to power in November 1986, his government proceeded to repeal this act. One may not approve of the imposition of governor's rule, yet the fact remains that political interference in the delivery of justice was considerably minimized during the period. And once the 'elected' government came to power, the old game of manipulation of personal and party patronage resumed with full vigour. An example will suffice. During the coalition government of National Conference and the Congress, 1,100 posts were created in the Agriculture Department in mid-1989. In regard to filling these posts, a dispute arose between the coalition partners—whether the posts would be divided between the National Conference and Congress in the ratio of 80:20 or 70:30 or 60:40.[209] This was the situation in 1989 when militancy had already started knocking at the doors of the state government.

Although the Public Service Commission was in existence for a long time, it had not been able to gain public credibility as political interference and misuse of power had ended up making this crucially important body, holding the public trust, a tool for legitimizing political and bureaucratic corruption. On top of this, successive governments did not refer the vacancies of gazetted positions in the civil services to the Public Service Commission

[208] Jagmohan, *My Frozen Turbulence in Kashmir*, 199.
[209] Jagmohan, *My Frozen Turbulence in Kashmir*, 198.

for more than 10 years between 1984 and 1995–6; this was deliberately done to fill up these vacancies through the back door. It
is reported that during the decade under reference, the political
bosses appointed a large number of persons to gazetted positions
in the civil services, bypassing the Public Service Commission.[210]

Referring to the unpopularity of Farooq Abdullah, Tavleen
Singh says, 'Rumours about the Begum's [Begum Sheikh Abdullah]
corrupt ways were also more and more widely believed.'[211] On 16
September 1989, Tourism Minister R. S. Chib resigned. Among the
reasons Chib gave for his resignation were the 'direction-lessness
of the government' and the 'unprecedented chaos and corruption
in the administration'. He further said 'all the three regions of the
State were in turmoil and methods of recruitment had made the
State Government infamous amongst the people.'[212]

Writing about the conditions prevailing during Farooq
Abdullah's period, the noted journalist Nikhil Chakravartty in his
political commentary of 11 March 1990, said:

> From personal experience borne out by two visits to the valley in
> that period, this correspondent gathered the very disturbing impres-
> sion that behind all the high-visibility political impetuosity and
> exhibitionism on the part of Farooq Abdullah, his ministry emerged
> as the symbol of utter corruption and mal administration. It was
> this very phase which saw the growing activity of the secessionist
> groups emerging out in the open.[213]

When D. D. Thakur, the deputy chief minister during the Shah
government, asked the leader of a delegation from Kargil as to why
the Kargilies are approaching the minister in so many delegations
when the same issues are repeated by each one of them, the answer
is revealing as it exemplifies the political corruption:

---

[210] Based on my fieldwork. A list of backdoor appointments to gazetted
positions was appended with a writ petition filed by KAS aspirants in the
High Court of Jammu and Kashmir in 1997.

[211] Singh, *Kashmir: A Tragedy of Errors*, 121.

[212] *Kashmir Times*, 17 September 1989.

[213] Cited in Jagmohan, *My Frozen Turbulence in Kashmir*, 340.

The answer was very interesting. They mentioned a Minister who was very poor and possessed only a small hut. In the first five years after he became an MLA, he built a palatial house for his sons and daughters. During his second term, he made all his relatives gazetted officers and in the third when again he was appointed as a minister, he had made provisions for every other relative in every possible form. They maintained that the politician was a role model for them and if they could acquire a membership of the assembly or ministership even after thirty years involvement in politics, it would make life worthwhile. 'It is therefore', they said, 'that every young man of the town is running a race for leadership by showing the people that they are the real representatives of the people who project their demands.' This statement appears to hold good not only for the people of Kargil but almost the entire country. Politics has become a lucrative profession, a gamble which can make a pauper a millionaire in no time.[214]

Corruption created a class war both open and passive. John Ray who was in Kashmir from 1970 to 1986 observed:

The signs of riches were all around us in Srinagar in luxurious houses and living. This growth of wealth increased the frustration of the semi-educated graduates who filled Government Service and those who tried to fight their way into it after the gates were closed. Obvious wealth and corruption was a prime source of envy from beyond the charmed circle, a factor fuelling the expectation that a religious radicalism would bring justice and freedom.[215]

Khem Lata Wakhloo, one of the 12 National Conference defectors of 1984, is partly right in saying that one of the basic reasons for the MUF wave was that the MUF promised employment for educated youth and an end to government corruption, the latter being an especially sensitive issue with Kashmiris, who say that most of the Rs 700,000 million given to the state as development aid by Delhi over the decades has been siphoned off by a nexus of corrupt politicians, bureaucrats, and businessmen.[216] Although

---

[214] Thakur, *My Life and Years in Kashmir Politics*, 394–5.

[215] Ray, 'Kashmir 1962 to 1986', 201.

[216] Khem Lata Wakhlu and O. N. Wakhlu, *Kashmir: Behind the White Curtain 1972–91* (New Delhi: Konark, 1992), 321.

the democratic expression of dissent against misgovernance was suppressed by the misuse of power, it resurfaced violently in 1989 by providing a mass base for armed revolt against Indian rule in Kashmir.

Radical Politics

Writing about the emergence of the underground campaign launched in 1988–9 by the radicalized youth of Kashmir, who associated themselves with the JKLF operating in PAK under the leadership of Amanullah Khan for the liberation of Kashmir, Navnita Chadha Behera rightly says that their 'understanding of Kashmiri nationhood and the right of self-determination was more or less in keeping with Sheikh Abdullah's tradition.' However, they differed with him both in the stability of their purpose and the means to achieve that purpose. They 'denounced his legacy for having surrendered the plebiscite demand for the prize of power in 1975'. And unlike Abdullah's espousal of political and constitutional means to achieve his political objective, the JKLF believed in the use of force to delegitimize all those political institutions of the state which stifled Kashmir's political aspirations: 'halt all political activities, paralyze the state apparatus and thereby transfer people's allegiance and loyalty to themselves'.[217] Clearly, Chadha argues that JKLF in Kashmir was born out of the debris of the Plebiscite Front, which was buried beneath the state power Abdullah opted for in 1975. The sections of youth who took it upon themselves to revive the cause represented by the erstwhile Plebiscite Front and reorient the line of action, constitute Toynbee's 'creative minority' who withdrew from the society during the deceptive peace that followed 1975 Accord, thought of an alternative means, and returned with it to mobilize the ever-present but silent sentiment around the new narrative.

Though the JLKF militants succeeded in making the underground campaign a durable movement, in which they were helped by circumstances both endogenous and exogenous, the employment of force for political purposes, as we saw in the preceding

[217] Behera, *Demystifying Kashmir*, 146.

chapter, was not an entirely new phenomenon in Kashmir politics. The new militancy, however, was born after the Indira–Abdullah Accord of 1975. Based on an interview with one of the early militants, Fayaz Parra says that 'border crossing activities and secret meetings with ISI officials were going on since 1983.'[218] In 1985, a number of bomb blasts took place at various places in Srinagar, namely, the office of *Srinagar Times*, Indian Coffee House, the residence of Justice Anand of J&K High Court, and near Allama Iqbal Library, University of Kashmir. According to police reports, the youth involved in these activities were all local and the devices were also locally assembled.[219] The real history of exfiltration for arms training and bringing back arms, explosives, and so on, however, began from early 1988. While in 1987, we have evidence of only two boys having crossed the Line of Control, from February 1988 to August 1988 seven groups went to PAK one after the other to receive training and return with arms.[220] It was one of these initial batches of PAK-trained Kashmiri youth who successfully carried out the bomb blast at the Central Telegraph Office and the Srinagar Club on 31 July 1988. Ajaz Dar was one among them who made an abortive bid on the life of the deputy inspector general (DIG) Kashmir, A. M. Watali, on the night of 17 September 1988.[221]

It is significant to note that almost all these radicalized youth were either campaigners or polling agents or counting agents for MUF candidates in the 1987 rigged elections.[222] It must be mentioned that the MUF was supported by emotionally surcharged youth who were against the existing National Conference rule, both for political and governance reasons. They were frustrated not only because of the fraud committed with the electoral process, but also for the treatment of criminals they met with as they

[218] Fayaz Parra, 'Political Unrest in Kashmir: Response of State and Society (1989–2010)', PhD thesis, Department of History, University of Kashmir, 2015, 7.

[219] Watali, *Kashmir Intifada*, 8.

[220] Watali, *Kashmir Intifada*, 19–21.

[221] Watali, *Kashmir Intifada*, 9.

[222] For details about the elections, see Watali, *Kashmir Intifada*, 483–92.

were hounded and many were imprisoned and severely tortured. With the option of expressing their dissent through the ballot having been closed, the other option of launching an armed struggle was inviting them with open arms, as the process of mobilizing the Kashmiri youth for this purpose had already begun. The youthful emotions of sections of the youth were thus appropriated by the idea that power flows through the barrel of the gun rather than through the democratic process.

The post-1987 militancy phase was signalled by an unprecedented intensification of bomb blasts. However, the militants raised the pitch during the second half of 1989 with selective killings of a block president of the National Conference, a BJP leader Tika Lal Taploo, and a former sessions judge, Neel Kant Ganjoo, who had sentenced to death Mohammad Maqbool Bhat, the founder of the Kashmir Liberation Front. For the first time in 1989, a complete shutdown was observed on 26 January. From now onwards, the militant leadership framed its own calendar according to which India's Red Letter Days were Black Days for Kashmir. Civil curfew was observed on 26 January (India's Republic Day), 15 August (India's Independence Day), and on 27 October, the day when Indian forces landed in Kashmir. On the contrary, from now onwards, the death of Maqbool Bhat was commemorated religiously. The Lok Sabha by-elections on November 1989 evoked a complete rejection. Neither the polling officials nor the voters were ready to go against the tehreek (movement).

To charge the emotions of the people in favour of the tehreek, the militants issued orders for the closure of wine shops and cinemas, which were obeyed without demur. The people in general were so well disposed towards a change that the sight of mujahid (the people called the gun-wielding youth 'mujahid'), especially of his gun, was considered no less than a *deedar* (vision of a holy relic). The kidnapping of Dr Rubaiya Sayeed, the daughter of the then home minister of India, Mufti Mohammad Sayeed, on 8 December 1989 and the subsequent release of five militants in lieu of her release gave a tremendous moral boost to the tehreek. There were jubilations everywhere in Kashmir. The first two months, especially the last fortnight of February 1990, are remarkable for the massive mass protests. There was no village or mohalla which did not take to the streets. Nor did any section of the Muslim community remain behind in joining the movement. *Chalo* (let us go) was the keyword for going in processions to important religious and political places/institutions for consolidating the dissent and bringing pressure to bear upon the government. At the end of February 1990, about 400,000 people gathered in Srinagar to present a memorandum to the headquarters of the UN Observer Group in India and Pakistan.[223] On 1 March 1990, more than one million people marched to Srinagar from different parts of the valley chanting slogans of azadi.[224] Along the highways, roads, and byroads, the processionists, who alternately visited different places, were provided with food and drink by the people of the localities with all fanfare.[225] The women folk composed wanwun (folk songs sung by women) in honour of the militants. People, especially youngsters, set their watches to Pakistani time.[226] The loudspeakers of mosques were used for playing freedom songs. There was such frenzied mass euphoria that freedom took precedence over survival. Azadi seemed just around the corner. Nobody was ready to listen to the logic that it would be a long-drawn battle. They were comparing their massive demonstrations with those nations where similar protests had yielded the desired results.

There was a complete collapse of administration. The local dailies prominently carried public notices called *izhar-i-la taluki* (no affiliation notice) through which the office bearers/workers belonging to mainstream parties declared their disassociation from mainstream politics. Now the badge of respectability had changed. All those who were associated with mainstream parties became traitors, and those who took up the gun against the Indian state became the mujahid (warriors for a sacred cause). And those who did not support the tehreek were despised as *mukhbir* (informers).

---

[223] Habibullah, *My Kashmir*, 87; *Guardian*, 24 February 1990, cited in Victoria Schofield, *Kashmir in Conflict: India, Pakistan and the Unending War* (London: I.B. Tauris, 2003), 150.

[224] Habibullah, *My Kashmir*, 87.

[225] Based on information gathered from my fieldwork.

[226] Singh, *Kashmir: A Tragedy of Errors*, 147; Watali, *Kashmir Intifada*, 63.

It was in this charged political environment based on proactive mass involvement in the struggle for azadi and the heroes' treatment given to militants that the youth scrambled to cross the Line of Control for arms training. While in 1988 and 1989, only a few groups crossed over to PAK for training and bringing back arms, in 1990 it became a mass movement.

Governance Lapses and Militancy

Civilizations die by suicide, not by murder. This is the lesson of history. And this is what happened to the Indian state in Kashmir. The youth were forced to take up the gun when the ballot was denied to them. No less important is the fact that both the Indian and the Kashmiri intelligence agencies showed culpable negligence till militancy took deep root in the valley. Why and how this happened is a debatable issue. Also, at crucial junctures, the government adopted completely flawed policies which converted a limited insurgency into a full-blown mass movement.

Although the Kashmir youth were crossing the Line of Control before 1988, and more so in 1989, surprisingly the intelligence agencies were seemingly completely ignorant about it. The then DIG, Kashmir, rues this sad state of affairs:

> Neither the IB, nor the Raw, or Army or BSF had informed the state government and the state police of any movement of Kashmir youth across the LOC. All agencies seemed to have been blissfully ignorant of the happenings along the LOC and the impending insurgency in the state. It is believed that the reason why the elite central agencies, especially Intelligence Bureau, were ignorant about movement of Kashmiri youth crossing LOC for receiving arms training is that the said agency has been all along misused by party in power at the Centre to promote their party interests, rather than allow them to perform their legitimate duty. To make their political bosses happy IB did not hesitate even to put senior officers of the state on their watch list, who would act independently in the interest of peace and public order.[227]

---

[227] Watali, *Kashmir Intifada*, 24–5.

On the basis of the information culled from arrested militants, the DIG Kashmir revealed in a high-level security-related meeting held in July 1989 under the chair of the governor that bribery at the Line of Control facilitated people in crossing in and out.[228] Watali says that in 1988, an MLA from Kashmir briefed Rajiv Gandhi about the training of Kashmiri youth in PAK and their subsequent infiltration into the valley. Mr Gandhi took cognizance of it, and passed orders to prepare the note and forward the same to the director, Intelligence Bureau. But the prime minister's note was taken 'non-seriously'; it became the victim of red tape, as happens with any routine government correspondence. 'This is the casual approach with which such an important information, which had a bearing on national security had been handled.'[229]

The kidnapping of Dr Rubaiya Sayeed on 8 December 1989, which proved a turning point in the escalation of militancy in Kashmir, is a typical case of pathetic negligence on the part of central and state agencies about a sensitive case. She was the daughter of the then home minister of India commuting around 8 kilometres daily between her home and the hospital. Though the situation in Kashmir was very grim, the security and intelligence agencies failed to appreciate the threat perception to the daughter of India's home minister, and did not make appropriate security arrangements for her.

The police bureaucracy at the top also made a mess of things. The director general of police, J. N. Saksena, was ill-advised by some 'over-ambitious Indian Police Service officers' who actually did not want Saksena to prove equal to the post. On his part, he did not trust Kashmiri officers. 'He openly accused them of being Pakistani agents.'[230] As a result, he failed to win the confidence of the local police, which was crucially needed to strengthen the grip on the fast deteriorating situation.[231]

Unable to understand the seriousness of the ground realities and tackle the situation with prudence and oversee sustained,

---

[228] Watali, *Kashmir Intifada*, 26.

[229] Watali, *Kashmir Intifada*, 29–30.

[230] Watali, *Kashmir Intifada*, 57.

[231] Watali, *Kashmir Intifada*, 57.

patient, and professional handling by the state police, the chief minister's

> response to the emerging situation indicated a sense of bravado rather than maturity. 'I will bury those people alive who are trying to exploit religious feelings,' he declared. Many other statements were in the same vein: 'I could break legs of my political detractors' ... 'I can send lakhs of people in jail. I have the backing of the Indian government...'. 'I will send them [arrested people] to Delhi where scorching heat will melt their fat.... Anybody seen carrying a gun will be shot dead....' 'I would throw out anti-national elements into Pakistan.' He threatened the militants that Batmaloo would be re-enacted.[232]

Referring to the conversation he had with Farooq Abdullah in 1989, Balraj Puri says: 'it was obvious that Farooq's main anxiety was to satisfy Delhi and not the people of the state. As V. N. Narayanan, editor of the Tribune, Chandigarh, observed: "The impression in Srinagar is that he cannot run the government without Delhi's orders, and paradoxically enough, he cannot run the government with Delhi's orders either."'[233]

---

[232] Puri, *Kashmir: Insurgency and After*, 62. Also see *Kashmir Times*, 4 April 1988 and 11 April 1989; and *Hindustan Times*, 7 May 1989.

[233] Puri, *Kashmir: Insurgency and After*, 63; *Tribune*, 22 and 23 June 1989. A. M. Watali also states that he had advised against the handing over of the city to the CRPF, but Chief Minister Farooq Abdullah was in no mood to accept any sane advice as he was under tremendous pressure from New Delhi. Rather than caring about the alienation of the people, he was more concerned about keeping Delhi happy. Watali, *Kashmir Intifada*, 41–4.

Conclusion

Governance is the function of a cluster of factors. The priorities of governance and their hierarchical order vary from place to place, depending on specific contexts. Jammu and Kashmir is a conflicted state with both exogenous and endogenous dimensions. There is a dispute over Kashmir, a dispute with the centre, and the dispute among the regions of the state. All cumulatively create permanent instability in Kashmir. The conflict began with the Partition and it continues to stay. In July 1952, Nehru stated in the Indian Parliament, 'If you go to Kashmir you will find normalcy and that the state is functioning adequately; but behind this normalcy is the constant tension because of the enemy trying to come in to create trouble and disturb.'[1] This statement was partially modified by Prem Nath Bazaz, Nehru's contemporary, saying this 'enemy' is not any outsider, but the victimized, suppressed, and wronged people of Kashmir.[2] Though decades have passed since Bazaz quoted Nehru (mediated by his own observation), the political condition of Kashmir has not shown any improvement; instead it has worsened further, making Kashmir, in the words of Haley Duschinski, the 'exceptional state' or the 'state of exception'.[3]

[1] Bazaz, *History of Struggle for Freedom in Kashmir*, 456.

[2] Bazaz, *History of Struggle for Freedom in Kashmir*, 456.

[3] Haley Duschinski, 'Reproducing Regimes of Impunity: Fake Encounters and the Informalization of Everyday Violence in Kashmir Valley', *Cultural Studies* 24, no. 1 (2010), 113.

Giorgio Agamben, the Italian philosopher, argues that the ordering principle of the modern state operates through the exercise of sovereign power over areas designated as 'exceptions', that constitute a particular 'juridical situation' in which inhabitants are stripped of their rights, reduced to 'bare life', and submitted to the sovereign power of the state.[4] 'Within the state of exception, the sovereign becomes absolute and unreferenced, requiring neither legitimacy nor legality for eternal justification.'[5] 'The state of exception', writes de la Durantaye, 'is the political point at which the jurisdiction stops, and a sovereign unaccountability begins; it is where the dam of individual liberties breaks and a society is flooded with the sovereign power of the state.'[6] Being a disputed/contested site on which to 'establish geographical boundaries of national territory as well as the metaphorical borderline of national identity through social control', the region has been designated as a *state of exception* and emergency. One important requirement of this site of emergency and exception is militarized governance to rescue the region from external aggression and internal revolts. It is, therefore, no wonder that Kashmir is the most militarized place in the world. Prem Nath Bazaz, the radical humanist, historian, and political activist, was pained to see the increasing number of army and paramilitary forces and state police and their involvement in maintaining peace in the mid-1960s:

> When Sheikh Abdullah and his colleagues invited the Indian Army in October 1947 to defend borders against aggression of tribesmen from Pakistan, it was believed that only a small number of Indian soldiers would be required to do the job on the Kashmir soil and none would be needed to assist the civil administration to maintain peace. But with the worsening of the political situation the number of army men increased year after year who did not remain confined to the border. Sometime later, Central Reserve Police was called in which was followed by squadrons of Punjab Armed Police

[4] Giorgio Agamben, *State of Exception* (Chicago: University of Chicago Press, 2005).

[5] Duschinski, 'Reproducing Regimes of Impunity', 114.

[6] L. de la Durantaye, 'The Exceptional Life of the State: Giorgio Agamben's *State of Exception*', *Genre* 38 (Spring/Summer 2005), 182.

(PAP), Bihar Armed Police (BAP), and Madhya Pradesh Armed Police (MPAP). An armed section of the Kashmir police also was raised and reinforced; the civil police was quadrupled. The National Militia and the Special Police Unit had been brought into existence at the commencement of the National Conference rule. In 1965, a force of ten thousand men called Home Guards was recruited. Wherever you go in the Valley, the non-Kashmiri armed police is conspicuous by its presence on both sides of the bridge, post, telegraph, and telephone offices and at the gates of many other public establishments; local policemen are seldom seen at any place of importance. The estates constructed for industrial development in the Valley are occupied by these non-Kashmiri armed personnel. It is derogatory to the national pride of a Kashmiri and a source of mortification to a patriot to see his homeland present the appearance of a police state. Perhaps under the prevailing conditions the rulers cannot help it, but to achieve Kashmir's emotional integration with India a political climate has to be generated where the need for these outside armed forces will disappear, at any rate for striking terror in the hearts of local population.[7]

With each passing year, militarized governance through security-related special legislations became more and more intensive. Besides intensive militarization and the regimes of impunity in post-1947 Kashmir, preventive detention—a legal and political tool that 'applies the logic of punishment as a method of pre-emptively controlling political and economic dangers to India's national sovereignty—has been a central component of governance'.[8] While at the time of the termination of the princely order, preventive detentions were made under one special act, namely,

---

[7] Bazaz, *Kashmir in Crucible*, 93–4.

[8] For details see Haley Duschinski and Bruce Hoffman, 'Everyday Violence, Institutional Denial and Struggles for Justice in Kashmir', *Race and Class* 52, no. 4 (2011), 44–70; Duschinski, 'Reproducing Regimes of Impunity', 110–32; Haley Duschinski and Shrimoyeen Nandini Ghosh, 'Constituting the Occupation: Preventive Detention and Permanent Emergency in Kashmir', *Journal of Legal Pluralism and Unofficial Law* (2017), 15–16, https://doi.org/10.1080/07329113.2017.1347850 (accessed 28 July 2017).

J&K Public Security Act, 1946, the post-1947 self-claimed 'popular governments' or awami hukumats (people's governments) added more draconian laws to the coercive apparatuses of the state to exact political silence. These include, for example, the Preventive Detention Act, 1954 (amended in 1958, 1964, and 1967), Jammu and Kashmir Prevention and Suppression of Sabotages Act of 1965, Unlawful Activities (Prevention) Act, 1967, Jammu and Kashmir Public Safety Act, 1978, and Terrorist and Disruptive Activities (Prevention) Act, 1985. Many more draconian Acts were imposed after the outbreak of militancy in 1989–90.

Significantly, the non-application of fundamental rights in Kashmir for a long time was also aimed at political containment through preventive detention, as it was feared that fundamental rights would put limitations on the state's authority in the exercise of its powers of preventive detention and other punitive measures to produce 'recognition of the state'. Nehru's note recording the discussion on preventive detention with Kashmir's delegation on 20 July 1952, *inter alia*, observes:

> It was further pointed out that, in view of the peculiar situation in the State because of the invasion of the State by Pakistan, subsequent war and ceasefire, very special precaution had to be taken against people infiltrating for espionage, sabotage, or to create trouble otherwise. If, by the full application of the Fundamental Rights in the Constitution, these persons could not be dealt with swiftly and effectively, the situation may well deteriorate and go out of hand. Therefore, the State Government required special powers to deal with this situation and the Fundamental Rights should not take away these powers. This principle was agreed to.[9]

Kashmir is an exceptional state also because it was not considered worthy of democracy and social liberties. Until the emergence of armed resistance, elections were a farce, a fraud played out with the people. The leaders openly declared that free and fair elections would hand over Kashmir to 'anti-national' forces. Governments were imposed by New Delhi; and through these

---

[9] Noorani, *Article 370*, 135.

client governments Kashmir's special position was eroded. The 1977 election was no doubt the only fair election, but the pressures which the Sheikh government faced from the interventionist prime minister Indira Gandhi was so overbearing that according to his principal secretary, 'the erstwhile lion became mortally afraid of Mrs Gandhi'.[10] Democratic institutions existed only in name. Sumantra Bose writes:

> Sheikh Abdullah's political career was merely the most immediate and obvious casualty of the Indian decision to impose hegemonic control. Indian-controlled Kashmir's democratic institutions and processes were subverted and permanently retarded by this policy. Elections were held to constitute Indian-controlled Kashmir's legislature at regular five-year intervals—in 1957, 1962, 1967, and 1972. All of these exercises made a cruel mockery of the principles and procedures of competitive democracy. They were fixed through a crude combination of fraud and intimidation to ensure the overwhelming victory of politicians favored by New Delhi. Sheikh Abdullah was himself no paragon of democracy.[11]

The policy of imposing governments and the denial of democracy was posited to make Kashmir restive, which could be controlled by turning the state, in the words of Bose, into 'a draconian police state in which civil rights and political liberties were virtually non-existent. Mass arrests, arbitrary detentions, and violence by hired thugs against political dissidents became a norm.'[12]

Revisionist political scientists contest the celebration of Kashmir's special position under Article 370, not because it has been hollowed through presidential orders across time, but because Kashmir's position has been rendered weaker than other states. According to Madhav Khosla, the features of Article 370 have enabled extensive revisions to J&K's status by presidential orders. As such, Article 370 has even emerged as 'asymmetric in the opposite sense to that intended as it has given Indian Union

[10] Ahmad, *My Years with Sheikh Abdullah*, 75.

[11] Sumantra Bose, *Contested Lands: Israel-Palestine, Kashmir, Bosnia, Cyprus, and Sri Lanka* (New Delhi: HarperCollins, 2007), 170–71.

[12] Sumantra Bose, *Contested Lands*.

greater powers over Kashmir than it has over other states.'[13] While the presidential order of 1954 extended fundamental rights to J&K, it however included the caveat that they could be suspended in the interest of 'security' and without judicial review,[14] meaning thereby that any law passed by the state legislature that violates the freedoms granted by the Indian constitution would be considered constitutional. It is significant to note that, 'prior to the passage of the J&K PSA [Public Safety Act] 1978, all preventive detention legislations in the state were immune from judicial review under this constitutional moratorium. These special legal conditions were justified through the association of J&K with danger, threat, and instability.'[15] Thus it has aptly been said: 'Designed to protect the state's autonomy it [Article 370] has been used systematically to destroy it.'[16] Tillin's study has also shown that the Supreme Court 'has effectively permitted the weakening of Article 370 regarding J&K over time, legitimizing ongoing revisions to the Article that have extended much of the constitution to the state.' 'The court', says Tillin, 'helped to strengthen the role of the President as guardian of the spirit of autonomy, rather than acting itself to protect Jammu and Kashmir's differential autonomy from political intervention on the basis of its distinctive constitutional settlement.'[17] On the basis of the provisions of Article 370 of the Indian constitution, A. G. Noorani maintains that all the powers which were given to the Indian Union over J&K state after 17 November 1956 are unconstitutional.[18] Moreover, as it has been frankly admitted

[13] Madhav Khosla, *The Indian Constitution* (Oxford London: Oxford University Press, 2002), 75.

[14] Bose, *Kashmir: Roots of Conflict*, 69; Khosla, *Indian Constitution*; Louise Tillin, 'Asymmetric Federalism', in *The Oxford Handbook of the Indian Constitution*, eds Sujit Choudhary, Madhav Khosla, and Pratap B. Mehta (Oxford: Oxford University Press, 2016), 546; Duschinski and Ghosh, 'Constituting the Occupation', 15–16.

[15] Duschinski and Ghosh, 'Constituting the Occupation', 15–16.

[16] Government of J&K, *Report of the State Autonomy Committee*, 129.

[17] Tillin, 'Asymmetric Federalism', 548.

[18] Elaborating his assertion, Noorani states that according to Article 370(1)(b), the Indian Parliament can make laws for J&K state and the

even by those prime ministers/chief ministers during whose time Article 370 was eroded or who were a party to such erosion, that elections were manipulated and the governments were made in Delhi,[19] the surrendering of sovereignty of the state in lieu of the undue favours they got is not only unethical and immoral but also daylight robbery of national property.

There is a general notion in Kashmir that the Indian intelligence agencies and the Hindu nationalist forces play an exceptionally greater role in the governance of Kashmir. There is substance in this folklore. The memoirs of B. N. Mullik, deputy director and later director in charge of internal affairs for India's Intelligence Bureau from 1948 until the 1960s, show how deeply

---

Indian constitution can be applied to it by presidential order, but subject to two conditions: one, that the jurisdiction cannot go beyond the subjects specified in the Instrument of Accession, and second, that it requires the concurrence of the state government. Similar concurrence was required beyond the agreed ones. 'But Article 370 (2)', says Noorani, 'stipulated clearly that if that concurrence is given before the Constituent Assembly . . . is convened, it shall be placed before such Assembly for such decision as it may take thereon.' Once J&K's Constituent Assembly was 'convened' on 31 October 1951, the state government lost all authority to accord any 'concurrence' to the union. With the assembly's dispersal on 17 November 1956 after adopting the Constitution of Jammu and Kashmir, the only authority vanished which alone could (a) provide more powers to the union; and (b) accept union institutions other than those specified in the Instrument of Accession. *All additions to union powers since then are unconstitutional.* This understanding informed decisions right until 1957. It was abandoned thereafter. Noorani, *Article 370*, 7–8.

[19] 'We had promised democracy to the people of the state. Although we gloat upon our electoral process we must admit that we have not been able to give our people an opportunity to choose their representatives freely and fearlessly. The State High Court found the role played by officials in the elections conducted by the Central Election Commission objectionable. We must remove our people's impression that their rulers are selected from somewhere else', Bakhshi Ghulam Mohammad stated in 1968. Abdullah, *The Blazing Chinar*, 527. '[T]he elections [of 1977], which for the first time in thirty years were truly fair and free, produced results contrary to their [ruling party at the centre's] expectations.' Qasim, *My Life and Times*, 155.

the Indian Intelligence Bureau shaped regional political developments.[20] Even after 1975, Sheikh told the prominent Indian journalist Kuldip Nayar that the 'bureaucracy and the intelligence agencies of government of India treat me like a *chaprasi* (peon)'.[21] The Central Bureau of Investigation was also directly involved in toppling the Farooq government by bribery.[22] The memoirs of A. S. Dulat,[23] who served as joint director, Intelligence Bureau, in Kashmir from 1988 to 1990 and then as head of Indian spy agency R&AW, also throws a flood of light on the role of Indian intelligence agencies in harnessing registers of consent and affirmation. Indeed, in Kashmir, the parties in power are extremely hamstrung, not able to do more than 'routine work' in the awful presence of central institutions, whose role in conflict zones, especially in J&K, becomes much more pronounced.

The centre's policies towards Kashmir have also been greatly influenced by the pressures that Hindu nationalist parties have been bringing to bear upon the government. Being a Muslim-majority state and the one where the Muslims fought against the princely order headed by a Hindu, pressurizing the government to be uncompromising towards Kashmiri Muslims satisfies the psychological needs of the Hindutva forces. Their emotional involvement in seeing Kashmiri Muslims oppressed is fuelled by 'disloyal' Kashmiri Muslims who are supported by the neighbouring Muslim country—the traditional enemy. Even though the Congress party ruled India for a pretty long time, it has been constantly under pressure from Hindu nationalist parties which, right from the time of transfer of power to the National Conference government in 1947, have extended their anti-Muslim agenda to Kashmir and played a proactive role in the separative and anti-Kashmiri politics of Jammu Hindu Dogras and Ladakhi Buddhists. The special position of Kashmir has always been an eyesore to them. In prompting Nehru to integrate Kashmir with India, they brought significant

[20] Mullik, *My Years with Nehru-Kashmir*, 14, 27, 31, 46, 58–9.

[21] Kuldip Nayar, 'Kashmir: No Ideal Solution', *Greater Kashmir*, 9 August 2010.

[22] Nehru, *Nice Guys Finish Second*, 627.

[23] Dulat, *Kashmir: The Vajpayee Years*.

pressure to bear upon him. It was largely because of the threat of Hindutva forces that Congress governments adopted a hard-line policy towards the Sheikh and his demands. The role they played during Sadiq's time against his policy of 'liberalization' is disapprovingly underlined by Prem Nath Bazaz. So was their provocative role in the Pandit agitation both within and outside Kashmir. It may be mentioned that the Hindu nationalists and Buddhists were not even in favour of the Indira–Abdullah Accord.[24] According to Mir Qasim, Indira Gandhi approved the participation of Jamat-i-Islami of J&K in the election process because she could not impose a ban on the RSS.[25]

And what is most disturbing is the fact that Kashmir is the only place in India which has so consistently complained about not being allowed to exercise choice over its political future, despite the repeated pledges made by the Indian prime minister on behalf of the Indian nation. This is the root cause of the Kashmir problem; from this has stemmed the politics of azadi, which is the dominant note of the political aspirations of Kashmiris. To be sure, Kashmiris were disappointed with the governance of Sheikh Abdullah, but he immediately regained ground when he took up cudgels with the Indian government over the autonomous position of Kashmir. His position was further enhanced when he sacrificed power and suffered repeated imprisonments for representing the collective political sentiment of Kashmiris. Bakhshi rendered memorable service to the economic and social development of Kashmir. Still, political sentiment took precedence over the material factor, which expressed itself during the moe-e-muqaddas agitation. Sadiq was a clean man. He institutionalized the system, and followed the policy of 'liberalization', but Kashmiris developed no love for him because he rode roughshod over their basic political sentiment. Following the Indira–Abdullah Accord, Sheikh Abdullah had to give false hopes to people to retain his popular base. But when the reality was exposed with the passage of time, the Muslims followed new leaders who represented their collective political sentiment.

---

[24] Qasim, *My Life and Times*, 141; Thakur, *My Life and Years in Kashmir Politics*, 192.

[25] Qasim, *My Life and Times*, 132.

Indeed, despite multiple means having been used by successive governments, including corruption and the paternalistic role of the state, towards the identification of the people with the Indian nation-state, the ground situation bears little resemblance to a carefully designed policy. State-led development brought substantial economic and social changes, but the results are a far cry from what policy makers had planned or intended. The disjuncture between the state leaders' will and the outcome is the dominant note of this book. Prem Nath Bazaz sounded a warning note in the mid-1960s, saying: 'I can never believe that Kashmir will be integrated with India by some trick, intimidation or draconian law. . . . It would be foolish to labour under the false hope that they would be afraid of it.'[26]

Sheikh Abdullah also had a similar understanding of the Kashmiri people: 'You can hold Kashmir in subjugation with bayonets only for a short while. You can cut Kashmiris to pieces but you cannot win their hearts by force.'[27] To be sure, the plebiscitary political sentiment, which itself is the product of geography, history, economy, and culture, is deeply embedded in the hearts and minds of Muslims of the state. It is so strong that no amount of oppression could silence it, and no dose of development could appease it. This is the reason, as Praveen Swami says, 'If some had abandoned the struggle to throw India out of Kashmir, though, another generation was readying itself to take up the baton.'[28]

However, Kashmir has proved a difficult state to govern. The divide between the political sentiments of the Muslims on the one hand, and those of the Hindus of Jammu and Buddhists of Ladakh on the other, has made it extremely difficult to strike a balance. A 'nationalist' in Kashmiri political discourse becomes 'anti-national' in the Hindu and Buddhist political discourses of Jammu and Ladakh and vice versa. It happened with the Sheikh, and it continues to happen today.

Besides conflict, governance in Kashmir is constrained by financial crisis. Indeed, there are some geographical constraints

---

[26] Bazaz, *Kashmir in Crucible*, 229.

[27] Abdullah, *The Testament of Sheikh Mohammad Abdullah*, 48.

[28] Swami, *India, Pakistan and the Secret Jihad*, 77.

that have impeded economic growth, but nature has not been altogether harsh to the state. It is bestowed with such locational advantages and resource endowments that they have not only the potential to compensate for the ecological limitations, but can surely make the state economically self-sufficient. The real cause of Kashmir's economic poverty is the geographical siege of the state, which snapped Kashmir's age-old ties with its neighbourhood, converted it into a citadel with a narrow opening, and denied it the opportunities of multidimensional interaction with a greater world. One may recall that the region had had close commercial relations with China, Russia, and Central Asia up to the beginning of the twentieth century. The Anglo-Russian rivalry resulted in the snapping of ties between Kashmir and these countries to suit the interests of British paramountcy. Then the Partition of India and the resultant hostilities between India and Pakistan culminated in the closure of the Baramulla route—the most important thoroughfare between Kashmir and the Indian subcontinent and other neighbouring countries. Although colonialism ended a long time ago, and the world has undergone revolutionary changes since 1947, the colonial legacy and the baggage of Partition continue to strangulate Kashmir. Soft borders—the buzzword of the contemporary global order—have yet to make any meaningful impact on Kashmir, as the historic linkages of Kashmir with its neighbourhood are yet to be restored, except for the recent half-hearted initiative of having limited trade with Pakistan. Clearly, Kashmir has suffered enormously by being deprived of its age-old position when it functioned as a trade entrepôt between different empires, and was famously called the place 'where three empires meet', or the 'Switzerland of Asia'. Even though J&K has suffered huge losses for 'national interests', the state has not been even marginally compensated. Worse, it has been deprived of its precious water resources without any recompense. Ultimately these man-made subversions of nature robbed Kashmir of its natural advantages with devastating effects on its economic health, which in turn fuelled the political discontent. Far from compensating the state for the losses it suffered for the Indian nation, it was not included among the 'special category' states until 1990, though the scheme had been introduced in 1969.

This work has tried to demonstrate that governance in Kashmir during the period under review does not constitute one single linear time, as governance is in essence diachronic assemblage, a composite result of different systems and structures with their own internal or imposed coherence moving at different speeds. Some are stable, some move slowly, and some wear themselves out more quickly depending on various forces and factors. We find rapid socio-economic progress and spectacular growth in education and exposure owing to institutional changes, introduction of new technologies, and infrastructural development. However, in certain important respects, there was continuity rather than change. These are political disputes, the stated positions of contending parties, the deficit of democracy, human rights violations, corruption, nepotism, financial crisis, and problem of unemployment (especially from the late 1970s), making the politically contested border state of Kashmir a volcano which exploded in 1989–90. As long as Kashmir continues to be a disputed territory, governance in Kashmir, especially in the valley and other Muslim-dominated districts of the state, will not mean more than containing the conflict by any means—more foul than fair. And well-wishers of humanity would continue exclaiming in pain: *What happened to governance in Kashmir!*

Bibliography

Archival Sources

'Bakhshi Number', *Sheeraza*, J&K Academy for Arts, Culture and Language, Srinagar.

Government of Jammu and Kashmir. *4 Years*. Archives Reference Library, Srinagar, Accession no. 501/GACC.

———. 'Aid from India', text of the speech by Bakhshi Ghulam Mohammad, Prime Minister, Jammu and Kashmir, in the State Assembly, 17 March 1955, Ministry of Information, *Kashmir Today* series II. Archives Reference Library, Srinagar, Accession no. 552/GACC.

———. *Ghulam Mohammad Sadiq, Prime Minister of Jammu and Kashmir*, Jammu and Kashmir Information Department. Archives Reference Library, Srinagar, Accession no. 456.

———. *Jammu and Kashmir (August 53–August 54): A Review of the Achievements of Bakhshi Government*, Directorate of Information and Broadcasting, 1954. Archives Reference Library, Srinagar, Accession no. 506.

———. *Land Reforms in Jammu and Kashmir*, Department of Information. Archives Reference Library, Srinagar, Accession no. 666/GACC.

———. *Message by G. M. Bakhshi*, 8 October 1956. Archives Reference Library, Srinagar, Accession no. 105/GACC.

———. *On the Road to New Kashmir*, Ministry of Information and Broadcasting (Jammu: Ranbir Government Press). Archives Reference Library, Srinagar, Accession no. 1241.

———. *Our Educational Policy: A Draft Statement*, 1955. Archives Reference Library, Srinagar, Accession no. 657/GACC.

Government of Jammu and Kashmir. Department of Education, File no. 1006-Edu-249-D/54/1954, Archives Reference Library, Srinagar.

———. Department of Education, File 1651 Ed-726-US/1948, Archives Reference Library, Srinagar.

———. Department of Education, File 1414,16/Schools/1949, File 359-EDU-490/C/54/206-3-55, and File no. 1410-nil/1955, Archives Reference Library, Srinagar.

Jammu and Kashmir 1947–50, Achievements of the Three Years of Sheikh Mohammad Abdullah's Government (Jammu: Government Ranbir Press, 1951). Archives Reference Library, Srinagar, Serial no. 293, Accession no. 149/GACC.

Jammu wa Kashmir 1947–50, Sher-i-Kashmir Sheikh Mohammad Abdullah ki hukumat kay teen salah kar kardagi (Urdu) (Jammu: Ranbir Government Press, 1951). Archives Reference Library, Srinagar, Serial no. 293, Accession no. 149/GACC.

Kashmir after August 9, 1953 (Srinagar: Lalla Rookh Publications, 1955). Archives Reference Library, Srinagar, Accession no. 554/GACC.

Kashmir Today, vol. 1, no. 1 (Srinagar: Lalla Rookh Publications, September 1956).

Nagar, Hakim Parshuram. *Is Abdullah Government Anti-Hindu?* (Jammu: National Publishing House). Archives Reference Library, Srinagar, Accession no. 533/GACC.

Sadiq, G. M. *Our Educational Policy*, Government of Jammu and Kashmir (Ranbir Press, 1955).

Reports/Documents

Crisis in Kashmir Explained. Srinagar: Lalla Rookh Publications, 1953.

Government of India. *Economic Development of Jammu and Kashmir,* Report of Working Group no. III (headed by Dr R. Rangarajan). Home Department, New Delhi, March 2007.

———. *Progress of Land Reforms.* Planning Commission of India, New Delhi, 1963.

Government of Jammu and Kashmir. *Address to the Joint Session of Jammu and Kashmir Legislature* by Sadr-e-Riyasat Karan Singh. Legislative Assembly Secretariat, Srinagar, 1963.

———. 'Administrative Re-organization', Order no. 2380-GD of 1976. General Administration Department. Government of Jammu and Kashmir, Srinagar/Jammu: 14 October 1976.

———. *Administrative Report of Jammu and Kashmir* (13 April 1949–12 April 1950). Srinagar/Jammu: Ranbir Government Press, 1952.

Government of Jammu and Kashmir. *Administrative Report of Jammu and Kashmir State*. Srinagar: General Administration Department, 1951.

———. *Administrative Report*, 1958–9. Srinagar/Jammu: General Administration Department, 1959.

———. *Administrative Report*, 1959–60. Srinagar/Jammu: General Administration Department, 1960.

———. *Budget Speech of Bakhshi Ghulam Mohammad*. J&K Legislative Assembly Secretariat, Srinagar/Jammu, 1960–61.

———. *Economic Review of Jammu and Kashmir, 1973–84*. Srinagar: Directorate of Economics & Statistics, Planning and Development Department.

———. *Educational Reorganizational Committee Report*. Srinagar, 1950.

———. *Fifth Five Year Plan*. Srinagar: Planning and Development Department.

———. *Fifty Years of Animal Husbandry in Kashmir, 1947–1998*. Srinagar: Animal Husbandry Department, Kashmir Division, 1998.

———. *First Five Year Plan Document*. Srinagar: Department of Planning and Development.

———. *Five Months*. Srinagar: Bureau of Information, 1953.

———. *In Ninety Days: A Brief Account of Agrarian Reforms Launched by Sheikh Mohammad Abdulla's Government in Kashmir*. Jammu: Land Reforms Officer, 1948.

———. *Jammu and Kashmir 50 Years*. Srinagar: Department of Information, 1998.

———. Jammu and Kashmir Agrarian Reforms Act, 1976. Department of Information, Srinagar, 1976.

———. Jammu and Kashmir Constituent Assembly, Opening Address by Sheikh Mohammad Abdullah the Hon'ble Prime Minister of Jammu and Kashmir, Srinagar, 5 November 1951.

———. 'Re-organization of District Administration', Order no. 2973-GD of 1976. General Administration Department, Srinagar/Jammu, 1976.

———. *Report of the Commission of Inquiry against Bakshi Ghulam Mohamed* (Ayyangar Commission), 30 June 1967.

———. *Report of the Commission to Enquire into Grievances and Complaints* (B. J. Glancy). Jammu: Ranbir Government Press, 1932.

———. *Report of the Committee on Economic Reforms for Jammu and Kashmir* (Godbole Report). Srinagar/Jammu: General Administration Department, 1998.

———. *Report of the Development Review Committee*, Part IV, Industrial Development. Srinagar: Industries & Commerce Department.

Government of Jammu and Kashmir. *Report of the Inquiry Committee Appointed to Examine the Working of Land Reforms, Price Control, etc.* (Wazir Committee Report). Srinagar, 1953.

———. *Report of the J&K State Finance Commission*, November 2010, vol. I.

———. *Report of the Jammu and Kashmir Commission of Inquiry* (Gajendragadkar Commission Report). Srinagar, December 1968.

———. *Report of the Kashmir Constitutional Reforms Conference* (Glancy Commission Report). Jammu: Ranbir Government Press, 1932.

———. *Report of the Land Commission*, 1968.

———. *Report of the Working Group on Good Governance in Jammu and Kashmir.* Srinagar, 2007.

———. *Republic Day Series (January 1956–8).* Srinagar: Press Information Bureau, n.d.

———. *Sixth Five Year Plan*, 1985–1990. Srinagar: Planning and Development Department.

———. *Some Basic Statistics.* Jammu: Ranbir Government Press, 1964.

———. State of Jammu and Kashmir Agrarian Reforms Act of 1972. Revenue Department, Srinagar, 1976.

———. *Unanimous Vote of Confidence in Bakhshi Ghulam Mohammad: An Account of the Proceedings of the State Legislature on the Motion of Confidence in Bakhshi Government (October 5, 1953).* Jammu: Naya Kashmir Orient Press, 1953.

Hamara Pyare Khalid-e-Kashmir. Lalla Rookh Publications, Srinagar, n.d.

Iyer, K. Gopal. *The Findings of the Empirical Study on Land Reforms in India, 1989–90*, Summary Report, Land Reforms Unit. LBS National Academy of Administration, Mussoorie, 1990.

Kashmir: An Estimate of Progress. Srinagar: Lalla Rookh Publications, n.d.

Kashmir: An Open Book. Srinagar: Lalla Rookh Publications, 1958.

Kashmiri Mussalmanun ki Siyasi Jadujahad, 1931–1939: Muntakhib Dastawaizat. Srinagar, 1991.

'Srinagar Dairy', *Kashmir Today*, vol. 1, no. 1. Srinagar: Lalla Rookh Publications, September 1956.

Dissertations/Theses

Aziz, Javid ul. 'Economic History of Modern Kashmir with Special Reference to Agriculture (1947–1989)'. PhD thesis, Department of History, University of Kashmir, Srinagar, 2010.

Bakshi, Shirin. 'Social Change in Kashmir with Special Reference to European Impact (1846–1947)'. PhD thesis, Department of History, University of Kashmir, 1992.

Dar, Ashiq Hussain. 'Inter-community Relations in Kashmir (Sixteenth to Twelfth Century)'. PhD thesis, Department of History, University of Kashmir, 2015.

John, Masarat, *Jammu and Kashmir under Bakhshi Ghulam Mohammad*, M.Phil Dissertation (unpublished), Department of History, University of Kashmir, Srinagar, 2012.

Kanjwal, Hafsa. 'Building a New Kashmir: Bakhshi Ghulam Mohammad and the Politics of State-Formation in a Disputed Territory'. PhD thesis, University of Michigan, 2017. https://deepblue.lib.umich. edu/bitstream/handle/2027.42/138699/hafsak_1.pdf?sequence= 1&isAllowed=y (accessed 21 October 2017).

Para, Altaf. 'Emergence of Modern Kashmir: A Study of Sheikh Mohammad Abdullah's Role'. PhD thesis, Department of Political Science, University of Kashmir, 2008.

Parra, Fayaz. 'Political Unrest in Kashmir: Response of State and Society (1989–2010)'. PhD thesis, Department of History, University of Kashmir, 2015.

Sheikh, Bashir Ahmad. 'Kashmir's Response to European Technology'. MPhil thesis, Department of History, University of Kashmir, 1984.

Books

Abdullah, Farooq. *My Dismissal*. New Delhi: Vikas, 1985.

Abdullah, Sheikh. *The Testament of Sheikh Abdullah*. New Delhi: Palit and Palit, 1974.

———. *Ātash-i-Chinar* (autobiography) (Urdu). Srinagar: Ali Mohammad and Sons, 1986.

———. *The Blazing Chinar* (English translation of Abdullah's autobiography, *Ātash-i-Chinar*, by Mohammad Amin). Srinagar: Gulshan Books, 2013.

Agamben, Giorgio. *State of Exception*. Chicago: University of Chicago Press, 2005.

Ahmad, Ghulam. *My Years with Sheikh Abdullah: Kashmir 1971–1987*. Srinagar: Gulshan Books, 2008.

Ahmad, Imtiyaz. *Ritual and Religion among Muslims in India*. New Delhi: Manohar, 1981.

Ahmad, Mirza Nazir. *Management of Tourism in Kashmir*. New Delhi: Dilpreet Publishing House, 2010.

Akbar, M. J. *India: The Siege Within*. New York: Viking, 1985.

———. *Kashmir behind the Vale*. New Delhi: Viking, 1991.

Alam, Jawaid, ed. *Kashmir and Beyond (1966–84): Selected Correspondence between Indira Gandhi and Karan Singh*. New Delhi: Penguin/Viking, 2011.

Ali, Agha Ashraf. *Kuch to Likhye ki Loag Kehte Hain*. Srinagar: Shalimar Art Press, 2010.

Amnesty International. *India: Torture and Deaths in Custody in Jammu and Kashmir*. London: Amnesty International, January 1995.

Anand, A. S. *The Constitution of Jammu and Kashmir: Its Development and Comments*. New Delhi: Universal Law Publishing, 1980.

Asia Watch, *Kashmir under Siege*. New York: Human Rights Watch, May 1991.

Asia Watch and Physicians for Human Rights. *The Human Rights Crisis in Kashmir: A Pattern of Impunity*. New York: Human Rights Watch, June 1993.

———. *Rape in Kashmir: A Crime of War*. New York: Human Rights Watch, June 1993.

Bamzai, Anana Koul. *The Kashmiri Pandit*. Delhi: Utpal, 1991.

Bardhan, P. K. *The Political Economy of Development in India*. New Delhi: Oxford University Press, 1984.

Baruah, Sanjib. *Durable Disorder: Understanding the Politics of Northeast India*. New Delhi: Oxford University Press, 2005.

Bashir, Khalid Ahmad. *Kashmir: Exposing the Myth behind the Narratives*. New Delhi: Sage, 2017.

Basu, Durga Das, ed. *Sardar Patel's Correspondence 1945–50*. Ahmedabad: Navajivan, 1971.

Bazaz, Prem Nath. *Azad Kashmir: A Democratic Socialist Conception*. Mirpur: Verinag, 1992.

———. *Democracy through Intimidation and Terror*. New Delhi: Heritage, 1978.

———. *The History of Struggle for Freedom in Kashmir*. Srinagar: Gulshan Books, 2009 (originally published by New Delhi: Kashmir Publishing Company, 1954).

———. *Kashmir in Crucible*. Srinagar: Gulshan Books, 2005 (originally published by Pamposh, 1967).

Beg, Mirza Mohammad Afzal. *On the Way to Golden Harvest: Agricultural Reforms in Kashmir*. Srinagar: Government of Jammu and Kashmir, 1951.

Behera, Navnita Chadha. *State, Identity and Violence: Jammu, Kashmir and Ladakh*. New Delhi: Manohar, 2000.

———. *Demystifying Kashmir*. New Delhi: Pearson/Longman, 2007.

Bell, Stephen and Andrew Hindmoor. *Rethinking Governance: The Centrality of the State in Modern Society*. Cambridge: Cambridge University Press, 2009.

Bevir, Mark, ed. *Public Governance*. London: Sage, 2007.

Bevir, Mark and Rod Rhodes. *Interpreting British Governance*. Basingstoke: Routledge, 2003.

Bhartiya Jana Sangh. *Kashmir Problem and Jammu Satyagraha*. Delhi, 1952.

Bhat, Mohammad Yousseff. *Prison Dairy: Kashmir Untold Story 1965–68*. Srinagar, 2017.

Bhattacharjea, Ajit. *Kashmir: The Wounded Valley*. New Delhi: UBS, 1994.

———. *Sheikh Mohammad Abdullah: Tragic Hero of Kashmir*. New Delhi: Roli Books, 2008.

Bhushan, Vidya. *State Politics and Government: Jammu and Kashmir*. Srinagar: Jaykay Book House, 1985.

Birdwood, C. B. *India and Pakistan: A Continent Decides*. New York: Praeger, 1954.

———. *Two Nations and Kashmir*. London: Robert Hale, 1956.

Bose, Sumantra. *The Challenge in Kashmir: Democracy, Self-Determination and Just Peace*. New Delhi: Sage, 1997.

———. *Kashmir: Roots of Conflict, Paths to Peace*. New Delhi: Vistaar, 2003.

———. *Contested Lands: Israel-Palestine, Kashmir, Bosnia, Cyprus, and Sri Lanka*. New Delhi: HarperCollins, 2007.

———. *Transforming India: Challenges to the World's Largest Democracy*. Cambridge: Harvard University Press, 2013.

Bourdieu, Pierre. *Distinction: A Social Critique of the Judgment of Taste*. London: Routledge/Kegan Paul, 1984.

Brecher, Michael. *The Struggle for Kashmir*. Toronto: Ryerson Press, 1953.

Butt, Sanaullah. *Kashmir in Flames*. Srinagar: Ali Mohammad & Sons, 1981.

Campbell, Johnson. *Mission with Mountbatten*. London: Robert Hale, 1951.

Chakrabarty, Bidyut and Mohit Bhattacharya, eds. *The Governance Discourse: A Reader*. New Delhi: Oxford University Press, 2008.

Choudhary, Zafar. *Kashmir Conflict and Muslims of Jammu*. Srinagar: Gulshan Books, 2015.

Chowdhary, Rekha. *Identity Politics in Jammu and Kashmir*. New Delhi: Vitasta, 2010.

———. *Jammu and Kashmir: Politics of Identity and Separatism*. New Delhi: Routledge, 2016.

Committee for Initiative on Kashmir. *Kashmir: A Land Ruled by Gun*. New Delhi, 1991.

Corbridge, Stuart and John Harriss. *Reinventing India: Liberalization, Hindu Nationalism and Popular Democracy*. London: Polity Press, 2000.

Crane, Robert I., ed. *Area Handbook on Jammu and Kashmir State*. Chicago: University of Chicago Press, 1956.

Dar, G. M. *Sheikh Abdullah: Important Speeches 1975–77*. Srinagar, 1981.

Das, Jyoti Bhusan. *Jammu and Kashmir*. The Hague: Martinus Nijhoff, 1968.

Devadas, David. *In Search of a Future: The Story of Kashmir*. New Delhi: Penguin/Viking, 2007.

Dhar, D. N. *Dynamics of Political Change in Kashmir: From Ancient to Modern Times*. New Delhi: Kanishka, 2001.

Dulat, A. S. *Kashmir: The Vajpayee Years*. New Delhi: HarperCollins, 2015.

Engineer, Asghar Ali. *Secular Crown on Fire: The Kashmir Problem*. Delhi: Ajanta, 1991.

Fazili, Manzoor. *Kashmir Government and Politics*. Srinagar: Gulshan Books, 1982.

Frankel, F. *India's Political Economy, 1947–1977: The Gradual Revolution*. Princeton: Princeton University Press, 1979.

Ganguly, Sumit. *The Crisis in Kashmir: Portents of War, Hopes of Peace*. London: Cambridge University Press, 1997.

Ganhar, J. N. and P. N. Ganhar. *Buddhism in Kashmir and Ladakh*. New Delhi: Ganhar and Ganhar, 1956.

Gauhar, G. N. *Hazratbal: The Centre Stage of Kashmir Politics*. Srinagar: Gulshan Books, 1998.

———. *Elections in Jammu and Kashmir*. New Delhi: Manas, 2002.

Geelani, Syed Ali Shah. *Wular Kinaray* (Urdu), vol. II. Srinagar: Meezan, 2012.

Gill, Lesley. *The School of the Americas: Military Training and Political Violence in the Americas*. Durham: Duke University Press, 2004.

Gramsci, Antonio. *Selections from the Prison Notebooks*. London: International Publishers, 1971.

———. *A Selection from the Cultural Writings*. London: Lawrence & Wishart, 1985.

Guha, Ramachandra. *India after Gandhi: The History of World's Largest Democracy*. New Delhi: Picador, 2008.

Gupta, Sisir. *Kashmir: A Study in India Pakistan Relations*. New Delhi: Asia Publishing House, 1966.

Habibullah, Wajahat. *My Kashmir: Conflict and the Prospects of Enduring Peace*. Washington, D.C.: United States Institute of Peace, 2008.

Haksar, Nandita. *The Many Faces of Kashmiri Nationalism*. New Delhi: Speaking Tiger, 2015.

Hansen, Thomas and Finn Stepputat. *States of Imagination: Ethnographic Explorations of the Postcolonial State*. Durham, NC: Duke University Press, 2002.

Hassnain, Fida Mohammad. *British Policy towards Kashmir (1846–1921)*. New Delhi: Sterling, 1974.

Huttenback, Robert A. *Kashmir and the British Raj*. New York: Oxford University Press, 2004.

Ishaq, Munshi Mohammad. *Nida-i-Haq* (autobiography). Srinagar: KBF Printers, 2014.

Jackson, Robert H. *Quasi-States: Sovereignty, International Relations and Third World*. Cambridge: Cambridge University Press, 1989.

Jagmohan. *My Frozen Turbulence in Kashmir*. New Delhi: Allied, 1991.

Kak, B. L. *Kashmir Problems and Politics*. New Delhi: Seema, 1981.

Kaul, Anand. *The Kashmiri Pandit*. Delhi: Utpal, 1991.

Kaul, Pyarelal. *Crisis in Kashmir*. New Delhi: Suman, 1991.

Kaul, Shridhar and H. N. Kaul. *Ladakh through the Ages: Towards New Identity*, 3rd edn. New Delhi: Indus, 2004.

Khan, Ghulam Hassan. *Freedom Movement in Kashmir 1931–1940*. Srinagar: Gulshan Books, 1980.

Khawaja, Ghulam Mohammad. *Sheikh Abdullah ki Wazarat kay Zawal kay Asbab* (The Causes of the Downfall of Sheikh Abdullah Government). Delhi, 1954.

Khohami, Hasan Shah. *Tarikh-i-Hasan*, vol. I, trans. Sharif Hussain Qasim. Srinagar: Ali Mohammad and Sons, 2013.

Khosla, Madhav. *The Indian Constitution*. Oxford: Oxford University Press, 2002.

Kjaer, Anne Mette. *Governance*. Cambridge: Polity Press, 2004.

Knight, E. F. *Where Three Empires Meet*, London: Longmans, Green, & Co., 1905.

Kohli, Atul. *Democracy and Discontent: India's Growing Crisis of Governability*. Cambridge: Cambridge University Press, 1990.

Koithara, Verghese. *Crafting Peace in Kashmir: Through Realist Lens*. New Delhi: Sage, 2004.

Korbel, Josef. *Danger in Kashmir*. Princeton: Princeton University Press, 1954.

Koul, M. L. *Kashmir, Past and Present: Unraveling the Mystique*. http://www.koausa.org/pastpresent/chapter11.html (accessed 9 November 2017).

Kumar, Sonali. *Un-Making Kashmir: A Bureaucrat Reveals*. New Delhi: Manhas, 2017.

Lakhanpal, P. L. *Essential Documents and Notes on Kashmir Dispute*. New Delhi: Transnational Books, 1965.

Lal, D. *The Hindu Equilibrium*. Oxford: Clarendon Press, 1988.

Lamb, Alaster. *Kashmir: A Disputed Legacy 1846–1990*. Karachi: Oxford University Press, 1993.

———. *The Incomplete Partition: The Genesis of the Kashmir Dispute (1947–1948)*. United Kingdom: Roxford Books, 1997.

Lawrence, Walter. *The Valley of Kashmir*. Srinagar: Gulshan Books, 2002 [1895].

Levy, Adriana and Scoot-Clark Cathy. *The Meadow: Kashmir 1995—Where the Terror Began*. London: HarperCollins, 2012.

Lijphart, Arend. *Democracy in Plural Societies: A Comparative Exploration*. New Haven, CT: Yale University Press, 1977.

Madhok, Balraj. *Kashmir: Centre of New Alignment*. New Delhi: Deepak Prakashan, 1963.

Mahjoor, Ghulam Ahmad. *Kalam-i-Majhoor*, vol. X. Srinagar, n.d.

Malaviya, H. D. *Land Reforms in India*. New Delhi: AICC, 1954.

Migdal, J. *State in Society: Studying How State and Societies Transform and Constitute One Another*. Cambridge: Cambridge University Press, 2004.

Migdal, J., Atul Kohli, and V. Shue, eds. *State Power and Social Forces: Domination and Transformation in the Third World*. Cambridge: Cambridge University Press.

Mir, Abdul Rehman. *Kashmir Mein Abpashi*. Srinagar: Shaheen, 1981.

Misra, Neelesh and Rahul Pandita. *The Absent State: Insurgency as an Excuse for Misgovernance*. Gurgaon: Hachette, 2010.

Mohammad, Zahid G. *Kashmir: Changing Shades*. Srinagar: Gulshan Books, 2013.

Moorcroft, William and Trebeck George. *Travels in Ladakh and Kashmir*, vol. I. Calcutta: Asiatic Society, 1841.

Mullik, B. N. *My Years with Nehru-Kashmir*. New Delhi: Allied, 1971.

Najar, G. R. *Kashmir Accord (1975): A Political Analysis*. Srinagar: Gulshan Books, 1988.

Nehru, B. K. *Nice Guys Finish Second*. New Delhi: Viking, 1997.

Nehru, Jawaharlal. *Letters to Chief Ministers 1947–1964*, vol. 3: *1952–1954*, gen. ed. G. Parthasarthi. Teen Murti, New Delhi, 1987. https://archive.org/stream/letterstochiefmi03nehr/letterstochief-mi03nehr_djvu.txt (accessed 23 February 2018).

Noorani, A. G. *Article 370: A Constitutional History of Jammu and Kashmir*. New Delhi: Oxford University Press, 2011.

Nordstrom, Carolyn. *Shadows of War: Violence, Power, and International Profiteering in the Twenty-first Century*. Berkeley: University of California Press, 2004.

Pampori, M. S. *Kashmir in Chains*. Srinagar: Pampori Publishing House, 1992.

Panikkar, K.M. *Gulab Singh (1792–1858): Founder of Kashmir*. Srinagar: Gulshan Books, 1989.

Parumu, R. K. *A History of Muslim Rule in Kashmir*. New Delhi: People's Publishing House, 1969.

Patel, Sardar Vallabhbhai. *Sardar Patel's Correspondence, 1945–1950*, vol. I. Ahmedabad: New Light on Kashmir, 1971.

Pierre, Jon and Guy Peters. *Governance, Politics and the State*. Basingstoke: Macmillan, 2000.

Poplai, S. L. *Selected Documents on Asian Affairs: India 1947–50*, vol. I. Bombay: Oxford University Press, 1959.

Puri, Balraj. *Jammu: A Clue to Kashmir Triangle*. New Delhi, 1966.

———. *Jammu and Kashmir: Triumph and Tragedy of Indian Federalisation*. New Delhi: Sterling, 1981.

———. *Simmering Volcano: Study of Jammu's Relations*. New Delhi: Sterling, 1983.

———. *Kashmir towards Insurgency*. New Delhi: Orient Longman, 1993.

———. *Jammu and Kashmir: Regional Autonomy (a Report)*. Jammu, 1999.

———. *Kashmir: Insurgency and After*. New Delhi: Orient Longman, 2008.

Qari, Saif-ud-Din. *Vadi Ya Purkhar*. Srinagar: Chinar, n.d.

Qasim, Mir. *My Life and Times*. New Delhi: Allied, 1992.

Rai, Mirdu. *Hindu Rulers Muslim Subjects*. New Delhi: Permanent Black, 2004.

Raju, C. Thomas. *Perspectives on Kashmir*. Boulder: Westview Press, 1992.

Rasool, G. and Minakshi Chopra. *Education in Jammu and Kashmir: Issues and Documents*. Jammu: Jay Kay Book House, 1986.

Rhodes, Rod. *Understanding Governance: Policy Networks, Governance, Reflexivity, and Accountability*. Buckingham: Open University Press, 1997.

Rizvi, Janet. *Ladakh: Crossroads of High Asia*. New Delhi: Oxford University Press, 1996.

———. *Trans-Himalayan Caravans: Merchant Princes and Peasant Traders in Ladakh*. New Delhi: Oxford University Press, 1999.

Roy, Srirupa. *Beyond Belief: India and the Politics of Postcolonial Nationalism*. Durham, NC: Duke University Press, 2007.

Rudolph, L. and S. H. Rudolph. *In Pursuit of Lakshmi*. New Delhi: Orient Longman, 1998.

Sahni, Sat Paul. *Jammu and Kashmir: Landmarks in State Public Administration (Post Independence)*. Jammu: Jay Kay Book House, 1997.

Saraf, Mohd Yousuf. *Kashmiris Fight for Freedom*, vol. I. Lahore, 1977.

———. *Kashmiris Fight for Freedom*, vol. II. Lahore: Ferozsons, 2005.

Saraf, Mulk Raj, ed. *Jammu & Kashmir Trade Guide (J&K Guide)*. Delhi: Universal Publications, 1969.

Schaffer, H. B. *The Limits of Influence: America's Role in Kashmir*. New Delhi: Penguin Viking, 2009.

Schofield, Victoria. *Kashmir in Conflict: India, Pakistan and the Unending War*. London: I.B. Tauris, 2003.

Seru, S. L. *History and Growth of Education in Jammu and Kashmir 1872–1973*. Srinagar: Ali Mohammad and Sons, 1977.

Shah, Ghulam and G. N. Reshi. *State Subjectship in Jammu and Kashmir*. Srinagar: Jupitor, 1988.

Sharma, Suresh K. and S. R. Bakhshi. *Encyclopedia of Kashmir*, series 4. New Delhi: Anmol, 1995.

Singh, Tavleen. *Kashmir: A Tragedy of Errors*. New Delhi: Penguin, 1996.

Sinha, Aditya. *Farooq Abdullah: Kashmir's Prodigal Son (A Biography)*. New Delhi: UBS, 1996.

Swami, Praveen. *India, Pakistan and the Secret Jihad: The Covert War in Kashmir, 1947–2004*. London: Routledge, 2007.

Teng, Mohan Krishan. *Kashmir: Article 370*. New Delhi: Anmol, 1990.

Thakur, D. D. *My Life and Years in Kashmir Politics*. Delhi: Konark, 2005.

Thomas, Raju, ed. *Perspectives on Kashmir: The Roots of Conflict in South Asia*. Boulder: Westview, 1992.

Thorner, Daniel. *The Agrarian Prospect in India*. Bombay: Allied, 1976.

———. *The Shaping of Modern India*. New Delhi: Allied, 1980.

Thorp, Robert. *Cashmere Misgovernment*. London: Longmans, Green & Co., 1870.

Toshkhani, S. S. and K. Warikoo. *Cultural Heritage of Kashmir Pandits*. New Delhi: Pentagon Press, 2009.

Vashisht, Satish. *Sheikh Abdullah: Then and Now*. Delhi: Maulit Sahitya Prakashan, 1968.

Vide, Steven Jones. *Antonio Gramsci*. London and New York: Routledge, 2006.

Wadeen, Lisa. *Ambiguities of Domination: Politics, Rhetoric, and Symbols in Contemporary Syria*. Chicago: University of Chicago Press, 1999.

Wakhlu, Khem Lata and O. N. Wakhlu. *Kashmir: Behind the White Curtain 1972–91*. New Delhi: Konark, 1992.

Walinsky, L. J. *Agrarian Reforms as Unfinished Business*. Oxford: Oxford University Press, 1952.

Warikoo, K. *Central Asia and Kashmir: A Study in the Context of Anglo-Russian Rivalry*. New Delhi: Gyan Publishing House, 1989.

Watali, A. M. *Kashmir Intifada: A Memoir*. Srinagar: Gulshan Books, 2016.

Winetrout, Kenneth. *Arnold Toynbee*. Boston: Twayne, 1975.

Younghusband, Sir Francis. *Kashmir*. London: Adam and Charles Black, 1911.

Articles

Aivalli, Veerana. 'Single Line Administration: An Administrative Experiment in Jammu & Kashmir'. *Indian Journal of Public Administration* 43, no. 3 (1998).

Akhtar, Shaheen. 'Elections in Indian-Held Kashmir, 1951–1999'. *Regional Studies* 18, no. 3 (2000).

Ali, Nisar. 'Structural Changes in Jammu and Kashmir Economy and Conflict'. In *Conflict and Politics of Jammu and Kashmir: Internal Dynamics*, edited by Avineet Prashar and Puvan Vivek. Jammu: Saksham Books International, 2007.

Amin, Shahid. 'Gandhi as Mahatma: Gorakhpur District, Eastern UP, 1921–22'. In *Selected Subaltern Studies*, edited by Ranajit Guha and Gayatri Chakravorty Spivak. New York: Oxford University Press, 1988.

Beg, Mirza Mohammad Afzal. 'Land Reforms in J&K'. *Mainstream* 15 (1976).

Bevir, Mark. 'Governance'. In *Encyclopedia of Governance*, edited by Mark Bevir. London: Sage, 2007.

Bhan, Mona. 'Refiguring Rights, Redefining Culture: Hill-Councils in Kargil, Jammu and Kashmir'. *Sociological Bullitin* 58, no. 1 (2009).

Bhatt, R. K. 'Kashmir: Politics of Integration'. In *Fourth General Election in India*, edited by S. P. Verma and Iqbal Narain. New Delhi: Orient Longman, 1968.

Choudhary, Shushma. 'Does the Bill Give Power to the People?' In *Panchayat Raj in Jammu and Kashmir*, edited by Mathew George. New Delhi: Institute of Social Sciences & Concept, 1990.

Copland, Ian. 'The Abdullah Factor: Kashmiri Muslims and the Crisis of 1947'. In *The Political Inheritance*, edited by D. A. Low. New York: St. Martin's, 1991.

Deane, James. 'Media and Communication in Governance: It's Time for a Rethink'. In *A Governance Practitioner's Notebook: Alternative Ideas and Approaches*, edited by Alan Whaites, Eduardo Fyson, and Graham Teskey. OCED, 2015.

Dinesh, Mohan, Gautam Navlakha, Sumanta Banerjee, and Tapan Bose. 'India's Kashmir War'. *Economic and Political Weekly* 25, no. 13 (1990).

Duschinski, Haley. 'Reproducing Regimes of Impunity: Fake Encounters and the Informalization of Everyday Violence in Kashmir Valley'. *Cultural Studies* 24, no. 1 (2010).

Duschinski, Haley and Shrimoyeen Nandini Ghosh. 'Constituting the Occupation: Preventive Detention and Permanent Emergency in Kashmir'. *Journal of Legal Pluralism and Unofficial Law* (2017): 15–16. https://doi.org/10.1080/07329113.2017.1347850 (accessed 28 July 2017).

Duschinski, Haley and Bruce Hoffman. 'Everyday Violence, Institutional Denial and Struggles for Justice in Kashmir'. *Race and Class* 52, no. 4 (2011).

Fazal, Shahab and Amin Arshid. 'Impact of Urban Land Transformation on Water Bodies in Srinagar City'. *Journal of Environmental Protection* 2, no. 2 (2011).

Gundevia, Y. D. 'On Sheikh Abdullah', foreword to Sheikh Abdullah, *The Testament of Sheikh Abdullah*. New Delhi: Palit and Palit, 1974.

International Crisis Group. 'India, Pakistan, and Kashmir: Stabilizing a Cold Peace'. *Asia Briefing*, no. 51 (June 2006).

Jamwal, Anuradha Bhasin. 'Auditing the Mainstream Media: The Case of Jammu and Kashmir'. In *Three Case Studies: Media Coverage on Forced Displacement in Contemporary India*, edited by Kumar Das Samir. Kolkata: Calcutta Research Group, 2004. http://www.mcrg.ac.in/mediareport2.htm (accessed 8 January 2018).

Ladejinksy, Wolf. 'Land Reforms: Observations in Kashmir'. In *Agrarian Reforms as Unfinished Business*, edited by L. J. Walinsky. Oxford: Oxford University Press, 1952.

Lipton, M. 'Agriculture, Rural People, the State and Supplies in Asian Countries'. In *Rural Transformations in Asia*, edited by J. Breman and S. Mundle. Oxford: Oxford University Press, 1991.

Madan, T. N. 'Ritual and Religion among Muslims in India'. In *Religious Ideology and Social Structure: The Muslims and Hindus of Kashmir*, edited by Imtiyaz Ahmad. New Delhi: Manohar, 1981.

Mahjoor, Ghulam Ahmad. *'Azadi* (Freedom)'. In *An Anthropology of Modern Kashmiri Verse (1930–1960)*, translated and edited by Trilokinath Raina. Pune: Sangam Press, 1972.

Migdal, J. 'The State in Society: An Approach to Struggles for Domination'. In *State Power and Social Forces: Domination and Transformation in the Third World*, edited by J. Migdal, Atul Kohli, and V. Shue. Cambridge: Cambridge University Press, 1994.

Navlakha, Gautam. 'Kashmir: Political Economy of Fiscal Autonomy'. *Economic and Political Weekly* 38, no. 40 (2003).

———. 'Achieving Fiscal Autonomy'. *Economic and Political Weekly* 39, no. 2 (2004).

Prakash, Siddhartha. 'The Political Economy of Kashmir since 1947'. *Economic and Political Weekly* 35, no. 24 (2000).

Punjabi, Riyaz. 'Corruption: A Factor in Kashmiri Alienation'. *Mainstream Weekly* 29 (1991).

Puri, Balraj. 'The Budget of Kashmir: What the Centre Means to the State'. *Economic Weekly* (18 April 1959).

———. 'Central Aid to Kashmir: Effects of Finance Commission's Recommendations'. *Economic Weekly* (19 May 1962).

———. 'The Era of Sheikh Abdullah'. *Economic and Political Weekly* 18, no. 6 (1983).

———. 'Jammu and Kashmir'. In *State Politics in India*, edited by Myron Weiner. Princeton: Princeton University Press, 1968.

Ray, John. 'Kashmir 1962 to 1986: A Footnote to History'. *Asian Affairs* 33, no. 2 (2002).

Siegel, Benjamin. 'Self Help Which Ennobles a Nation: Development, Citizenship, and the Obligations of Eating in India's Austerity Years'. *Modern Asian Studies* 50, no. 3 (2016).

Srinivasan, T. N. 'Neoclassical Political Economy, the State and Development'. *Asian Development Review* 3, no. 2 (1985).

Staniland, Paul. 'Kashmir since 2003: Counterinsurgency and the Paradox of Normalcy'. *Asian Survey* 53, no. 5 (2013).

Thorner, Daniel. 'The Kashmir Land Reforms: Some Personal Impressions'. *Economic and Political Weekly* no. 37 (1953).

Tillin, Louise. 'Asymmetric Federalism'. In *The Oxford Handbook of the Indian Constitution*, edited by Sujit Choudhary, Madhav Khosla, and Pratap B. Mehta. Oxford: Oxford University Press, 2016.

Tremblay, Reeta Chowdhari. 'Jammu: Autonomy within and Autonomous Kashmir?' In *Perspectives on Kashmir: Roots of Conflict in South Asia*, edited by C. Raju Thomas. Boulder: Westview Press, 1992.

———. 'Kashmir's Secessionist Movement Resurfaces: Ethnic Identity, Community Competition, and the State'. *Asian Survey* 49, no. 6 (2009).

Trissal, S. S. 'Kashmiri Pandit: At the Crossroads of History'. In *The Story of Kashmir: Yesterday and Today*, vol. 2, edited by Grover Virender. Delhi: Deep and Deep, 1991.

Van Beek, Martijn. 'Beyond Identity Fetishism: "Communal" Conflict in Ladakh and the Limits of Autonomy'. *Cultural Anthropology* 15, no. 4 (2000).

———. 'Dangerous Liaisons: Hindu Nationalism and Buddhist Radicalism in Ladakh'. In *Religious Radicalism and Security in South Asia*, edited by Satu P. Kimaye, Robert G. Wirsing, and Mohan Malik. Asia-Pacific Centre for Security Studies, 2004. http://apcss.org/Publications/Edited%20Volumes/ReligiousRadicalism/ReligiousRadicalismandSecurityinSouthAsia.pdfs (accessed 2 October 2017).

Wani, Mohammad Ashraf. 'Religion, Economy and Political Crisis in Kashmir'. In *Identity Politics in Jammu and Kashmir*, edited by Rekha Chowdhary. New Delhi: Vitasta, 2010.

———. 'In Search of an Authentic Text of Sheikh Sheikh ul Alam's Poetry'. *Alamdar* (Centre for Shaikh ul Alam Studies, University of Kashmir) 5, no. 5 (2012).

Warikoo, K. 'Trade Relations between Central Asia and Kashmir Himalayas during the Dogra Period (1846–1947)'. *Cahiers d'Asie centrale (En ligne)* (1 February 1996).

Index

About the Author

Aijaz Ashraf Wani is senior assistant professor in the Department of Political Science, University of Kashmir, Srinagar, J&K, where he has been teaching for more than 10 years. His research interests include governance, politics of J&K, Indian politics, and peace and conflict studies. He has published a number of research papers and essays with reputed international publishers. He has participated and presented papers in a number of conferences both within and outside India (Sri Lanka, USA, Pakistan). Wani has received funding for research projects from ICSSR, New Delhi, and CSDS, New Delhi, India.